6G: The Next Horizon

The first book on 6G wireless presents an overall vision for 6G – an era of intelligence-of-everything – with drivers, key capabilities, use cases, KPIs, and the technology innovations that will shape it. These innovations include immersive human-centric communication, sensing, localization, and imaging, connected machine learning and networked AI, Industry 4.0 and beyond, with connected intelligence, smart cities and life, and the satellite mega-constellation for 3D full-Earth wireless coverage. Also covered are new air interface and networking technologies, integrated sensing and communications, and integrated terrestrial and non-terrestrial networks. In addition, novel network architectures to enable networked AI, user-centric networks, and native trustworthiness are discussed. Essential reading for researchers in academia and industry working on B5G wireless communications.

Wen Tong is the CTO of Huawei Wireless and is the Huawei 5G chief scientist. Dr. Tong is also an IEEE Fellow and a Fellow of the Canadian Academy of Engineering. He was the recipient of the IEEE Communications Society Distinguished Industry Leader Award, and the R. A. Fessenden Medal.

Peiying Zhu is Senior Vice President of Wireless Research at Huawei and is a Huawei Fellow. She is also an IEEE Fellow and a Fellow of the Canadian Academy of Engineering.

"This is the first comprehensive book on 5G, beyond 5G, and 6G written by experts in the field. It elaborates very clearly the potential of 5G and future enhancements before outlining the 6G vision and technical challenges. I highly recommend this book to all in academia and industry who are interested in advanced research."

Rahim Tafazolli, University of Surrey, UK

6G: The Next Horizon

From Connected People and Things to Connected Intelligence

Edited by

WEN TONG

Huawei Technologies Co., Ltd., Canada

PEIYING ZHU

Huawei Technologies Co., Ltd., Canada

CAMBRIDGE
UNIVERSITY PRESS

University Printing House, Cambridge CB2 8BS, United Kingdom

One Liberty Plaza, 20th Floor, New York, NY 10006, USA

477 Williamstown Road, Port Melbourne, VIC 3207, Australia

314–321, 3rd Floor, Plot 3, Splendor Forum, Jasola District Centre, New Delhi – 110025, India

79 Anson Road, #06–04/06, Singapore 079906

Cambridge University Press is part of the University of Cambridge.

It furthers the University's mission by disseminating knowledge in the pursuit of
education, learning, and research at the highest international levels of excellence.

www.cambridge.org
Information on this title: www.cambridge.org/9781108839327
DOI: 10.1017/9781108989817

First published 2021

Printed in the United Kingdom by TJ Books Limited, Padstow Cornwall

A catalogue record for this publication is available from the British Library.

ISBN 978-1-108-83932-7 Hardback

Contents

Contributors

Note: Contributors are listed alphabetically by surname.

Arashmid Akhavain
Huawei Technologies Co., Ltd., Canada

Xueli An
Huawei Technologies Co., Ltd., Germany

Hadi Baligh
Huawei Technologies Co., Ltd., Canada

Alireza Bayesteh
Huawei Technologies Co., Ltd., Canada

Jean-Claude Belfiore
Huawei Technologies Co., Ltd., France

Xiaoyan Bi
Huawei Technologies Co., Ltd., China

Yan Chen
Huawei Technologies Co., Ltd., Canada

Wenshuan Dang
Huawei Technologies Co., Ltd., China

Merouane Debbah
Huawei Technologies Co., Ltd., France

Yiqun Ge
Huawei Technologies Co., Ltd., Canada

Huanhuan Gu
Huawei Technologies Co., Ltd., Canada

Maxime Guillaud
Huawei Technologies Co., Ltd., France

Gaoning He
Huawei Technologies Co., Ltd., China

Jia He
Huawei Technologies Co., Ltd., China

Artur Hecker
Huawei Technologies Co., Ltd., Germany

Huang Huang
Huawei Technologies Co., Ltd., China

Rong Li
Huawei Technologies Co., Ltd., China

Xu Li
Huawei Technologies Co., Ltd., Canada

Yanchun Li
Huawei Technologies Co., Ltd., France

Zhongfeng Li
Huawei Technologies Co., Ltd., China

Hui Lin
Huawei Technologies Co., Ltd., China

Yingpei Lin
Huawei Technologies Co., Ltd., Germany

Fei Liu
Huawei Technologies Co., Ltd., Singapore

Yong Liu
Huawei Technologies Co., Ltd., China

Jianmin Lu
Huawei Technologies Co., Ltd., China

Hejia Luo
Huawei Technologies Co., Ltd., China

Jiajin Luo
Huawei Technologies Co., Ltd., China

Yongxia Lv
Huawei Technologies Co., Ltd., China

Jianglei Ma
Huawei Technologies Co., Ltd., Canada

Mengyao Ma
Huawei Technologies Co., Ltd., China

Amine Maaref
Huawei Technologies Co., Ltd., Canada

Michael Mayer
Huawei Technologies Co., Ltd., Canada

Rui Ni
Huawei Technologies Co., Ltd., China

Chenghui Peng
Huawei Technologies Co., Ltd., China

Morris Repeta
Huawei Technologies Co., Ltd., Canada

Xueliang Shi
Huawei Technologies Co., Ltd., China

Huan Sun
Huawei Technologies Co., Ltd., China

Rob Sun
Huawei Technologies Co., Ltd., Canada

Danny Kai Pin Tan
Huawei Technologies Co., Ltd., China

Hao Tang
Huawei Technologies Co., Ltd., China

Wei Tan
Huawei Technologies Co., Ltd., China

Wen Tong
Huawei Technologies Co., Ltd., Canada

Guangjian Wang
Huawei Technologies Co., Ltd., China

Jian Wang
Huawei Technologies Co., Ltd., China

Jun Wang
Huawei Technologies Co., Ltd.,China

Lei Wang
Huawei Technologies Co., Ltd., China

David Wessel
Huawei Technologies Co., Ltd., Canada

Jianjun Wu
Huawei Technologies Co., Ltd., China

Xun Xiao
Huawei Technologies Co., Ltd., Germany

Xiuqiang Xu
Huawei Technologies Co., Ltd., China

Xueqiang Yan
Huawei Technologies Co., Ltd., China

Chenchen Yang
Huawei Technologies Co., Ltd., China

Xun Yang
Huawei Technologies Co., Ltd., China

Ziming Yu
Huawei Technologies Co., Ltd., China

Kun Zeng
Huawei Technologies Co., Ltd., China

Chunqing Zhang
Huawei Technologies Co., Ltd., China

Hang Zhang
Huawei Technologies Co., Ltd., Canada

Huazi Zhang
Huawei Technologies Co., Ltd., China

Liqing Zhang
Huawei Technologies Co., Ltd., Canada

Mingyu Zhao
Huawei Technologies Co., Ltd., China

Peiying Zhu
Huawei Technologies Co., Ltd., Canada

Preface

Wireless revolutions certainly necessitate the development of innovative, disruptive technologies and groundbreaking applications. When these two forces meet, a new generation of wireless technology emerges. This is exactly what happened when mobile voice and digitalized radio transmission converged, and when the mobile Internet came together with highly spectral efficient radio technology tailored towards the IP protocol. The recent 5G wireless network seeks to achieve wireless connectivity for both massive and ultra-high reliable links, ultimately connecting everything as we know it, and accelerating the digital transformation of every business. Building on the steps of 5G, the 6G wireless network has set a ubiquitous intelligence revolution as the goal. In actuality, 6G will serve as our society's neural network, and as the link between the physical and cyber worlds. Artificial intelligence (AI) based on machine learning will power 6G, and within this realm, our future will fully transform from connected people, connected things to connected intelligence. Phrased differently, the 6G wireless network aims to deliver intelligence to every person, home, and business, which in turn will bring the Intelligence of Everything into existence. From the wireless technology perspective, we are presented with the opportunity to sense the environment and things by exploiting wireless communication radio waves. As such, in addition to transferring bits, the 6G wireless network could serve as networked sensors that extract real-time knowledge and big data from the physical world. This extracted information will not only go a long way in enhancing data transmission, but will also facilitate machine learning for AI services. Another novel aspect, which is definitely worth remembering, is the expansion of VLEO satellites, which orbit the earth at very low altitude in very large constellations, ultimately creating 6G wireless networks "in the sky." As a result, wireless services and applications that are covered anywhere and everywhere on our planet will not be beyond imagination. Without doubt, this vision is immensely ambitious and will impact both our society and economy substantially. On top of that, the creation of the 6G wireless network will seize the opportunities offered by technological innovations in communications, computing, materials, and algorithms. That said, this decade-long journey will not be accomplished overnight.

In this book, we provide a comprehensive view of the 6G wireless network through a technology lens. Our motivation is to introduce some of the initial research results and critical thinking on the 6G wireless network. Not only do we review cutting-edge wireless services and technologies; we also explore the requirements, capabilities, and applications of this network, with an emphasis on the new radio air interface and network architecture. This book serves as a collective result of our research team's quest to define 6G. With that in mind, it should be viewed as a starting point, especially due to the fact that innovation never stops. Along those lines, 6G's trajectory will eventually be shaped by experts around the globe, as we hold firmly that open innovation and a single, globally unified standard are the foundation upon which 6G's success will be built. Just like its predecessors, the success of the 6G wireless network will translate into the success of the open and global ecosystem.

Finally, the wireless revolution we are familiar with today has been in progress for over four decades, with a widespread impact that continues to surpass all expectations. As such, we can neither underestimate nor overestimate the potential of the wireless future. With this spirit, let's recall that Guglielmo Marconi once asserted in 1932 that "it is dangerous to put limits on wireless."

Abbreviations

3GPP	3rd Generation Partnership Project
5G	5th generation of mobile communication
5GAA	5G automative association
5G-ACIA	5G alliance for connected industries and automation
ACLR	Adjacent channel leakage ratio
ADC	Analog-to-digital converter
AE	Auto encoder
AGV	Automated guided vehicle
AI	Artificial intelligence
AIaaS	AI as a service
AMC	Adaptive modulation and coding
AMI	Antagonist myoneural interface
AMP	Approximate message passing
ANC	Analog network coding
ANN	Artificial neural network
AoA	Angle of arrival
AoD	Angle of departure
APSK	Amplitude phase shift keying
AR	Augmented reality
A-SIC	Application-specific integrated circuits
ATIS	Alliance for telecommunications industry solutions
AWGN	Additive white Gaussian noise
B2B	Business-to-business
BCH	Bose–Chaudhuri–Hocquenghem
BCI	Brain–computer interface
BER	Bit error rate
BICM	Bit-interleaved coded modulation
BLAST	Bell Laboratories layered space time
BLER	Block error rate
BN	Bayesian network
BP	Belief propagation
BPSK	Binary phase shift keying
CA	Carrier aggregation
CAGR	Compound annual growth rate

CAPEX	Capital expenditure
CAPL	Concurrent passive and active localization
CC	Component carrier
CCSA	China communications standards association
CDMA	Code division multiple access
CEM	Computational electromagnetic
CJT	Coherent joint transmission
CN	Core network
CNN	Convolutional neural network
CNT	Carbon nanotube
CoMP	Coordinated multipoint
CP	Cyclic prefix
CS	Compressed sensing
CSI	Channel state information
CT	Computed tomography
CU	Control unit
CWS	Continuous wave spectroscopy
D2D	Device-to-device
DAC	Digital-to-analog converter
DBN	Dynamic Bayesian network
DC	Dual connectivity
DFT	Discrete Fourier transform
DL	Downlink
DLT	Distributed ledger technology
DMRS	Demodulation reference signals
DNN	Deep neural network
DoF	Degrees of freedom
DoU	Data of usage
DPC	Dirty paper coding
DRL	Deep reinforcement learning
DRX	Discontinuous reception
DSS	Dynamic spectrum sharing
E2E	End-to-end
EEG	Electroencephalography
EIT	Electromagnetic information theory
ELAA	Extremely large aperture arrays
eMBB	Enhanced mobile broadband
EMG	Electromyography
ESPRIT	Estimation of signal parameters via rotational invariance techniques
EVM	Error vector magnitude
FBMC	Filter bank multi-carrier
FDD	Frequency division duplex
FDMA	Frequency division multiple access
FEC	Forward error correction

FFT	Fast Fourier transform
FinFET	Fin field-effect transistor
FMCW	Frequency modulated continuous wave
f-OFDM	Filtered OFDM
FOV	Field of view
FPGA	Field programmable gate arrays
FPS	Frames per second
GaAs	Gallium arsenide
GAN	Generative adversarial network
GB	Grant-based
GBSM	Geometry-based stochastic model
GDP	Gross domestic product
GDPR	General Data Protection Regulation
GEM	General expectation maximization
GEO	Geostationary earth orbit
GF	Grant-free
GFDM	Generalized frequency-division multiplexing
GNSS	Global navigation satellite system
GPS	Global positioning system
GPU	Graphics processing unit
GQD-PD	Graphene photodetector sensitized with semiconducting quantum dots photodetectors
GSMA	Global System for Mobile Communications Association
HAP	High-altitude platform
HAPSs	High-altitude platform stations
HARQ	Hybrid automatic repeat request
HBT	Heterojunction bipolar transistors
HEMT	High electron mobility transistors
HMI	Human–machine interface
HMM	Hidden Markov model
Holo-MIMO	Holographic MIMO
IaaS	Infrastructure as a service
IAB	Integrated access and backhaul
IB	Information bottleneck
ICI	Inter-carrier interference
ICT	Information and communications technology
IDFT	Inverse discrete Fourier transform
IDMA	Interleave-division multiple access
IFFT	Inverse fast Fourier transform
IGMA	Interleave-grid multiple access
IMT	International mobile telecommunication
InH	Indoor hotspot
InP	Indium phosphide
IoE	Internet of everything

IRS	Intelligent reflecting surface
ISAC	Integrated sensing and communication
JSCC	Joint source and channel coding
KPI	Key performance indicator
LDPC	Low-density parity-check
LED	Light emitting diode
LEO	Low-earth-orbit
LIDAR	Light detection and ranging
LIS	Large intelligent surface
LLR	Log likelihood ratio
LNA	Low-noise amplifier
Log-MAP	Logarithmic maximum a posteriori
LOS	Line-of-sight
LSTM	Long short-term memory
LTE	Long-term evolution
LTE-A	LTE-advanced
LVDM	Lagrange–Vandermonde division multiplexing
M2M	Machine-to-machine
MA	Multiple access
MAC	Media access control
MARL	Multi-agent reinforcement learning
MCS	Modulation and coding scheme
MDP	Markov decision process
MEC	Mobile edge computing
MEO	Medium earth orbit
ML	Machine learning
MMSE	Minimum mean square error
mMTC	Massive machine type of communication
mmWave	Millimeter wave
MNO	Mobile network operator
MOSFET	Metal-oxide–semiconductor field-effect transistor
MP	Matching pursuit
MR	Mixed reality
MRI	Magnetic resonance imaging
MTP	Motion-to-photon
MUD	Multi-user detection
MUSA	Multi-user shared access
MUSIC	Multiple signal classification
MUST	Multi-user superposition transmission
NASA	National aeronautics and space administration
NB-IoT	Narrowband Internet of things
NCJT	Non-coherent joint transmission
NEP	Noise-equivalent power
NFI	Near-field imaging

NIST	National Institute of Standards and Technology
NLP	Natural language processing
NMR	Nuclear magnetic resonance
NOMA	Non-orthogonal multiple access
NR	New radio
NTN	Non-terrestrial network
OA&M	Operations, administration, and maintenance
OAM	Orbital angular momentum
OAMP	Orthogonal approximate message passing
OFDM	Orthogonal frequency-division multiplexing
OFDMA	Orthogonal frequency-division multiple access
OMA	Orthogonal multiple access
OMP	Orthogonal matching pursuit
OOBE	Out-of-band emission
OPEX	Operational expenditure
OQAM	Offset QAM
OSS	Operation support system
OT	Operational technology
OTFS	Orthogonal time–frequency space
OTP	One-time pad
OTT	Over-the-top
OWC	Optical wireless communication
P2P	Point-to-point
PA	Power amplifier
PaaS	Platform as a service
PAE	Power-added efficiency
PAPR	Peak-to-average power ratio
PAR	Packet arrival rate
PC	Photonic crystal
PCA	Photoconductive antenna
PCE	Photo conversion efficiency
PDMA	Pattern division multiple access
PDR	Packet drop rate
PDSCH	Physical downlink shared channel
PHY	Physical layer
PLE	Path loss exponent
PNC	Physical-layer network coding
PPD	Pixel per angular degree
PQC	Post quantum cryptography
QAM	Quadrature amplitude modulation
QD	Quantum dot
QHA	Quadrifilar helix antenna
QoS	Quality of service
RACH	Random access channel

RAN	Radio access network
RCN	Radio computing node
RE	Resource element
ReLU	Rectified linear unit
RF	Radio frequency
RIS	Reconfigurable intelligent surface
RL	Reinforcement learning
RM	Reed–Muller
RNN	Recurrent neural network
RRC	Radio resource control
RRH	Remote radio head
RS	Reference signal
RSMA	Resource spread multiple access
RSRP	Reference signal received power
RTT	Round-trip time
SAE	Society of Automotive Engineers
SCL	Successive cancellation list
SCMA	Sparse code multiple access
SDG	Sustainable development goal
SDMA	Space division multiple access
SEEG	Stereoelectroencephalography
SeGW	Security gateway
SIC	Successive interference cancellation
SiN	Silicon neuron
SLAM	Simultaneous localization and mapping
SLM	Spatial light modulator
SNR	Signal-to-noise ratio
SP-OFDM	Spectrally precoded OFDM
SPP	Surface plasmon polariton
SRR	Split-ring-resonator
SRS	Sounding reference signal
SSB	Synchronization signal blocks
STBC	Space–time block codes
SWC	Surface wave communication
TCA	Tightly coupled array
TDD	Time division duplex
TDL-C	Tapped-delay line channel
TDMA	Time division multiple access
TDS	Time domain spectroscopy
THz	Terahertz
TN	Transport network
TPC	Terminal-pipe-cloud
TR	Time reversal
TTI	Transmission time interval

UAV	Unmanned aerial vehicle
UCN	User-centric network
UCNC	User-centric no cell
UFMC	Universal filtered multi-carrier
UN	United Nations
URLLC	Ultra-reliable low-latency communication
UWB	Ultra-wideband
V2I	Vehicle-to-infrastructure
V2V	Vehicle-to-vehicle
V2X	Vehicle-to-everything
VAE	Variational auto encoder
VLC	Visible light communication
VLEO	Very low earth orbit
VLSI	Very-large-scale integration
VR	Virtual reality
W-OFDM	Windowed OFDM
WRC	World Radiocommunication Conference
XOR	Exclusive OR

Part I

Introduction

1 Mobile Communications Towards 2030 and Beyond

We are in the midst of a great digital wave that is bringing a continuous stream of innovations, flexibility, and new opportunities to every person, family, car, and industry in many countries, redefining how we live, work, learn, and stay healthy. Today, as the global rollout of 5G picks up pace and unleashes unimaginable possibilities, we are witnessing 5G transform every aspect of our lives, industry, and society. Looking ahead to 2030 and beyond, what might we expect from the next generation of mobile communications?

1.1 Evolution of Mobile Communications

Mobile communication systems have evolved dramatically since the 1980s, with a new generation emerging every 10 years or so. At the same time, the mainstream services provided by mobile networks and the application of new frequency bands usually take two generations – or 20 years – to mature. As shown in Figure 1.1, each new generation brings significant capability improvements compared with its predecessor by introducing new technologies, new design principles, and new architectures in the radio access networks and core networks.

For the 2G and 3G networks, the main drivers were mobile subscriptions primarily focusing on voice communication services. As the penetration rate of mobile phones and usage of voice service reached saturation point, this subscription-based business model began to plateau.

From 3G to 4G, the data service grew rapidly and mobile broadband became the dominant service for 4G. Over the last 10 years, major advancements in mobile communications have had a profound impact on people's way of life. For example, smartphones carrying all kinds of applications have become deeply ingrained in every aspect of the lives of many people. 4G network operators therefore mainly rely on traffic volume rather than subscriptions for revenue, and the growth of traffic consumption per capita drives business growth.

4G has had a dramatic impact on our lives – the technology capabilities it brought have led to numerous innovations in mobile-oriented applications that have revolutionized our daily lives. One primary example of this in China is the shift from using cash to now using online payment methods. Today, young and old alike favor using online payment methods such as AliPay and WeChat Pay, finding it more convenient

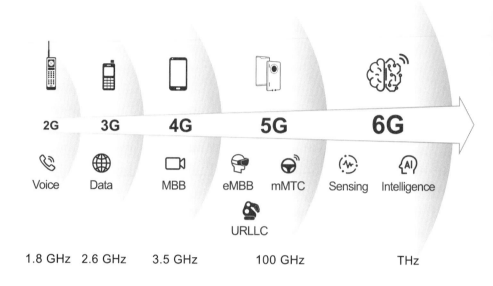

Figure 1.1 Evolution of mobile networks.

to pay for anything from grocery shopping to car parking without carrying cash with them. Another example is the rise of social media. Anyone can now share pictures and videos – effectively becoming a news anchor – to anyone else, anywhere, and at any time from smartphones, speeding up how information is spread.

This trend continues in 5G, as more and more bandwidth-demanding applications continue to emerge. Such applications include high-definition videos and immersive media applications such as augmented reality (AR), virtual reality (VR), and mixed reality (MR). At present, there are approximately 3.8 billion smartphones in use around the world. We expect this number to reach 8 billion by 2025, at which time there will be more than 6.5 billion mobile Internet users with 80% of them having mobile broadband. On top of that, there will be 440 million AR/VR users, and 40% of cars will be connected.

As narrowband Internet of Things (NB-IoT), industry IoT, and vehicle to every-thing (V2X) have become standardized, mobile networks have shifted some of their focus from connecting people with enhanced mobile broadband (eMBB) to connecting things with ultra-reliable low-latency communication (URLLC) and massive machine type of communication (mMTC). This in turn should enable every business to achieve digitalization for the next wave of economic growth. The 5G commercial deployment initially focused on consumer services, but later releases of 3GPP 5G standardization (e.g., Release 16 and Release 17) have evolved with the aim of driving the maturity of vertical applications such as V2X and industry IoT. To enable different levels of autonomous driving and Industry 4.0 across a vast range of businesses and industries, the mobile industry is working closely with verticals at various consortiums such as 5G-ACIA [1] and 5GAA [2] to accelerate the application of mobile technologies. It is expected that level-4 autonomous driving will be available around 2024, and improvements in transportation efficiency are widely anticipated due to the proliferation of V2X. Optimized business process and production efficiency will become the key drivers for future gross domestic product (GDP) growth.

While 5G opens the door for the Internet of Everything (IoE), we predict that 6G – the successor to 5G – will be the platform for connected intelligence, where the mobile network connects vast amounts of intelligent devices and connects them intelligently. The next wave of digitalization is expected to create more innovations that will meet every aspect of our needs. Through artificial intelligence (AI) and machine learning (ML), we will be able to build a real-time connection between the physical and digital worlds, allowing us to capture, retrieve, and access larger amounts of information and knowledge in real time and thus make the connected world a connected intelligence. Furthermore, sensing and distributed computing, together with advanced and integrated non-terrestrial network (NTN) and short-distance wireless communication technologies, will lay the foundation to build intelligent mobile communication networks in the future.

1.2   Key Drivers

As illustrated in Figure 1.2, we predict three key drivers calling for a new generation of connected intelligence. We explain each of these drivers below.

Driver 1: New Applications and Businesses

Today, business revenue is driven by the increase of traffic consumption per subscription. Figure 1.3 shows how the average global mobile traffic per subscription per

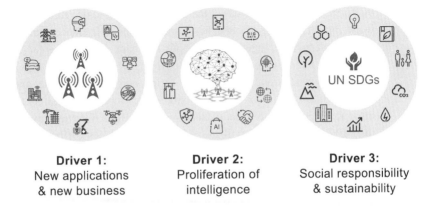

Driver 1:
New applications
& new business

Driver 2:
Proliferation of
intelligence

Driver 3:
Social responsibility
& sustainability

Figure 1.2 Key drivers for 6G.

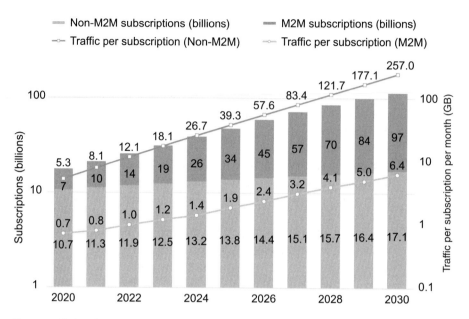

Figure 1.3 Estimation of subscriptions and mobile traffic from 2020 to 2030.

month (solid lines) and the number of subscriptions (bars) are expected to grow from 2020 to 2030 for machine-to-machine (M2M) and non-M2M devices. The data used in the figure were obtained from the ITU-R report M.2370 [3]. From the figure, we can see that the growth of smartphone subscriptions is already saturated in 2020; the compound annual growth rate (CAGR) from 2020 to 2030 is about 6%. In addition, the Global System for Mobile Communications Association (GSMA) expects that the penetration rate of unique mobile subscribers will increase only three percentage points, i.e., from 67% to 70%, from 2019 to 2025 [4]. Yet despite this, the mobile data traffic per MBB subscription is expected to increase 50-fold over the 10-year period, from 5.3 GB per month at 2020 to 257 GB per month at 2030.

A wide range of bandwidth-hungry applications have been supported on the 5G platform, which have increased the traffic volume and driven up the demand for network capacity. In the 6G era, more applications will emerge, and extended reality (XR) cloud services together with haptic feedback and holographic display are likely to become mainstream applications, covering 360 degree VR movies, AR-assisted remote services, virtual 3D educational trips, haptic telemedicine, and remote teleoperation. Huawei's global industry vision (GIV) report [5] predicts that there will be more than 337 million users of head-mounted VR/AR devices by 2025, while more than 10% of enterprises will use AR/VR technologies for business operations, and these numbers are certain to increase by 2030. As cloud XR applications increase both in number and popularity, and as the resolution, size, and refresh rate increase, the bandwidth and latency requirements may exceed what 5G evolution can offer. The exponential increase in the traffic demand per device, together with strict latency and reliability requirements, will become a major challenge for 6G network design in terms of the huge capacity needed. Furthermore, the unlimited data plans that many operators offer have become a key business model and will also contribute to the potential rise of data consumption.

We can also see from Figure 1.3 that there will be about 13 times more M2M devices in 2030 than in 2020, and both enterprise and consumer IoT connections will continue to increase. In its mobile economy report 2020 [4], GSMA predicts that enterprise IoT will overtake consumer IoT by 2024. AI will therefore become an engine for all kinds of automation and will use large amounts of data to convert real-time situational awareness into real-time decision-making. Massive numbers of wideband sensors will be deployed in scenarios such as smart home, smart health, smart car, smart city, smart building, and smart factory to obtain the huge amount of data needed by AI. Big data is foundational to the success of machine learning, and this becomes a major driver for the order of magnitude increase in 6G network throughput. In addition, new capabilities such as networks-as-sensors and non-terrestrial communication will become an integrated part of 6G mobile systems, enabling environmental monitoring and imaging over large areas in real time with an even larger number of sensors.

In addition to this, high-performance industrial IoT applications have demanding requirements on wireless performance in terms of deterministic latency and jitter, and they expect guaranteed availability and reliability. For example, high performance is needed for time-sensitive command and control, as well as multi-robot movement coordination and collaboration. Such use cases also drive the extreme and diverse performance needed for 6G.

Driver 2: Proliferation of Intelligence

In the coming decades, the digital economy will continue to be a major driver of economic growth worldwide, growing much faster than the global economy. In 2019, the digital economy grew 3.5 times faster than the global economy and reached US$15.60 trillion, accounting for 19.7% of the global economy. This is expected to reach 24.3% by 2025. In terms of investment leverage, analysis shows that every $1 increase in digital investment over the past 30 years has led to a $20 increase in GDP, compared to an average of 1 : 3 for non-digital investment [6].

As one of the most dynamic sectors within the information and communications technology (ICT) industry, the mobile industry has had a profound impact on people's lives, helped to mitigate the digital divide, and contributed significantly to society's overall productivity and economic growth. By 2024, mobile technologies and services are expected to generate 4.9% of the global GDP (approaching US$5 trillion), with more industries benefiting from the improvements in productivity and efficiency brought by the increased adoption of mobile services [4].

We believe this trend will continue into 2030 and beyond. In particular, as pervasive intelligence becomes the key enabler of business and economic models in the future, paradigm shifts in radio technology and network architecture will be driven by four critical factors, as illustrated in Figure 1.4.

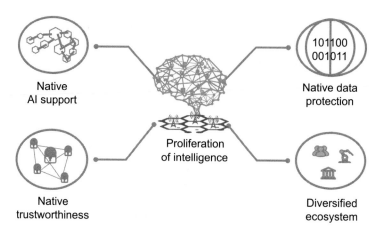

Figure 1.4 Business drivers brought by pervasive intelligence and big data.

- **Native AI support:** Although the core network design in 5G supports intelligence by introducing a new type of network function (i.e., network data analytics), there is limited scope for usage in network operations and management. Instead, 5G delivers AI as an over-the-top (OTT) service. Conversely, for 6G, end-to-end (E2E) mobile communication systems are designed with optimal support for AI and machine learning – not only as a basic functionality, but also for optimal efficiency. From the architecture perspective, running distributed AI at the edge could achieve ultimate performance while also addressing the data ownership concerns of individuals and enterprises as well as meeting regional and national regulatory constraints. "Native" AI support in 6G aims to deliver AI services anywhere and at any time, and will continuously improve system performance and user experience through continual optimization. Consequently, truly pervasive intelligence in combination with deeply converged ICT systems, featuring diverse connectivity, computing, and storage resources at the edge, will become a native trait. The corresponding capabilities (such as algorithms, neural networks, databases, and application programming interface (APIs)) must be integrated into the 6G system as part of the network realization. The 6G network architecture with native AI support will bring "Networked AI," moving away from today's centralized "Cloud AI."
- **Native data protection:** In addition to the security capabilities developed in 5G and earlier generations, privacy will be a critical design requirement and principle for 6G. The protection of privacy in every aspect of 6G networking and data will be essential. On the one hand, the key driver behind this is data ownership and the right to access data, posing a challenge in how the network architecture can ensure privacy protection. On the other hand, native AI requires the capability to process and access data in a distributed manner. Instead of relying on network and application service providers to ensure data protection, we expect that users – which might be people or machines – will be empowered as data subjects with control and

operation rights. The design of the next-generation system should make privacy-guarantee the top priority instead of only a side feature. Such a design should also ensure the proper rights of data subjects, enable data control and processing, and integrate support for policies like the General Data Protection Regulation (GDPR) in order to establish fundamental guidelines for technology design and usage in the future.

- **Native trustworthiness:** To support a diverse range of use cases and markets, it is essential to have customized, verifiable, and measurable trustworthiness. The nomothetic network ownership and operation for the current and previous generations of networks will evolve into many-parties, many-players, and many-actors. This business driver will promote a trustworthiness architecture that includes many factors. An inclusive multi-lateral trust model will be more vital than the single-trust model. In addition to being future-oriented, the trustworthiness architecture should include security, privacy, resilience, safety, and reliability.

- **Diversified ecosystem:** The three fundamental elements for AI are data, algorithms, and computing. However, individual businesses may not possess full capabilities in these aspects to achieve digital transformation with pervasive intelligence and fast technological innovation. Consequently, it is essential to establish an open, sustainable, and multi-party collaborative ecosystem in order to achieve business success.

Furthermore, as 5G capabilities gradually expand, the vertical wireless market is expected to ramp up throughout the 2020s. Players in both the ICT and operational technology (OT) sectors are exploring how to collaborate in order to generate new sources of revenue. As we approach the 6G era, it would be beneficial if there were a universal ICT framework that could offer an overarching perspective for all industries and thereby accelerate the collaboration and convergence of the ICT and OT sectors. The first wave of 6G commercial use is likely to boost both the consumer and vertical markets.

Driver 3: Social Responsibility and Sustainability

In terms of social responsibility and sustainability, let's take the COVID-19 pandemic as an example. This global crisis has had an impact on almost everyone worldwide, during which time the ICT industry stepped up and played a crucial role in saving human lives. Wireless communication and positioning technologies were used to trace infected patients and monitor the spread of the disease, while minimizing the exposure of medical personnel to it has given rise to significant innovations in 5G for healthcare automation. To limit large gatherings of people while also keeping the economy active, many countries used various remote applications over wireless networks, including applications such as telemedicine, tele-education, telecommuting, industrial automation, and e-commerce. As the mobile industry supported different sectors of the global economy and society during this pandemic, the resulting use cases that emerged from it have contributed to the future technological evolution of the mobile industry.

Mobile networks have the potential to transform business, education, government, healthcare, agriculture, manufacturing, and the environment, as well as the way we

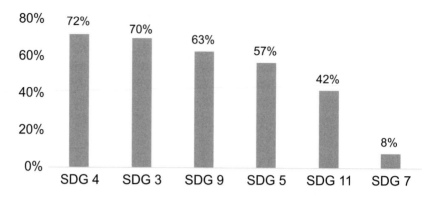

Figure 1.5 Top correlations between individual SDGs and ICT development. From a joint report of Huawei and UN [7]. SDG 4: Quality education; SDG 3: Good health and well-being; SDG 9: Industry, innovation, and infrastructure; SDG 5: Gender equality; SDG 11: Sustainable cities and communities; SDG 7: Affordable and clean energy.

interact with others. They have become one of the key drivers of social evolution and may redefine our very existence. According to the GSMA [4], mobile communication is central to the sustainable development goals (SDGs) set by the United Nations (UN) in 2015 to transform our world, and is a powerful tool for achieving these SDGs. The mobile industry has had a significant impact across all 17 SDGs – an impact that continues to grow, providing a solid foundation for the digital economy and acting as a catalyst for a diverse and innovative range of services.

Huawei and the UN worked together on the ICT SDG Benchmark [7] to measure the degree to which ICT development enables progress on the SDGs in a quantitative manner. The assessment in 2019 showed a strong correlation between ICT maturity and the progress on the SDGs, reaching $R^2 = 0.86$. Among them, as shown in Figure 1.5, SDG 3 (Good health and well-being) and SDG 4 (Quality education) showed the strongest correlation with ICT, signaling the areas where digital technology has the highest potential to accelerate a country's performance.

Regarding environmental sustainability, the evolution of ICTs is critical to achieving SDG 7 (Affordable and clean energy) and SDG 11 (Sustainable cities and communities). The world is becoming increasingly urbanized, with projections showing that 5 billion people will live in cities by 2030 and occupy just 3% of the inhabitable land while accounting for 60%–80% of energy consumption worldwide [8]. ICTs will become more carbon neutral as they achieve higher and higher energy efficiency per bit, and ICT-enabled solutions (e.g., smart grids, smart logistics, and smart industry) will help transform the world towards a more sustainable and energy-efficient future [9].

As of the year 2020, about 90% of people worldwide have access to 3G or 4G networks. For the remaining 10%, both 5G and 6G will aim to connect them by using technologies such as satellite communications. As an example, 5G has already attempted to integrate non-terrestrial access technologies into 5G New Radio (NR) technologies, while ambitious plans for very low earth orbit (VLEO) – constellations of tens of thousands of VLEO satellites – might come to fruition in the 6G era. Services, applications, and contents enabled by mobile networks are helping to expand financial and social inclusion, while emerging technologies such as IoT, big data, AI, and machine learning are being integrated into the network infrastructure, demonstrating their potential to significantly transform society and the environment. The strong correlations between ICT and SDGs mean that we must consider all aspects of the SDGs when designing 6G communication systems and networks.

1.3   Overall Vision

Mobile communication, over the span of only 40 years, has completely revolutionized the world. Today, we depend heavily on wireless connectivity for both work and life; it has become a key enabler for the digital transformation of every business. As 5G – the current generation of wireless connectivity – starts to take hold, everything, in addition to every person, will be connected; with this hyper-connectivity, we will be able to automate every aspect of society. Moreover, the momentum for wireless innovations is accelerating. As was pointed out in [10]. *Science, the endless frontier*, wireless is indeed the endless frontier.

As wireless technology innovations continue over the next 10 years, the rise of machine-learning-based AI and the creation of digital twins (i.e., representing the physical world as a cyber one) are the two major catalysts fueling more technology breakthroughs. The resulting 6G will be a game-changer in terms of both the economy and society – it will lay a solid foundation for the future Intelligence of Everything.

6G will be the next generation of wireless communication, transiting from an era of connected people and connected things to one of connected intelligence. As society moves towards the Intelligence of Everything, 6G will be the key to proliferating AI, delivering intelligence to every person, home, car, and business.

Functioning like a distributed neural network with communication links, 6G will fuse the physical and cyber worlds. It will not be simply a pipe carrying bits, but rather a network for everything sensed, everything connected, and everything intelligent. As such, 6G will be a network of sensors and machine learning, where data centers will become neural centers with machine learning spread over the entire network. This is the blueprint of the cyber world for the future Intelligence of Everything.

6G will be a key enabler in achieving the full-scale digital transformation of all vertical businesses. Offering extreme performance, such as multi-Tbit/s data rates, sub-millisecond latency, and seven nines (99.99999%) reliability, 6G will realize major improvements in terms of key performance indicators (KPIs) – more than an order of magnitude higher in some cases – compared with 5G. It will provide universal, high-performance connectivity comparable with fibers in terms of speed and reliability, except that it will do so wirelessly. Free from functional and performance limitations, 6G will be a generic platform that supports the creation of any service and any application, reaching an ultimate "connectivity supremacy"!

The disruptive technologies used in 6G – along with major innovations – will set it apart from previous generations. Here we describe just some of 6G's key highlights, which will have a profound effect for decades to come.

- 6G will be designed from the ground up with native AI capabilities, and the network architecture will include machine learning capabilities, in particular, distributed machine learning. To put it simply, 6G is a network designed to support AI, or a network for AI, where many network elements will perform AI and machine learning functions.
- Wireless sensing, a natural characteristic of radio wave propagation, will be a key disruptive technology in 6G, using radio transmission and echoes to detect (or sense) the physical world. Previous generations of wireless systems primarily carried information over radio waves. However, in order to support AI and machine learning, we need to collect extremely large amounts of data from the physical world; 6G radios can function as sensors for this purpose. In particular, by using higher frequencies, such as millimeter wave (mmWave) and the terahertz (THz) spectrum, 6G will enable high-resolution sensing.
- The integration of VLEO satellite constellations with terrestrial networks will also be a major differentiator in 6G. Densely deployed small satellites will enable a "wireless network in the sky" for full-earth coverage. This has been made economically viable with the breakthroughs achieved by Space-X in advanced satellite launching technologies, significantly lowering the cost to construct massive satellite constellations. This novel non-terrestrial wireless infrastructure will be complementary to the existing terrestrial-based cellular system with a completely integrated design, which will be a key enabler for 6G.
- The network architecture used in 6G will differ significantly from earlier generations. 6G centers on data and the intelligence and knowledge derived from it. The network architecture will be designed to enable native trustworthiness with advances in security technologies, privacy preservation, and data governance. This will require a basic network re-architecture in 6G to reflect the Intelligence of Everything. In addition, 6G will employ new data ownership, trust models, and security designs resistant to quantum-computing-based attacks.
- Sustainability is a central topic for 6G, particularly in terms of energy consumption across the entire network and associated ICT infrastructure and devices. The design of 6G must meet stringent requirements in this regard. Specifically, the total power consumption of 6G infrastructure must be much lower than previous generations, and realizing an E2E architecture that is both sustainable and energy-efficient must be prioritized. As a global ICT infrastructure, 6G will be designed with the ultimate goal of social, environmental, and economic sustainability. The future of intelligence must align with our common goal of making our planet a better place in which to live.

In summary, *shifting from connected people, connected things to connected intelligence is the driving force for 6G*. It is the guiding principle behind 6G use cases, network designs, and technologies. Artificial intelligence, the fusion of the physical and cyber worlds, and connectivity supremacy are the new pillars on which we aim to build a society with Intelligence of Everything.

We envisage that 6G technologies will offer the following six new capabilities in order to address the potential challenges faced in the 6G era:

6G will reach the ultimate level of connectivity supremacy; employing all radio frequencies up to terahertz or even visible light.

Traffic growth typically drives the need for additional wireless spectrum, while the cellular network infrastructure favors the lower-frequency spectrum in order to achieve ubiquitous coverage. As several generations of wireless networks have evolved, increasing amounts of spectra have been allocated for network upgrades. In addition to the mmWave spectrum, 6G will utilize terahertz or even visible light for the first time, meaning that all potential spectra will be employed to achieve extreme connectivity. Emerging coverage solutions such as VLEO satellites, which provide new coverage infrastructure, and on-demand high-altitude platform stations (HAPSs), which provide temporary coverage infrastructure, require innovative approaches to aggregate the available spectra for 6G capacity and network solutions. This includes defining spectrum ownership and usage models such as sharing, flexible spectrum allocation, and duplexing. In this way, 6G will offer near-infinite capacity with an unprecedented wireless connection speed. With such capability, the new 6G air interface can unify eMBB, URLLC, and mMTC physical-layer technologies, as capacity and latency will no longer be bottlenecks hampering the design of a truly customized wireless connection for every user, service, application, and scenario.

6G will support AI natively, connecting intelligent things and connecting things intelligently.

One of the primary objectives for 6G is to support ubiquitous AI, where 6G will be an E2E system designed to support AI-based services and applications. This capability is neither an add-on nor an over-the-top (OTT) feature; instead, the 6G system itself should be the most efficient platform for AI. However, this presents new challenges in terms of realizing minimal costs for both communication and computation, each of which is a KPI for future study. For minimal communication costs, it is necessary to design a 6G system capable of transferring massive amounts of big data for AI training while using minimal capacity. In terms of minimizing computation costs, it is necessary to optimally distribute computing resources across the networks at strategic locations where we can best leverage mobile edge computing. To support machine learning, 6G will need to enable the collection of massive data from the physical world in order to create a cyber world. This is a significant increase in data, posing a major challenge for 6G. As such, achieving effective compression of training data based on information and learning theories will be a new and essential area of research. Another challenge is reducing the computational load involved in AI training by using collaborative learning. At the network level, data splitting and model splitting will be incorporated into the 6G architecture, where the distributed and federated learning will not only be used to optimize the computing resource, the local learning, and the global learning, but also to conform to new data local governance requirements. In terms of the network architecture, the core network functions will be pushed towards a deep-edge network, while software cloudification will shift towards machine learning. In addition, the 6G RAN will shift from downlink-centric radio access to uplink-centric radio access, because the massive training data involved in machine learning require significantly higher throughput in the uplink. Moreover, the 6G air interface can be designed with new machine learning capabilities to achieve intelligent communication.

6G will be networked AI, redefining networking and computing.

AI will be ubiquitous in 6G and drive new architectures for networking and computing. For example, the cloud-based data centers in use today will evolve into native AI neural centers. This involves a shift from CPU-based computing to graphics processing unit (GPU)-based computing. In most cases, AI-specific computing hardware must be co-designed and optimized with the AI algorithms. However, the rise of AI brings a significant challenge in terms of computing. On average, the human brain achieves data rates of 20,000 Tbit/s and can store 200 TB of information while consuming only 20 watts. Conversely, today, the computing power of AI is doubling every two months, far in excess of Moore's law. To achieve the same capabilities as the human brain, a neural center will consume 1000 times more power than the human brain at a point in time near the end of Moore's law. In order for neural centers to replace data centers and fully leverage the potential of AI, it is imperative to use significantly advanced machine learning technologies that facilitate sustainable AI-based 6G [11]. A standardized approach to implementing a neural center computing architecture and software orchestration is critical to enabling the 6G network to be an open platform and an open ecosystem.

6G will function as a networked sensor, enabling the fusion of cyber, physical, and biological worlds.

Sensing is a new and foundational feature for 6G – it is a new channel through which we can link the physical and biological worlds to a cyber world. In order to create a parallel cyber universe that is a true replica of the physical one, specifically, a digital twin, we need real-time sensing. 6G radio wave will be used to realize sensing functionality across all radio access nodes and devices, including base stations and mobile devices. The sensing data collected by the network and devices can be used for two purposes: to enhance communications, especially for beam-based mmWave and THz frequency bands; and to facilitate machine learning and AI. In both cases, sensing data contain real-time information and knowledge about the physical and biological worlds. As such, we can consider 6G to be a networked sensor, differing significantly compared with previous generations of wireless systems, which simply transport information. Network- and device-based sensing can realize global and local sensing, respectively. With such functionality, 6G will bring real-time AI and machine learning to the next level.

6G with integrated terrestrial and non-terrestrial networks will deliver complete full-earth coverage, eliminating digital divide.

Integrating non-terrestrial networks, especially VLEO satellite mega constellations, into 6G is a very attractive feature. A VLEO satellite system, in addition to delivering full-earth coverage, offers a number of new capabilities and advantages. For example, it eliminates the issue with communication latency inherent in conventional geostationary earth orbit (GEO) and medium earth orbit (MEO) satellite systems. It can also provide coverage to areas uncovered by terrestrial networks, offering complementary radio access. One of the VLEO satellite system's unique advantages is that it provides a low-latency global communication link, which is essential for mission-critical applications such as frequent stock trading. VLEO satellite systems can also provide more accurate positioning, which is critical for autonomous driving and important for earth sensing and imaging. In addition to satellite communications, new radio nodes such as drones, unmanned aerial vehicles (UAVs), and HAPSs will be an integral part of 6G, functioning as either mobile terminals or temporary infrastructure nodes.

6G will support a prosumer-centric instead of operator-centric network architecture, embracing an inclusive open ecosystem.

6G will bring about a paradigm shift as it drives economic and social changes with advances in virtualization and AI. 6G networks will have intelligence at their foundation, enabling a participatory approach to networking and service provisioning. This will redefine the intelligent connectivity infrastructure as a dynamic pool consisting of all participating users' resources. It is a radical paradigm shift from the conventional operator-centric view to an inclusive prosumer-centric view (a prosumer both produces and consumes a particular commodity). Through a collaborative model bringing together many networks, key aspects such as multi-lateral ownership, data ownership and privacy, and trust models of involved players must be designed as built-in features rather than built-after ones. Furthermore, in order to achieve local data governance and network sovereignty, 6G will adopt new trust-model and security technologies.

In an inclusive prosumer-centric model, every system participant can both contribute and consume resources and services. Employing AI and machine learning technologies, 6G networks will be fully autonomous, requiring no manual intervention. In this regard, 6G networks will be tailored rather than proprietary, giving rise to the concept of "my network."

1.3.1 Key Technology Trends

Based on the overall vision of 6G discussed in the previous section, we can conclude the following points.

- 6G will enhance human communication, providing the ultimate immersive experience with true human perception anywhere.

- 6G will enable novel machine communication, redefining intelligent communication for efficient machine-oriented access and connectivity. 6G will fully integrate both machine learning and AI.
- 6G will expand beyond just communication. It will integrate new functions such as sensing and computing, enabling new services and leveraging enhanced knowledge of the environment for machine learning.
- 6G will usher in a new and distinctive wireless generation to support the future Intelligence of Everything. As part of this, AI, trustworthiness, and energy efficiency will be native features of 6G.

In what follows, we discuss six major technology trends.

Trend 1 – New spectrum up to THz and optical wireless communications for extremely high data rates.

To enable new applications such as AR/VR/MR and holographic communications, ultra-high data rates up to tens of Tbit/s are needed. The mmWave sub-THz and bands will be the key spectrum in 6G cellular networks, while the lower-THz band (0.3–1.0 THz) will be a prime candidate for short-distance transmission, e.g., for indoor or for device-to-device (D2D) communication. The THz band offers ultra-wide bandwidth exceeding tens of GHz. The THz will enable a wide range of data-hungry and delay-sensitive applications; in addition to this the THz can be used for wireless sensing.

THz communications is a new wireless technology that involves numerous challenges. Research is currently exploring the design of high-power devices, new materials for antennas, radio frequency power transistors, THz transceiver on-die architecture, channel modeling, and array signal processing. Whether THz technology is successfully adopted in 6G depends on the engineering breakthroughs in THz-related components such as electronic, photonic, and hybrid transceivers and on-die antenna arrays.

Communication through visible light is a potential radiation-free transmission technology that enables connectivity without significant electromagnetic field exposure. However, a large-scale micro-LED array technology will be required to attain data rates reaching tens of Tbit/s for short-distance communications with low power consumption, small form factors, and low-cost devices. In addition, visible light communication (VLC) can access large amounts of unlicensed spectrum, but whether VLC can be successfully utilized in 6G hinges on several challenges in terms of uplink transmission, mobility management, and high-performance transceivers.

Trend 2 – Integrated sensing and communication (ISAC) for new services and enhanced wireless communications.

Traditionally, sensing is a stand-alone function with a set of dedicated devices and equipment, such as radar, lidar, computed tomography (CT), and magnetic resonance imaging (MRI). Mobile phone positioning in mobile systems, assisted by air interface signaling and device-based measurements, is an elementary sensing-like capability. However, by utilizing the mmWave and THz bands, which offer wider bandwidth and smaller wavelength, 6G will make it possible to integrate the sensing function into the communication system, which is especially relevant for the mmWave and terahertz bands. In a full ISAC system, the sensing and communication functions will complement each other to offer the following key benefits.

- **Cellular as a sensor:** Communication signals will be used for new sensing functions, such as high-accuracy localization, gesture and activity recognition, object detection and tracking, imaging, and environmental object reconstruction.
- **Sensing-assisted communication:** Sensing assists and improves quality of service (QoS) and performance for communication, including path selection, channel prediction, and beam alignment.

Integrated sensing and communications make it possible for sensing services in 6G to move beyond simple positioning. Instead, they will be new services that offer additional sensing features with enhanced accuracy (which describes the difference between sensed values and real values in range, angle, velocity, etc.) and sensing resolution (which describes the capability to separate between multiple objects in range, angle, velocity, etc.). Chapter 3 discusses this in greater depth.

Compared with traditional radar sensing, 6G sensing, utilizing the broadband spectrum and larger antenna arrays, will enable technological innovations such as large-scale cooperation between base stations and user devices, joint design of communication and sensing waveforms, advanced techniques for interference cancellation, and sensing-assisted AI. Sensing will potentially be one of the most disruptive services available in 6G, upon which numerous real-time machine learning and AI applications can be created.

A much higher sensing accuracy and resolution can be achieved with THz sensing, due to the ultra-wide bandwidth. Given the range of wavelength and properties of molecular vibration, THz sensing can perform spectrogram analysis to identify the constituent parts of different types of food, medicine, and air pollution. Due to its compact form factor and non-ionizing safety, THz sensing can be integrated into mobile devices and even wearables to identify the number of calories in food and help detect hidden objects. 6G sensing devices will become a gateway for realizing numerous innovative AI applications.

Trend 3 – AI as both a service and a feature in the 6G communication system to intelligently connect intelligent devices.

The key design challenge in 6G is to combine wireless and AI technologies at the beginning rather than designing a wireless system first and then applying AI. Leveraging AI to enhance the 6G wireless system creates opportunities for post-Shannon communications theory research and innovations in wireless technologies.

There are two types of designs involved in 6G: (1) *AI for network* uses AI applications as tools to optimize the network. (2) *Network for AI* tailors a network to support and optimize AI applications. The network also assists in providing AI functionality or even performing such functionality itself; for example, the network can handle inferencing and machine computations. Of course, AI can be applied as a generic tool to optimize and facilitate efficient operations, as is the case for 5G and its evolution.

The AI for network and network for AI concepts are described below.

- **AI for network:** AI technologies, which are inherently data driven, can be integrated with the classic model-driven communication system design to cope with use cases where the model-driven design is complex or cannot achieve high enough accuracy. In AI for network, it is possible to create an intelligent communication link that adapts to a dynamic E2E transmission environment. Furthermore, by fully integrating signal processing and data analysis, we can simplify and unify the computing and inference architecture while also transforming the network from one relying on dynamic processing and response to one where proactive prediction and decision-making are possible. Although this is an area full of promise, it also poses many challenges. For example, system design is hampered by a lack of robust analytical tools and a universal neural network architecture, making it difficult to find the optimal balance among system parameters. Consequently, obtaining a solid theoretical understanding of AI is an important aspect requiring further research. In order to achieve ultimate AI support in 6G networks, it may require a more disruptive approach – one where we revisit the fundamental aspect of how the communication system transmits intelligence. A great deal more theoretical research is needed for a deeper understanding of post-Shannon communications.

- **Network for AI:** The 6G network will develop towards a more distributed architecture with embedded MEC capabilities for local data collection, training, reasoning, and inference together with global training and inference for enhanced privacy protection, lower latency, and reduced bandwidth consumption. One candidate technology that will enable such features is federated learning. During the initial design phase of 6G radio interfaces and network architecture, it is important to factor in distributed AI learning and inference in order to realize efficient large-scale intelligence. A more disruptive approach is to study the theoretical information bottleneck in the context of AI – a new direction of 6G research. This would help us compress the huge amounts of training data sent over the network using minimal resources such as bandwidth and memory.

Trend 4 – 6G native trustworthiness based on a multi-lateral trust model and new cryptographic technologies.

In 6G, mobile devices will be a portal to a cyber world that is a true replica of the physical world. As we come to depend more heavily on 6G and the services it offers, both network and service trustworthiness will be critical. As part of the network architecture design, a robust security system is the basis for establishing trusted relationships between different entities in the network. Security and network services should be jointly designed and dovetailed to meet the service requirements from both individual and business customers. The following discusses two aspects of native trustworthiness.

- **Multi-lateral trust model:** Any security architecture, regardless of whether it is centralized or decentralized, has both advantages and disadvantages. The centralized architecture employs a set of very strong security mechanisms that leverage strict security policies, but in terms of security dialog it involves higher complexity during roaming, handover, or re-login operations performed in the network. Consequently, the probability of being attacked is higher due to the fact that the more interfaces there are, the more vulnerable the system becomes. The decentralized architecture, which features flexible and customizable mechanisms, supports various requirements and can serve nearby services. If subjected to an attack, this architecture will contain it to within a small localized area. In cases where not all stakeholders are trustworthy, the multi-lateral trust model can be used to implement resilience so that other stakeholders remain trusted. Compared with the centralized architecture, the distributed one lacks efficient synchronization of unified security policies. A unified security architecture with multiple security attributes is therefore needed to accommodate the requirements involved in both centralized and decentralized architectures.

A multi-lateral trust model, one that is more inclusive, will serve as the foundation of future security systems. This multi-lateral trust model will help establish a resilient and native trustworthiness architecture that covers the entire 6G lifecycle. It will also flexibly implement centralized security policies, consensus-based distributed mechanisms, and verified third-party reference and verification.

- **New cryptographic algorithms:** As quantum computing continues to develop, challenges arise with regard to classical cryptography, which is based on mathematical problems such as large-prime factorization and discrete algorithms. Key generation and exchange algorithms are two indispensable elements involved in cryptography. In 6G, one-time pad (OTP) encryption can be used for full-duplex bidirectional communications at the physical layer in order to safeguard against quantum-computing-based attacks. One important aspect in terms of cryptography is that 6G requires cryptographic algorithms to complete operations within microseconds in order to ensure ultra-low latency. When quantum computing becomes a reality, quantum communication technologies are expected to be more secure due to quantum entanglement. In addition, lightweight cryptographic algorithms and privacy-compliance-related algorithms are just some of the possibilities that deserve further research in 6G.

Some of the key questions involved in implementing the preceding native trustworthiness mechanisms are as follows. (1) How will new technologies be integrated with traditional security mechanisms? (2) How will decentralized technologies be integrated with wireless network architecture? (3) How will open and transparent data security and privacy protection standards be realized? All these issues will require further study.

Trend 5 – Integrated terrestrial and non-terrestrial networks for full-earth ubiquitous access.

Today, even in developed countries, many rural and remote areas still lack high-speed Internet connections. The situation in developing countries is even worse. In fact, more than 3 billion people around the world are still without Internet access, creating a serious digital divide between the connected and the unconnected [4].

Currently, the main barrier to achieving seamless global coverage is due to economic factors rather than technical ones. To overcome this barrier and provide seamless coverage with high-speed mobile Internet services regardless of geographical constraints, the integration of terrestrial and non-terrestrial networks is expected to be a cost-effective solution.

As the cost to manufacture and launch satellites decreases, huge fleets of small low-earth-orbit (LEO) or VLEO satellites will become a reality. Furthermore, the use of UAVs and HAPSs will mean that the coverage provided by the future mobile system will no longer be only horizontal or two-dimensional. Instead, a three-dimensional hybrid network architecture comprising multiple layers and numerous moving access points will enable communication and navigation services anywhere and at any time. This means a radical shift in terms of cell planning, cell acquisition and handover, and wireless backhaul.

At present, UAVs and HAPSs are designed and operated separately, but in the future 6G networks their functions and operations along with their resource and mobility management are expected to be tightly integrated. Such an integrated system will identify each user device with a unique ID, unify billing processes, and continuously provide high-quality services via optimal access points.

In order to seamlessly integrate a new UAV or LEO satellite without the need for manual configuration, the integrated network requires self-organization. With intelligent air interface design, the addition and deletion of an access point would be transparent to user devices, with respect to the physical-layer procedures (such as beamforming, measurement, and feedback) associated with the access point. Given that the deployment, maintenance, and energy source of satellites differ completely from those of terrestrial networks, it is expected that new operating and business models will emerge.

Trend 6 – Green and sustainable networking for low total cost of ownership (TCO) and sustainable development worldwide.

The increasing number of connected devices, base stations, and network nodes will not only lead to a massive surge of data traffic, but also result in a substantial increase in energy consumption across all parts of the network. Energy efficiency, defined as bits/joule, has long been a focal design target. In the 6G network design, it will become an even more important requirement – it will no longer be just a nice-to-have feature; rather, it will be a make-or-break requirement for 6G mobile networks.

As of today, ICT produces about 2% of the global greenhouse gas emissions (of which, mobile networks represent about 0.2%) [12]. This percentage is expected to increase year-over-year. For 6G, in addition to energy efficiency, it will be important to reduce the energy consumption of networks. This is necessary to not only cut the electricity bill but also reduce greenhouse gas emissions, an important social commitment. At the same time, however, it will also be necessary to consider both capital expenditure (CAPEX) and operational expenditure (OPEX). While the design of cost-effective and energy-efficient networks moves the ICT industry closer to sustainable development, the ICT industry as a whole can play an important role in reducing global CO_2 emissions for a cleaner and healthier living environment. It is expected that ICT can achieve a 20% reduction of global CO_2 emissions by 2030 compared with 2015 levels [13]. Meanwhile, the 6G communication system should support new business use cases and application scenarios to facilitate other industries while also enabling sustainable social development.

The so-called green radio network is a vast research discipline. The potential energy-efficiency technologies span architectures, materials, hardware components, algorithms, software, and protocols. Dense network deployment (leading to shorter propagation distance), centralized RAN architecture (resulting in fewer cell sites and higher resource efficiency), energy-aware protocol design, and cooperation between users and base stations are some factors that need to be carefully considered in order to achieve an energy-efficient 6G communication system. Another challenge is the need for innovative ways to deal with the reduced power amplifier (PA) efficiency as we move towards using higher and higher frequencies. In addition, renewable energy and radio frequency (RF) energy harvesting technologies should also be considered.

On the other hand, as AI capabilities become pervasive across data centers, edge nodes, and even mobile devices, the energy consumption involved in AI learning and training becomes a key issue that must be addressed. It was pointed out in [14] that training a single AI model emits as much carbon as five cars produce throughout their lifetime. Data centers alone consumed more than 2% of the world's electricity in 2018, and this percentage is expected to increase as more AI-enabled edge nodes and devices will emerge by 2030. Some recent research shows that training a once-for-all network that supports diverse architectural settings would significantly reduce CO_2 emissions compared with finding a specialized neural network and training it from scratch for each case by using neural architecture search [15].

1.3.2 Typical Use Cases

As new technologies are increasingly adopted in wireless communications systems, many aspects of our daily lives will be augmented by ultra-high-speed wireless connections, AI, and advanced sensor technologies. Simply put, the way we communicate and interact with technology will change as we know it.

In addition to broadband services, 5G has taken a leap towards low-latency and highly reliable wireless access, thereby enabling a set of vertical and IoT applications. ITU-R identified three types of usage scenarios for 5G applications (eMBB, URLLC, and mMTC) in the IMT-2020 vision document [16]. As intelligence and sensing capabilities are introduced in 6G, and coverage is extended beyond terrestrial, the next-generation networks will create new applications and improve existing applications. Some of these applications may have already been discussed in the 5G vision, even though they may not have been included in 5G deployments due to technological limitations or to an immature market. We'll focus on the use cases that the 5G network cannot support (such as sensing), as well as the use cases that were discussed but not extensively adopted in 5G such as intelligence and enhancements to the three 5G applications (eMBB+, URLLC+, and mMTC+). We identify six categories for the potential use cases of 6G, as shown in Figure 1.6.

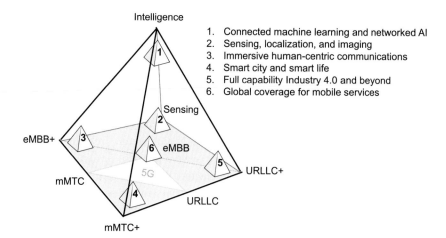

Figure 1.6 Overview of 6G use cases.

- **Immersive human-centric communications:** The pursuit of a better communication experience never stops. In order to provide immersive experience in remote presence and human-centric applications such as AR, VR, and MR as well as holographic communications, we need to continuously push the display resolution towards the human perception limit. As such, an ultra-high data rate of up to Tbit/s is required, and today's 5G network is incapable of achieving this. To avoid motion sickness (dizziness and fatigue) and obtain real-time haptic feedback from teleoperation, extremely low E2E network latency is another key requirement approaching the limits of human senses and perception.
- **Sensing, localization, and imaging:** In addition to communication, the use of higher-frequency bands (THz and mmWave) delivers other capabilities such as sensing, imaging, and localization. As a result, various value-added innovative applications are introduced, such as high-precision navigation, gesture recognition, mapping, and image reconstruction. Compared with communication, sensing, localization, and imaging have different requirements, such as sensing resolution and accuracy for range, angle, or velocity. They also include a new set of performance metrics, such as the probabilities of misdetection and false alarm.
- **Full-capability Industry 4.0 and beyond:** The industry use cases of 5G have been extensively studied [1]. Although 5G has been designed with low latency and high reliability, some of these cases pose extremely high requirements (such as precise motion control) that exceed 5G's capabilities. That said, 6G will enable these use cases through technologies that support ultra-high reliability, extreme low latency, and the deterministic communication capability. Moreover, as new AI-based human–machine interaction methods become viable, future automated manufacturing systems will be centered on collaborative robots, cobots, or even cyborgs. Real-time intelligence interaction between robots and humans requires even lower latency and higher reliability in comparison with 5G.

- **Smart city and life:** A massive number of sensors will be deployed for smart transportation, building, health, car, city, and factory. These sensors will collect a huge amount of data for AI algorithms, which are then used to provide AI as a service. It is also believed that we will live in a world where physical reality is accompanied by a digital twin in the cyber world. Phrased differently, automation and intelligence will be created within the cyber world and delivered to the physical world through 6G wireless networks. To achieve this, it is imperative that we develop extensive sensing capabilities for retrieving the big data used to train deep neural networks (DNNs). This, in turn, demands a very-high-throughput wireless link for collecting sensing data in real time. As such, a massive number of reliable and secure connections are required for the 6G-enabled smart urban city and life use cases.

- **Global coverage for mobile services:** In order to provide seamless mobile services everywhere on earth, 6G aims for integrated terrestrial and non-terrestrial communications. In such an integrated system, a mobile user with a single device can access mobile broadband services in both urban and rural areas, or even on moving planes and ships. In these scenarios, the best links for terrestrial and non-terrestrial networks are dynamically optimized without interrupting ongoing services. Self-driving enthusiasts will also benefit from the integrated system with seamless high-precision navigation in all types of geographical areas. Some of the other potential use cases involve a wide range of IoT connections for real-time environmental protection and precise agriculture.

- **Connected machine learning and networked AI:** Full-scenario use of AI capabilities is a fundamental use case for 6G. Basically, on the one hand, AI capabilities can be augmented and integrated into most of the functions, features, and capabilities of 6G. On the other hand, almost all 6G applications will be AI-based, and AI can also be applied to all the preceding use cases to achieve different levels of automation. Put differently, there are challenges and limitations involved with providing AI as an OTT service. The first challenge is the fact that machine learning requires the transfer of huge amounts of data to data centers, especially for customized AI services. The second challenge is directly related to the local data governance requirement. More specifically, the transfer of data to overseas data centers is not permitted. The third challenge involves the interaction of different AI agents (even if separated remotely) through the 6G network. As one of the most important use cases, distributed machine learning agents will be fully connected through the 6G network to achieve networked intelligence while also allowing for better data privacy protection. In light of this, connected machine learning and networked AI will essentially involve the following aspects: the 6G design for maximizing the machine learning capability; network architecture that supports distributed and AI-at-network-edge capabilities for real-time AI services; high-capacity, low-latency, and highly reliable AI inference and action. Furthermore, AI will serve as a native feature of 6G, thereby facilitating the design of future transmission schemes, intelligent control and resource management, as well as "zero-touch" network operations.

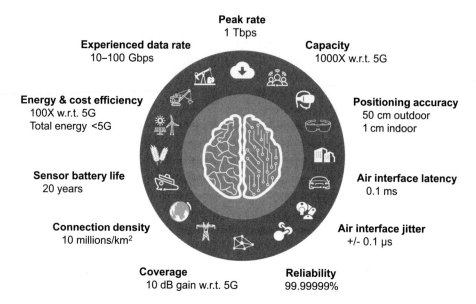

Figure 1.7 Summary of KPIs for radio access networks.

1.3.3   Target KPIs

In order to achieve ultimate user experience in all use cases, 6G must realize significant improvements in terms of key capabilities. Based on the upgrade trends of earlier mobile generations, the improvements in 6G are expected to increase by 10 to 100 times compared with 5G. Our preliminary expectations are shown in Figure 1.7 and elaborated as follows.

Very high data rate and spectrum efficiency.

To realize human-centric communication with immersive experience will pose very high demand on the bandwidth. The transmission of 300 degree AR/VR and holographic information may require a data rate from several Gbit/s to a few Tbit/s, depending on factors such as the resolution, size, and refresh rate of images.

In IMT-2020 (5G), the ITU-R minimum requirements for the peak and user-experienced data rates are 10–20 Gbit/s and 100 Mbit/s, respectively [17]. In 6G, the peak and user-experienced data rates should be 1 Tbit/s and 10–100 Gbit/s, respectively. In addition, 6G is expected to further utilize spectrum, improving the peak spectrum efficiency by 5 to 10 times compared with 5G.

Very high capacity and ultra-massive connectivity.

The area traffic capacity is the total traffic throughput per geographic area. It is the product of the area's connection density (the total number of devices per unit area) and the average data rate provided for users. In 5G, the ITU-R minimum requirement for connection density is 1,000,000 devices per km^2 [17]. In 6G, where we must support use cases such as Industry 4.0 with connected intelligence and smart city in the next 10 years and beyond, the connection density should be increased by about 10 to 100 times, resulting in up to 100 million/km^2. Such a massive number of connections should be able to accommodate diversified types of services with different characteristics, (e.g., different throughput, latency, and QoS). In this regard, the 6G system capacity should be 1000 times larger than 5G in order to provide high-quality services for a large number of connections.

Very low latency and jitter, and ultra-high reliability.

In some IoT use cases such as autonomous driving and industry automation, it is critical that data are delivered in time (with low latency and jitter). 6G air interfaces will achieve a latency of 0.1 ms and a jitter of +/−0.1 s. Taking into account the requirements of remote XR presenting services, the total E2E round-trip latency should be 1–10 ms. In addition to low latency, IoT applications also require reliability (i.e., the correct transmission of information). ITU-R requires that reliability in 5G be 99.999% for URLLC services. In 6G, where various vertical applications are expected to be more prevalent, a ten-fold increase in reliability is needed, reaching 99.99999%.

Very high localization, sensing accuracy, and resolution.

Sensing, localization, and imaging are new functions in 6G, marking a significant step forward for connected intelligence, as will be discussed in Chapter 9. With the increased resolution of the frequency range – up to THz – and advanced sensing technologies, 6G is expected to provide an ultra-high localization accuracy of 50 cm for outdoor scenarios and 1 cm accuracy for indoor scenarios. For other sensing services, as will be described in Chapter 5, the extreme sensing accuracy and resolution can be 1 mm and 1 cm, respectively.

Very wide coverage and very high mobility.

To provide high-quality mobile Internet services with wider coverage, the link budget for 6G air interfaces should be increased by at least 10 dB compared with 5G. This applies to MBB services with a guaranteed data rate in addition to NB-IoT services. The 6G coverage should not be defined only by link budget. By integrating both terrestrial and non-terrestrial networks, 6G should achieve 100% coverage of the earth's surface and population, connecting unconnected areas and people.

In addition, 6G will support coverage for aircrafts traveling at speeds around 1000 km/hr, much higher than 5G (500 km/hr, mainly for high-speed trains).

Very high energy and cost efficiency.

Energy consumption is a challenging aspect in 6G systems. On the one hand, this is due to transmission at very high frequency bands, very large bandwidth, and very large numbers of antennas. The lower PA efficiency and increased number of RF chains are the two key aspects to address. On the other hand, with the convergence of communication and computing, and the support of native AI, AI training and inference in 6G networks will consume more energy. This means that the energy consumption per bit in 6G should be at least 100 times lower than that in 5G in order to achieve comparable levels of total energy consumption. From the perspective of devices, with increased data rates, the energy efficiency for signal processing should be increased accordingly. Furthermore, sensing devices should support a battery life of up to 20 years for use in smart city, smart building, smart home, and smart health scenarios.

Native AI.

As discussed in Section 1.3.1, the native AI support in 6G mobile communications systems includes two aspects: AI for network and network for AI.

- **AI for network** enables an intelligent framework for designing air interfaces and network functions, supporting E2E dynamic transmission, truly zero-touch network operation, and automatic creation of specialized network slices for diverse services and diverse enterprises.
- **Network for AI** demands a more distributed architecture with embedded mobile edge computing capabilities to combine local data collection, training, and reasoning/inference with global training and inference for better privacy protection and lower latency/bandwidth consumption.

Native trustworthiness.

6G will strengthen the connection between the physical and digital worlds, becoming an integral part of our life. Trustworthiness is a fundamental aspect for any network service and covers topics including security, privacy, resilience, safety, and reliability [18].

- **Security** is a condition that results from the establishment and maintenance of protective measures that enable an organization to perform its mission or critical functions despite risks posed by threats to its use of systems, as defined in [19]. Protective measures may involve a combination of deterrence, avoidance, prevention, detection, recovery, and correction that should form part of the organization's risk management approach.
- **Privacy** is freedom from intrusion into the private life or affairs of an individual when that intrusion results from undue or illegal gathering and use of data about that individual, as defined in [20].
- **Resilience** is the ability to quickly adapt and recover from any known or unknown changes to the environment through holistic implementation of risk management, contingency, and continuity planning, as defined in [21].
- **Safety** is the freedom from conditions that can cause death, injury, occupational illness, damage to or loss of equipment or property, or damage to the environment, as defined in [22].
- **Reliability** is the ability of a system or component to function under stated conditions for a specified period of time, as defined in [23].

1.4   Structure of the Book

This book is composed of seven parts that cover specific topics about 6G. Part I (Chapter 1) describes the evolution of mobile communications from 2G to 6G and the overall vision of 6G, identifying six technology trends that are central to the three fundamental drivers for the future of connected intelligence.

In Part II, we discuss potential use cases for 6G and analyze the key performance requirements. The use cases range from evolutionary 5G ones that will gain popularity and mature in 6G due to its larger capacity, lower latency, and higher reliability, to brand new ones made possible by the new functions and capabilities that 6G offers. Part II organizes the typical use cases into six categories, describing each one in separate chapters. They cover human-centric communications with extremely immersive experience (Chapter 2); highly accurate sensing, localization, and imaging plus enhanced human sensing (Chapter 3); full-capability Industry 4.0 with connected intelligence (Chapter 4); smart city and smart life (Chapter 5); global 3D coverage for mobile services with integrated terrestrial and non-terrestrial communications (Chapter 6); and native AI support in all use cases (Chapter 7).

Part III explores the scope and boundaries of 6G design, discussing the theoretical foundations for 6G radio technologies and network technologies before examining a range of enabling technologies with the potential to achieve key performance indicators (KPIs).The introduced content covers theoretical foundations for native AI and machine learning (Chapter 8), theoretical foundations for massive capacity and connectivity (Chapter 9), theoretical foundations for future machine type communications (Chapter 10), as well as theoretical foundations for energy-efficient systems (Chapter 11).

In Part IV, we analyze the future International Mobile Telecommunications (IMT) spectrum from the perspective of communication and sensing (Chapter 12), and the corresponding channel modeling methodologies and some example channel measurements (Chapter 13). Then, to provide a broad and comprehensive understanding of how 6G will mature over the next 10 years, we describe potential new materials for hardware production (Chapter 14), new antenna structures for ultra-massive multiple-input multiple-output (MIMO) systems (Chapter 15), new radio frequency components to support the use of the THz band (Chapter 16), computing evolution following the end of Moore's law (Chapter 17), and new demands and features for terminal devices (Chapter 18).

Part V focuses on the overall design principles and potential enabling technologies for 6G air interfaces. We describe the paradigm shifts in designing air interfaces compared with 5G and earlier generations in the introduction to the Part, and then discuss a range of enabling technologies. These potential technologies are intelligent air interface (Chapter 19), integrated non-terrestrial and terrestrial communication (Chapter 20), integrated sensing and communication (Chapter 21), new waveform and modulation (Chapter 22), new coding (Chapter 23), new multiple access (Chapter 24), ultra-massive MIMO (Chapter 25), and super short-range communication (Chapter 26). For each technology, we describe the background and motivation for its use, examine the existing solutions, clarify the new design expectations, and highlight the potential research problems and directions for future study.

Similarly, Part VI focuses on the design principles and potential enabling technologies for 6G network architecture design. It also starts with the paradigm shifts in designing the network architecture in the introduction to the part. Following that, it delves into several of the new major features and technologies involved in 6G network architectures. They are architecture technologies for network AI (Chapter 27), user-centric network (Chapter 28), native trustworthiness (Chapter 29), data governance (Chapter 30), multi-player ecosystems (Chapter 31), and integrated non-terrestrial networking (Chapter 32).

Part VII (Chapter 33) concludes the book by describing the current status of the 6G ecosystem globally, covering research projects, platforms, workshops, and papers on 6G, and then predicts the potential roadmap towards 2030.

References

[1] 5G-ACIA. [Online]. Available: https://www.5g-acia.org/
[2] 5GAA. [Online]. Available: https://5gaa.org/
[3] ITU-R, "IMT traffic estimates for the years 2020 to 2030," Report ITU-R M.2370-0, July 2015.
[4] "The mobile economy 2020," Intelligence, GSMA, 2020.
[5] "Touching an intelligent world," Sept. 2019. [Online]. Available: https://www.huawei.com/minisite/giv/Files/whitepaper_en_2019.pdf
[6] "Terahertz spectroscopic system TAS7400 product specification." [Online]. Available: https://www.advantest.com/documents/11348/146157/spec_TAS7400_EN.pdf
[7] "2019 ICT Sustainable development goals benchmark," 2019. [Online]. Available: https://www.huawei.com/minisite/giv/Files/whitepaper_en_2019.pdf
[8] United Nations Development Programme, "Goal 11: Sustainable cities and communities." [Online]. Available: https://www.undp.org/content/undp/en/home/sustainable-development-goals/goal-11-sustainable-cities-and-communities.html
[9] "ICTs for a sustainable world #ICT4SDG," ITU, 2019. [Online]. Available: https://www.itu.int/en/sustainable-world/Pages/default.aspx
[10] V. Bush, *Science, the endless frontier.* Ayer Company Publishers, 1995.
[11] N. C. Thompson, K. Greenewald, K. Lee, and G. F. Manso, "The computational limits of deep learning," *arXiv preprint arXiv:2007.05558*, 2020.
[12] "ICT sector helping to tackle climate change," Dec. 2016. [Online]. Available: https://unfccc.int/news/ict-sector-helping-to-tackle-climate-change
[13] "SMARTer2030: ICT solutions for 21st century challenges," The Global eSustainability Initiative (GeSI), Brussels, Brussels-Capital Region, Belgium, Technical Report, 2015.
[14] E. Strubell, A. Ganesh, and A. McCallum, "Energy and policy considerations for deep learning in NLP," *arXiv preprint arXiv:1906.02243*, 2019.
[15] H. Cai, C. Gan, T. Wang, Z. Zhang, and S. Han, "Once-for-all: Train one network and specialize it for efficient deployment," *arXiv preprint arXiv:1908.09791*, 2019.
[16] ITU-R, "IMT Vision – framework and overall objectives of the future development of IMT for 2020 and beyond," Recommendation ITU-R M.2083-0, Sept. 2015.
[17] ITU-R, "Minimum requirements related to technical performance for IMT-2020 radio interface(s)," Report ITU-R M.2410-07, Nov. 2017.
[18] E. R. Griffor, C. Greer, D. A. Wollman, and M. J. Burns, "Framework for cyber-physical systems: Volume 1, overview, Version 1.0," NIST Special Publication, 2017.

[19] C. Dukes, "Committee on national security systems (CNSS) glossary," CNSSI, Fort Meade, MD, USA, Technical Report, vol. 4009, 2015.

[20] S. L. Garfinkel, "De-identification of personal information," National Institute of Standards and Technology (US Department of Commerce), 2015.

[21] M. Swanson, *Contingency planning guide for federal information systems.* DIANE Publishing, vol. 800, 2011.

[22] "System safety," Department of Defense Standard Practice, May 11, 2012. [Online]. Available: https://e-hazard.com/wp-content/uploads/2020/08/department-of-defense-standard-practice-system-safety.pdf

[23] IEEE Standards Coordinating Committee, "IEEE standard glossary of software engineering terminology (IEEE Std 610.12-1990). Los Alamitos, CA," CA: IEEE Computer Society, vol. 169, 1990.

Part II

Use Cases and Target KPIs

Introduction to Part II

Following the overview of use cases in Chapter 1, Part II discusses the typical use cases of 6G in greater depth, covering various aspects of how they affect our life and work in the future. By 2030, because services take at least two generations to mature, some services defined in 5G will reach maturity in 6G. Furthermore, some new services will exceed the capabilities of 5G. In the following chapters, we introduce six categories of use cases together with the key requirements of radio transmission and network architectures. The first category includes use cases that require extremely immersive human-centric experience. XR video, haptic and multi-sensory information, and 3D holographic images will be used to provide users with an immersive experience that transcends distance limitations in the physical world. The second category is sensing, localization, and imaging. These new capabilities will bring enhancements to many fields and even to new regions that were never covered by previous mobile communication systems. The next two categories are industry automation, and smart city and smart life. Highly reliable and ultra-low-latency wireless communication networks, together with advancements in the IoT and ICT domains, will help implement a much greater degree of automation in future factories. Various types of robots will be used in both industrial and daily life scenarios, while cities and the environment will become smart. The fifth category is global coverage for mobile services. By integrating terrestrial and non-terrestrial networks, 6G will provide coverage to every corner of the earth. In addition to wireless broadband and IoT communication services, this three-dimensional full coverage brings new momentum for use cases such as navigation and earth observation. The last category is connected machine learning and networked AI. In this category, machine learning and AI are shown to be the key enablers for better 6G network performance and operations. At the same time, the ultimate performance and the converged sensing, communication, and computing capabilities of 6G become the key enabler for the proliferation of native AI services with connected intelligence.

2 Extremely Immersive Human-Centric Experience

From 1G to 4G, as shown in Figure 2.1, wireless networks were mainly used to meet our fundamental needs for remote and instant information sharing, such as in voice calls, text messages, and video streaming or clips. The large-scale deployment of 5G is unleashing the potentials of XR applications, including VR, AR, and MR. However, in order to achieve satisfactory results in terms of true immersion into remote scenarios, high-definition audiovisual senses and the haptic perception of people and objects in target environments must be achieved. We believe that the 6G wireless network will support an extremely immersive user experience, which will evolve in the following three key directions:

- 360° full-view extremely immersive XR with a very high resolution and video frame rate close to the human perception limit
- Interactive haptic and multi-sensory communications for teleoperation with new human–machine interfaces
- Glass-free 3D and holographic display integrated with XR

In the following section, we will delve into more details regarding the potential use cases and their requirements.

2.1   Ultimate Immersive Cloud VR

Example 2.1 Kate loves playing football. However, she doesn't have much time to play as she often has to go on business trips. But now, with 360° VR devices, she can play football virtually with her friends anywhere and at any time. With the cloud VR techniques, which provide an optimal immersive visual experience and extremely low interactive latency, Kate feels like she is actually playing football in the stadium, and can do so for extended periods without feeling dizzy, as shown in Figure 2.2.

In addition to playing virtual football games, Kate likes watching live football matches with the 360° VR devices. They enable her to view the matches from the referee's perspective, which makes her feel like a participant on the field and surpasses the experience of being a spectator in a stadium.

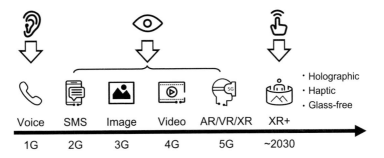

Figure 2.1 Evolution of human-centric applications in wireless communications.

Figure 2.2 Interactive gaming experience with 360° VR.

3GPP has already started the XR study of the current 5G network [1], in which numerous use cases have been identified such as immersive online gaming, 3D messaging, immersive six degrees of freedom (DoF) streaming, real-time 3D communication, remote assistance in industry, and online shopping. Estimates indicate that an increasing number of people will subscribe as more advanced XR services with extremely immersive experience emerge over the next decade. In the following section, we will discuss the requirements that extreme experience poses on mobile transmission throughput and latency.

2.1.1 Transmission Latency Requirements

Requirements on transmission latency mainly arise from the need to avoid motion sickness, which is caused by disparity between what our vestibular senses expect to experience after movement, and the actual images we see. This disparity is usually caused by the latency between the user interacting with a cloud VR system and the system responding to that user interaction. 3GPP refers to this type of interactive latency as motion-to-photon (MTP) latency [2].

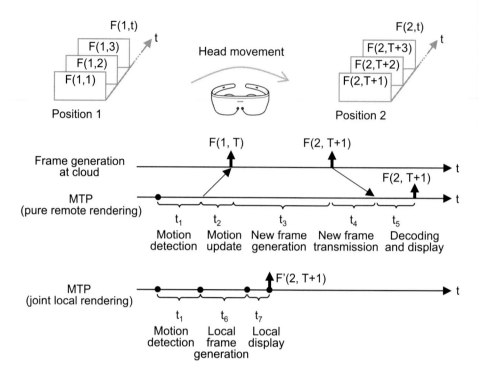

Figure 2.3 Illustration of MTP with and without local rendering.

To achieve optimal immersive experience, the target MTP latency should be close to the human perception limit, which is reported to be approximately 10 milliseconds (ms) [3, 4]. This implies that if the update of the correct visual image is within 10 ms, most people would have a comfortable and smooth experience.

However, there are numerous signal processing and transmission steps between motion and photon (the reception of an image on a screen), as illustrated in Figure 2.3. The MTP latency would be different depending on whether local processing occurs on the VR device side to generate predicted frames before receiving updated frames from the cloud.

XR System with Pure Remote Rendering

For a cloud XR system with pure cloud-based rendering, the following steps are needed after the motion is performed and before the updated image is displayed.

- **Motion detection:** Captures the movements, e.g., change in head posture, which is done on the device side, takes a time t_1, and is related to the sensitivity of the device sensors and detection algorithms.
- **Motion update:** Sends signals from the device to the cloud for motion update, which takes t_2 and is related to the network latency.
- **New frame generation at cloud:** Generates (including rendering and encoding) new video frames upon receiving the updated motion information, which takes t_3

and is related to the processing efficiency at the cloud and video frame rate (frames per second, FPS) of the source.

- **New frame transmission:** Transmits the updated video frame from the cloud to the device, which takes t_4 and is related to the network latency.
- **Decoding and display:** Decodes and displays the updated video frame on the device, which takes t_5 and is related to the hardware efficiency.

Therefore, the MTP latency with pure remote rendering can be obtained from the following formula:

$$\text{MTP}_{remote} = t_1 + t_2 + t_3 + t_4 + t_5 = (t_1 + t_5) + t_3 + (t_2 + t_4)$$

which can be categorized into three parts:

- $t_1 + t_5$: device processing latency.
- t_3: remote processing latency.
- $t_2 + t_4$: round-trip time (RTT), i.e., network transmission latency, including RTT in the radio access and core networks.

With a moderate video frame rate at the cloud, say 60 FPS, only t_3 is at the $1/60 = 16.7$ ms level, which is already larger than the 10 ms target MTP. An increase in the video frame rate, say from 60 FPS to 120 FPS, would help reduce MTP_{remote} by 8.3 ms, though it would be at the cost of an increased throughput requirement. In the current 3GPP standard [2], the target MTP latency has been relaxed to 20 ms, leaving more time for processing and transmission. Furthermore, owing to the higher processing capability of today's XR devices, joint local rendering in the cloud XR system on the device side is a popular technique used to exclude the RTT from the MTP latency calculation.

As devices diversify in the future, dedicated market space might still exist for pure remote rendering-based VR, at least for very light wearable devices that have particularly limited processing capabilities. Given this, to achieve the ultimate goal of 10 ms MTP latency, all the three parts in MTP_{remote} need to be significantly reduced. Specifically:

- Device processing latency ($t_1 + t_5$) can be reduced by simply following Moore's law. New materials and computing architectures may help accelerate the process, which will be discussed in Part IV of the book.
- Remote processing latency (t_3) can be reduced by reducing the compression ratio, taking advantage of the greatly increased data rate predicted for 6G.
- RTT ($t_2 + t_4$) can be reduced through distributed edge computing architectures and extremely low-latency air interfaces in 6G.

Cloud XR System with Joint Local Rendering

For a cloud XR system with joint local rendering and frame generation, the steps between motion and photon have been simplified considerably as follows:

Table 2.1 Throughput and latency requirements for ultimate 360° VR experience.

Parameter	Current VR	Ultimate VR1	Ultimate VR2
Video frame resolution	4K	24K	48K
FPS	60	120	120
Color depth (bits per pixel)	24	36	36
Raw rate	10.62 Gbit/s	1146.62 Gbit/s	2293.24 Gbit/s
Compression ratio	100 : 1	100 : 1	20 : 1
Throughput after compression	0.1 Gbit/s	11.5 Gbit/s	114.66 Gbit/s
MTP latency	20 ms [2]	10 ms [2]	10 ms
RTT latency	< 20 ms [8]	< 8 ms [8]	< 2 ms

Current/Ultimate VR1: Local rendering, heavy compression
Ultimate VR2: Remote rendering, light compression

- **Motion detection:** Same as in the preceding case and takes t_1.
- **Local frame generation:** Locally generates the frame to be updated based on the local rendering algorithms and takes t_6.
- **Local display:** Displays the locally updated frame without decoding signals as in the preceding case and takes a time t_7 less than t_5.

Therefore, the MTP latency with joint local rendering can be calculated through the following formula:

$$MTP_{local} = t_1 + t_6 + t_7$$

Needless to say, the quality of experience depends heavily on the quality of local rendering, and this is not viable in the long term. As such, frames should still be updated at the cloud in parallel. But the round-trip latency, remote rendering time, and compression and decompression latency can be relaxed significantly. For instance, $t_2 + t_4$ is not included in the MTP_{local} calculation for optimal XR experience with joint local rendering; therefore, the RTT requirement can be relaxed from less than 2 ms to approximately 8 ms, as shown in Table 2.1.

2.1.2   Throughput Requirements

Resolution, color depth, and video frame rate are the three key factors that determine the throughput requirement of a cloud VR system. We can calculate the overall throughput by simply multiplying the three values as follows:

$$\text{Throughput} = \text{Video frame resolution} \times \text{Color depth} \times \text{Video frame rate}$$

To achieve optimal VR experience, the three dimensions all need to approach the human perception limit. Specifically they are as follows:

- **Video frame resolution:** The resolution is the pixel count for digital image display. The resolution requirement is mainly driven by the pixel density requirement in a unit area. The latter requirement is usually measured in pixels per angular degree (PPD). Early research [5] claimed that the human eye has limitations on the minimum degree of resolution at the 0.3 arcminute level, where 1 arcminute is 1/60 of a degree. Therefore, the human perception limit will be $1/[(1/60) \times 0.3] = 200$ PPD. On the other hand, as reported in [2], the common horizontal field of view (FOV) for the ordinary human eye is approximately 120 degrees, so the maximum resolution for our visual perception is approximately 120 degrees $\times$ 200 PPD = 24K pixels. This implies that images with a resolution higher than 24K would be a waste for most cases. Note that for a VR system with pure remote processing, the number of frames or the size of transmitted video frames needs to be doubled compared with that of the joint local processing. This is so because the video frames are rendered and transmitted for binocular display (two eyes) directly, which doubles the size of the video frames for monocular display (one eye). In the VR system with joint local rendering, the transmitted video frames are intended only for monocular display, and binocular images can be rendered locally.

- **Color depth:** This is defined as the number of bits used to indicate the color of a single pixel. Eight-bit color depth can provide $2^8 = 256$ different colors, while 24-bit color depth (eight bits per color component in the RGB color framework) can provide approximately 16 million color variations. The latter depth is referred to as true color and is widely applied in today's VR systems. Even though in some cases it seems as though true color already covers more details than the human eye can perceive (as was pointed out in [6] that people can only distinguish up to 10 million colors), there is a risk of producing abrupt changes between shades of the same color, known as the effect of color banding [7]. We can mitigate this issue by further increasing the color depth, e.g., by using 36-bit color depth with 12 bits per color component.

- **Video frame rate:** This is the rate at which video frames are generated, counted in FPS. If there is no interactive motion during a VR experience (e.g., watching a movie), 120 FPS (which equals 8.3 ms between two frames) is sufficiently high for immersive experience since it approaches the human perception limit (approximately 10 ms). However, as discussed in Section 2.1.1, a higher video frame rate can help reduce MTP in interactive scenarios, at the cost of higher throughput. For instance, a video frame of 4K resolution, 60 FPS, and 24-bit color depth would result in a raw data rate of 10.62 Gbit/s. This number increases to 1.15 Tbit/s if the parameters are all set to values approaching human limits, e.g., 24K resolution, 120 FPS, and 36-bit color depth.

2.1.3 Summary of Main Requirements for Ultimate VR

On the basis of the analysis provided in the previous sections, the throughput and latency requirements for the ultimate immersive VR experience are summarized in Table 2.1, in comparison with the current VR system. As we can see, increases in the resolution, video frame rate, and color depth correlate to an over 100-fold increase in the raw data rate, thereby exceeding 1 Tbit/s. To reduce the burden on network transmission, a high compression ratio (e.g., $133:1$ as used in [8]) is usually applied when local rendering is feasible. Subsequently, a data transmission rate above 10 Gbit/s needs to be guaranteed. Also, when the MTP target is set to the 10 ms limit, the requirement on the RTT should be reduced proportionally. For example, a value less than 8 ms is required in the future VR systems with joint local rendering [8].

On the other hand, for devices with strict power and weight constraints, the pure remote rendering architecture is more suitable. In this case, to achieve the goal of $MTP_{remote} = 10$ ms, if the compression ratio is too high, there will be no time left for transmission. The light compression ratio (e.g., 20:1 in Table 2.1) is therefore assumed to reduce $t_3 + t_5$, but the RTT ($t_2 + t_4$) still needs to meet a very stringent bound close to 1 ms (less than 2 ms is estimated in Table 2.1). Meanwhile, owing to the low compression ratio, the guaranteed data rate in this case is above 100 Gbit/s.

2.2   Haptic and Multi-Sensory Communication

Example 2.2 Kate purchases haptic clothing to wear while playing virtual football. This clothing provides multi-sensory communication with the game server to make the game more realistic, letting Kate feel the texture, weight, and pressure of the virtual ball when she touches, holds, and kicks it.

Kate is away on a business trip. Although she cannot meet her husband and children in person during the trip, the 6G network lets her meet her family virtually and even hug them via haptic clothing – almost as if she were there with them.

In addition to audio and video, haptics is a new dimension of human-centric content that will be transmitted over mobile networks. Haptic communication involves real-time haptic information, regarding surface texture, touch, actuation, motion, vibration, and force, all of which is transmitted over the network along with audiovisual information. These capabilities will allow people to remotely control machines or robots, or even project a holographic image of themselves via mobile networks, in order to perform complex, real-time tasks that would otherwise be too dangerous or expensive to perform in person, or simply to hug a family member living in another city.

In some applications that offer immersive virtual experience, multi-sensory feedback, such as video, audio, and haptics, might occur simultaneously. For example, in a 360° VR gaming application, gamers might see the environment, hear spoken dialogs, and feel sensations simultaneously. In order to deliver the ultimate human-centric experience – one that smoothly connects the physical and virtual worlds – multi-sensory communication will therefore be essential.

The human brain is a complex and powerful system that is adept at building a mental model of the physical world based on multiple sources of sensory information, as shown in Figure 2.4. The overall round-trip latency can be divided into two components: the transmission time $T_{transmission}$, and the processing time $T_{processing}$ (which includes transmission through the nervous system and processing by the brain). Because $T_{processing}$ is normally the same for both local and remote operations and predictions can be used to compensate for environment changes, the target for haptic

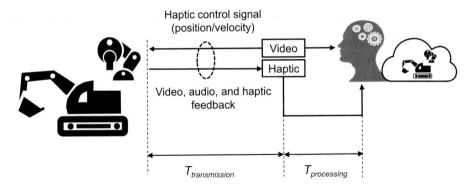

Figure 2.4 Haptic communication with multi-sensory feedback.

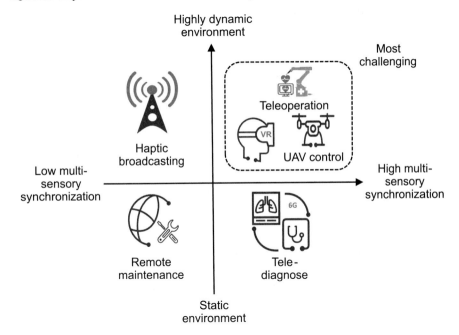

Figure 2.5 Typical applications with haptic communications.

and multi-sensory communication is to reduce $T_{transmission}$ in order to achieve a local operation experience.

2.2.1   Teleoperation in Highly Dynamic Environments

As shown in Figure 2.5, applications such as interactive VR with haptic feedback and remote control of UAVs – in addition to teleoperation – involve frequent interactions with the target objects and require synchronized multi-sensory feedback, representing a highly dynamic environment. These haptic applications are the most challenging, because in order to provide smooth and reliable connections for them, today's wireless networks need to be optimized or even redesigned to account for the behavior of human brains.

The haptic feedback in teleoperation use cases, such as tele-surgery, tele-diagnosis, and tele-motion-control, is extremely important for remote control. This is so because haptic feedback, such as pressure and texture information, can stimulate the human brain and help users adjust their operation time, stress, gestures, and so on.

For haptic control and interactive haptic applications in highly dynamic environments, very low latency is usually required. Many operations performed remotely, such as tele-surgery and industrial-level control, require extremely precise motion control. In such cases, movement accuracy that is not controlled within a 1 mm tolerance may cause the operation to fail, severely lowering the work efficiency or even posing a risk to human life. Assuming that we are moving at a speed of 1 m/s relative to an object and want to control operations within 5 mm tolerance, we would require an overall latency of less than 5 ms. Although we can predict and compensate for transmission latency to some extent, the maximum overall latency should still not exceed 10 ms so that we can effectively handle changes in highly dynamic and interactive environments [9]. It is worth pointing out that this suggested latency is the overall latency, including the motion detection, transmission RTT through the network, and interactive feedback time. In some cases such as interactive teleoperation, the RTT requirement for air interface transmission may be less than 1 ms or even 0.1 ms.

Multi-sensory feedback, including haptics, is important in teleoperation use cases. For example, in order to control a remote machine accurately and quickly, an operator needs to perceive audiovisual and haptic information simultaneously. This means that the relative transmission latency between audio, video, and haptic information must be within the natural perception thresholds that humans expect. However, the perception thresholds for different senses are not the same. Furthermore, it is not fully understood how the relative perception thresholds – defined as the relative latency between different senses – affect the simultaneous reception of multi-sensory information. This is still an open research topic, but an early study in [10] drew the following conclusions:

- Individual differences in perception thresholds appear to be significant.
- The relationship between the average audio, visual, and haptic information perception thresholds is haptic < audio ≈ visual, indicating that haptic information is the most latency-sensitive threshold among the three.

2.2.2   Main Requirements for Highly Dynamic Teleoperation

The requirements for teleoperation in a highly dynamic environment, the main haptic communication scenario for 6G, are as follows and summarized in Table 2.2.

Table 2.2 Summary of teleoperation requirements in a highly dynamic environment.

Parameter	Haptic information	Other sensory information
Overall latency	≤ 1–10 ms	≤ 10–20 ms
Throughput	1000–4000 packets/s	≤ 100 Mbit/s
Reliability	≥ 99.999%	≥ 99.999%

- **Latency:** As discussed in the previous section, the overall latency requirement for highly dynamic teleoperation is between 1 to 10 ms. To achieve this while ensuring stable remote control, we require an extreme high packet transmission rate. Sending 1000 packets per second corresponds to a delay of 1 ms between two successive packets, i.e., in order to satisfy the 1 ms overall latency requirement, the transmission rate needs to exceed 1000 (ideally up to 4000) packets per second [9].
- **Relative latency:** The synchronization between video, audio, and haptic feedback is important. Some experiments [10] show that a relative latency below 10–20 ms between any two types of sensory information is needed to ensure a good user experience.
- **Throughput:** For haptic information, the throughput is related to the packet transmission rate and packet size. In most cases, haptic information comprises several bytes for each DoF. This means that the packet size is linearly related to the number of DoFs (the number of touch spots). For example, the size of a packet containing six DoFs is normally 12–48 bytes. Assuming that the transmission rate is 1000 to 4000 packets per second, the haptic throughput will be from 96 kbit/s to 1.5 Mbit/s. If the number of DoFs is increased to 100, the packet size will increase to 200 to 800 bytes, meaning that the haptic throughput will be 1.6 to 25.6 Mbit/s. For other sensory information, the throughput depends on the audio and video quality. For instance, around 100 Mbit/s is needed to transmit a video in 4K, while a much higher throughput is needed for the ultimate VR use case discussed in Section 2.1.
- **Reliability:** Transmission failure may affect the quality of experience in teleoperation or even pose a risk to human life, meaning that a transmission reliability of 99.999% is required in some highly reliable teleoperation scenarios [9].

2.3 Glass-Free 3D and Holographic Displays

Example 2.3 Jack visits a client in a mountainous city. The downtown area of the city has many high-rise buildings and overpasses. The road network is like a maze, and people may easily get lost there if they are not familiar with the area. However, it is not the case for Jack this time. Jack installed a 3D navigation application on his smartphone prior to his journey. The application uses 3D display techniques to show all the buildings and roads virtually, in real time. With the help of these 3D display techniques, Jack can easily navigate the complex road network and arrive at his destination on time.

2.3.1 Background of Glass-Free 3D Displays

Although VR delivers a similar visual experience to the real world, it cannot simulate the way humans' eyes perceive depth (i.e., when eyes focus on near objects, distant objects become blurred, and vice versa). In VR, users always focus on the VR screen, regardless of whether the object is near or distant. This creates a conflict of depth perception, causing users to experience dizziness or other unwanted effects. To eliminate such effects, glass-free 3D displays based on visual accommodation rather than psychological perception or motion parallax are expected to be the next generation of immersive XR.

The concept of glass-free 3D displays has been around for a long time. Various types have been developed using light field and holographic display techniques [11, 12]. Today, demonstrations of such displays mainly focus on local scenarios; for example, pre-recorded holographic concerts [13]. These scenarios do not currently require mobile communication; therefore, they are not closely related to mobile networks. In the future, however, as new applications such as mobile 3D navigation emerge, it is expected that 6G networks will need to transmit 3D images or holograms. This poses extremely high requirements on the performance of 6G networks, including factors such as bandwidth and latency. In this section, we focus mainly on the light field and holographic display techniques used in glass-free 3D displays [12].

2.3.2   Glass-Free 3D Image Reconstruction Techniques

- **Light field displays** collect light field information from the source, typically using camera arrays, and transmit the captured information to a device screen, where 3D images are reconstructed. The reconstruction process relies on the projection of separate 2D images with continuous parallax information on multiple viewpoints. This process is usually based on techniques that can separate light rays from the screen on an angular basis, such as lenticular lens arrays, micro-lens arrays, and ray guiding optics.

  Figure 2.6(a) shows how users view different parallax images at different viewpoints in order to reconstruct 3D images. Each viewpoint receives a collection of light rays emitted from the screen plane at different angles.

- **Holographic displays** are based on the wavefront reconstruction process, in which wave optics or spatial light modulators (SLMs) are typically used as active control elements. As shown in Figure 2.6(b), a holographic display is normally composed of SLMs and a light source such as a laser or light emitting diode (LED) (single

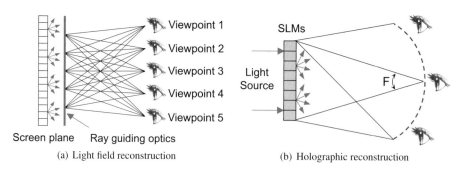

(a) Light field reconstruction   (b) Holographic reconstruction

Figure 2.6 Glass-free 3D image reconstruction based on (a) directional rays from the screen plane and (b) spatial light modulators (SLMs).

color or RGB colors) [14]. Changing the phase distributions on the SLMs enables users to reconstruct and perceive holographic images.

2.3.3 Resolution and Latency Requirements

Using light field displays, users can reconstruct 3D images at different viewpoints by using a collection of light rays emitted from the screen plane. Because a larger number of viewpoints are needed to achieve a good 3D image perception experience, the pixel density on the screen plane can be extremely high; therefore, displaying a high-resolution image requires a large amount of data to be transmitted. The resolution of a light field display can be calculated on the basis of the number of display elements on the screen plane and the number of rays per element [12]. It can also be calculated from the number of viewpoints and the resolution of each viewpoint. Take a 6-inch (133 mm$\times$75 mm) light field display image as an example: if the number of viewpoints is 144 (i.e., 12$\times$12) and the resolution of each viewpoint is 1080 pixels (1920$\times$1080), the pixel size is $\sim$5.8 μm and the total resolution is 3.0×10^8 pixels.

For wavefront-based holographic displays, the pixel size is calculated using the diffraction angle of the holographic display's FOV. A pixel size of 1 μm is required for an FOV of $\sim$30$°$ [15]. Take a 10-inch (200 mm$\times$150 mm) holographic display as an example: as calculated in [12], if the pixel size of the 3D image is 1 μm, the resolution is (200 mm/1 μm)$\times$(150 mm/1 μm) = 3.0×10^{10} pixels.

Because the development of glass-free 3D displays is still in its infancy, in-depth studies have not yet been conducted on how the image refresh rate affects the display quality. A good starting point is to use an image refresh rate of 60 Hz for light field displays [16], and use 30 Hz for holographic displays because of the complex and time-consuming process involved in generating holograms [12]. However, in order to achieve the ultimate immersive experience approaching the human perception limit, as discussed in previous sections, a refresh rate of 30 or 60 Hz may be insufficient.

2.3.4 Main Requirements for Glass-Free 3D Displays

On the basis of the analysis provided in previous sections, Table 2.3 lists the typical raw data rate requirements for different 3D display techniques. The raw data rates

Table 2.3 Raw data rate requirements for different types of glass-free 3D displays.

Parameter	Light field	Holographic	Holographic
Image size (inch)	6	10	50
Pixel size (μm)	5.8	1	1
Image refresh rate (Hz)	60	30	30
Pixel rate (pixels/s)	1.79×10^{10}	9.0×10^{11}	2.3×10^{13}
Color	Full color	Single color	Single color
Bits per pixel	24	8	8
Raw data rate (Tbit/s)	0.4	7.2	184

for glass-free 3D displays are extremely high: about 0.4 Tbit/s for a cellphone-sized light field display, and 184 Tbit/s for a 50-inch holographic display. Compressing this huge amount of data efficiently is a major challenge. In [11], the authors suggested a method to transmit compressed 3D object data rather than 3D hologram data. This method aims to reduce the bandwidth requirement, but it is yet to be explored.

References

[1] 3GPP, "Extended reality (XR) in 5G," 3rd Generation Partnership Project (3GPP), Technical Report (TR) 26.928, 03 2020, version 16.0.0. [Online]. Available: https://portal.3gpp .org/desktopmodules/Specifications/SpecificationDetails.aspx?specificationId=3534

[2] 3GPP, "Virtual reality (VR) media services over 3GPP," 3rd Generation Partnership Project (3GPP), Technical Report (TR) 26.918, 03 2020, version 16.0.0. [Online]. Available: https://portal.3gpp.org/desktopmodules/Specifications/SpecificationDetails .aspx?specificationId=3053

[3] P. Jombik and V. Bahỳl, "Short latency disconjugate vestibulo-ocular responses to transient stimuli in the audio frequency range," *Journal of Neurology, Neurosurgery & Psychiatry*, vol. 76, no. 10, pp. 1398–1402, 2005.

[4] M. S. Amin, "Vestibuloocular reflex testing," Medscape Article Number 1836134, 2016. [Online]. Available: https://emedicine.medscape.com/article/1836134-overview

[5] H. R. Blackwell, "Contrast thresholds of the human eye," *Journal of the Optical Society of America*, vol. 36, no. 11, pp. 624–643, 1946.

[6] D. Judd, *Color in business, science, and industry*, 1975.

[7] Wikipedia, "Colour banding," 2020, accessed Sept. 2020. [Online]. Available: https://en .wikipedia.org/wiki/Colour_banding

[8] Huawei Technologies Co., Ltd., Boe Technology Group Co., Ltd., and CAICT, "Ubiquitous display: Visual experience improvement drives explosive data growth," 2020, accessed Sept. 2020. [Online]. Available: https://www.huawei.com/minisite/static/ Visual_Experience_White_Paper_en.pdf

[9] O. Holland, E. Steinbach, R. V. Prasad, Q. Liu, Z. Dawy, A. Aijaz, N. Pappas, K. Chandra, V. S. Rao, S. Oteafy *et al.*, "The IEEE 1918.1 tactile internet standards working group and its standards," *Proceedings of the IEEE*, vol. 107, no. 2, pp. 256–279, 2019.

[10] D. L. Woods, J. M. Wyma, E. W. Yund, T. J. Herron, and B. Reed, "Factors influencing the latency of simple reaction time," *Frontiers in Human Neuroscience*, vol. 9, p. 131, 2015.

[11] X. Xu, Y. Pan, P. P. M. Y. Lwin, and X. Liang, "3D holographic display and its data transmission requirement," in *Proc. 2011 International Conference on Information Photonics and Optical Communications*. IEEE, 2011, pp. 1–4.

[12] M. Yamaguchi, "Light-field and holographic three-dimensional displays," *Journal of the Optical Society of America*, vol. 33, no. 12, pp. 2348–2364, 2016.

[13] P. Gallo, "Michael Jackson hologram rocks billboard music awards: Watch & go behind the scenes," 2014. [Online]. Available: https://www.billboard.com/articles/news/6092040/ michael-jackson-hologram-billboard-music-awards/

[14] L. Onural, F. Yaraş, and H. Kang, "Digital holographic three-dimensional video displays," *Proceedings of the IEEE*, vol. 99, no. 4, pp. 576–589, 2011.

[15] Q. Jiang, G. Jin, and L. Cao, "When metasurface meets hologram: Principle and advances," *Advances in Optics and Photonics*, vol. 11, no. 3, pp. 518–576, 2019.

[16] L. G. Factory, "Holographic displays," accessed Sept. 20, 2020. [Online]. Available: https://lookingglassfactory.com

3 Sensing, Localization, and Imaging

As stated in Chapter 1, sensing will become a new function integrated in the 6G communication system. The new sensing capability is based on measuring and analyzing wireless signals, and this is different from wireless sensor networks based on sensor data collection and transmission. This capability will open up brand new services for 6G, which are currently served by various dedicated sensing equipment, such as radar, light detection and ranging (LIDAR), and professional CT and MRI.

In the ISAC system, the sensing capability can turn 6G base stations, 6G devices, or even the whole 6G network into sensors. Some of the new sensing applications in 6G may include localization and gesture and activity recognition, as well as imaging and mapping.

For the 6G network, higher-frequency bands (mmWave and THz), wider bandwidths, and massive antenna arrays will enable sensing and imaging solutions with a very high resolution and accuracy. This will play a vital role in providing enhanced solutions in many fields, such as public safety and critical asset protection, health and wellness monitoring, intelligent transportation, smart homes and factories, gesture and activity recognition, air quality measurement, and gas/toxicity sensing. Within this context, several new KPIs are introduced for the sensing capability, including the following.

- **Sensing accuracy:** The difference between sensed and real values in range, angle, velocity, etc.
- **Sensing resolution:** The separation between multiple objects in range, angle, velocity, etc.
- **Detection/False alarm probabilities:** Probability that an object will be detected when one is present/not present.

In the following sections, four categories of integrated sensing and communication use cases are described and the corresponding KPI requirement analyzed.

3.1 High-Accuracy Localization

Example 3.1 Jack is the co-founder of a flying-car manufacturing startup. A 6G-enabled high-accuracy 3D localization system is deployed within his factory.

The system provides both communication and sensing including advanced localization and navigation, as well as mapping services to fully automate robots. This type of system serves as a cyber-physical bridge for the physical facilities, materials, processed parts, and digital manufacturing workflow. Within this system, Jack and his business partners can monitor, diagnose, and optimize the manufacturing process anywhere and at any time. Small parts are stored in the 3D stacked warehouse to save space. Drone robots can easily retrieve goods from the warehouse and load them onto automated guided vehicle (AGV) robots with a very-high-precision soft vertical landing, as shown in Figure 3.1. This involves accurate 3D localization, which is essential to avoid damaging the lifted parts. The AGV robots can efficiently transport parts at high speed and therefore need to avoid colliding with each other and obstacles, using real-time localization. Millisecond-level latency and millimeter-level accuracy make it possible to achieve real-time dynamic path planning. To lift a turbofan engine from a cargo bay and mount it on the target location, the AGV robots will need to closely engage with fixed robots while maintaining relative distance and arm gestures to prevent inadvertently damaging the engine. High-precision operations are required for lifting, carrying, mounting, and soldering, using a fusion of 6G wireless regional localization and other close range sensors.

The 6G network will provide services for 6G device-based localization and 6G device-free object localization. For 6G device-based localization, the location information of devices is derived from the received reference signals or the measurement feedback from the target devices. On the other hand, localization for 6G device-free objects' functions is more like radar detection. This involves processing the delay, Doppler, and angle spectrum information (used to depict the distance, velocity, and angle of environment objects) from the scattered and reflected wireless signals. We can process the signals further to extract coordinates, orientation velocity, and other geometric information in a physical 3D space. Note that highly accurate ranging can

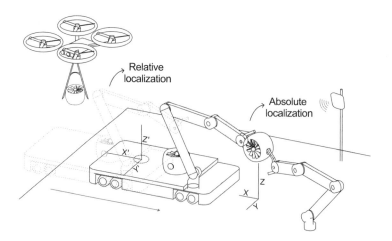

Figure 3.1 High-accuracy localization in a future flying-car factory.

also enable precise clock distribution and synchronization, which is important for industrial networks.

The target performance for outdoor deployment of the current 5G network includes horizontal and vertical localization errors of less than 10 meters and 3 meters, respectively, and 80% availability for both. For indoor deployment, the aim is for the horizontal and vertical localization errors (both with 80% availability) to be less than 3 meters [1]. In favorable conditions, the 6G network will provide higher-accuracy localization down to the centimeter level for outdoor cases with extremely large antenna array, and for indoor cases in near field. In general, the indoor environment is more challenging and complex since obstacles, i.e., walls, furniture, equipment, people, etc., greatly influence the propagation of 6G signals. Thanks to higher bandwidth, multi-spectrum operation, and increased antenna array aperture, the ISAC system in 6G can provide excellent resolving capabilities to separate multipaths and exploit multipath information for better localization and tracking performance.

3.1.1 Absolute Localization

From factories to warehouses, hospitals to retail shops, and agriculture to mining, high-accuracy localization and tracking enables meaningful association between cyber information and the locations of physical entities. 6G-based sensing is not affected by unfavorable light conditions, noise, and mechanical vibration, which means that this technology brings additional benefits and robustness to localization performance, especially when compared with the traditional time-of-flight-based light sensors or ultra-sound sensors. It goes a step further by enabling work process and workflow and layout optimization, a real-time view of current work, valet parking or convenient parking, etc. To add on to that, low-latency, high-accuracy localization can enable navigation without collision for AGVs and forklifts. The 10 cm level of accuracy enables device-level placement, while the 1 cm level of accuracy can further enable module-level installation and placement in tight spaces, allowing for the efficient storage of components with a small form factor, such as integrated chips and small metallic parts [2].

3.1.2 Relative Localization

Relative localization is defined as occurring when two or more entities exist and one or more approach each other, or the entities have coordinated moving direction and speed. This is of utmost importance to automatic docking and cooperative multi-robot applications. Relative localization could also be utilized by a team of navigating robots in situations where it is imperative that we know a vessel's relative location compared with its neighbors. Relative localization is necessary as a viable alternative for close-in maneuvering owing to the fact that complexity, physical limitations, and external infrastructure are mission critical for each robot to accurately determine its location in relation to a common datum. Let's take a mechanical part with a complex shape and length of one meter as an example. To avoid losing weight balance during collaborative lifting and carrying, given that the maximum tilt allowed is three degrees, a relative localization error of approximately 5 cm or even less between two moveable entities can ensure the object is moved at the right orientation and balance. Relative localization can also be applied to new scenarios, for example, when a drone docks onto a moving vehicle with an extremely small margin for landing, due to the moving vehicle cargo platform's limited area.

3.1.3   Semantic Localization

Empowered by pervasive AI, future systems can provide the semantic localization capability with context awareness. Legacy systems can perform simple dispatchable localization by providing the floor level, building number, street name of user, when serving people. To support future smart home/shopping mall/restaurant/hotel, and automatic factory applications, objects and parts need to have dispatchable localization, by providing information such as shelf level, seat number, table number, cube number, etc. Beyond these fine granularity addresses, pervasive AI can also support dynamic address registration according to service context. As illustrated by Figure 3.2, a restaurant uses drones as robotic waiters. With AI, these robots can understand the context of semantic instructions (such as delivering food to a table). Then, with the semantic localization capability, they can locate guests and deliver food. The actions of these drones and droids will be very similar to those of human waiters. Through integrated AI and localization, the robots even go a step further and set different goals according to task characteristics, e.g., fragile and rigid objects will be treated with different levels of care, including location and velocity accuracy, during transportation.

Figure 3.2 Semantic localization in a future restaurant with drones as robotic waiters.

3.2   Simultaneous Imaging, Mapping, and Localization

Example 3.2 Jack's flying-car manufacturing startup is located downtown in the futuristic city where he lives. An important feature of the futuristic city is real-time image reconstruction, where real-time information on the outdoor environment can be obtained to achieve the goal of a virtual city, and smart network, as shown in Figure 3.3. With this technology, he is able to notice the fine details of the city, thereby improving his way of life and ease of living. For instance, glass curtain walls and billboards that are being replaced, as well as crowd flows can all be sensed and imaged. Even seasonal changes in foliage are embedded in the background information. At times, when Jack is returning home from work, he avoids certain routes due to his allergic reaction to the willows in certain locations. This can all be imaged and mapped in real time. One night, he is in a fully autonomous vehicle driving along a street with limited lighting, and a deer suddenly dashes across the road. The autonomous vehicle immediately senses the situation and acts accordingly to avoid injuring the deer and passenger.

In the preceding example, sensing capabilities from three perspectives are mutually enhanced. In particular, the imaging function is used to capture the images of the surrounding environment, and the localization function is used to obtain the locations of surrounding objects. These images and/or locations are then used by the mapping function to construct a map. Finally, the map helps the localization function improve the inference of locations.

Compared with the traditional imaging radar, ISAC in 6G would leverage advanced algorithms, edge computing, and AI to produce super-resolution and highly

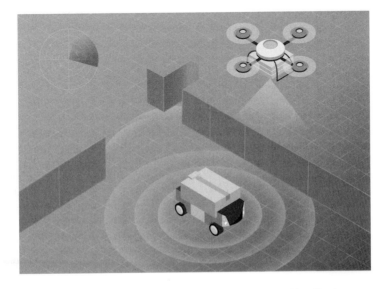

Figure 3.3 Real-time virtual environmental image reconstruction and localization.

recognizable images and maps. The imaging function of the 6G network will provide a fully functional all-day solution, which is in contrast to the current LIDAR and optical devices, which are highly affected by light and atmospheric conditions (such as fog or cloud). With the vast network of vehicles, base stations, etc., which will act as sensors, the imaging area can be remarkably extended. Moreover, performance will significantly improve owing to the fusion of imaging results that are shared globally through the network with cloud-based services. Different use cases with such sensing capabilities are discussed below.

3.2.1   Simultaneous Localization and Mapping

Simultaneous localization and mapping (SLAM) applications in the mmWave or THz band enable 3D maps of the surroundings in unknown environments to be reconstructed. SLAM is based on the concept that a sensing device, moving in an unknown environment, recognizes the surrounding objects (landmarks) and subsequently reconstructs a 2D or 3D map of the environment to further improve localization accuracy.

The current, accurate SLAM (for automatic creation of indoor and outdoor maps) requires both high-resolution distance estimation (ranging) and very high angular resolutions, which are traditionally achieved through laser and optics technologies. Although the LIDAR system and optical cameras can provide high resolution, they cannot operate in adverse weather conditions (fog, rain, cloud, etc.) and low ambient light environments.

We can perform SLAM using 6G wireless signals to address the preceding issues and simultaneously avoid having to incur equipment costs for a separate LIDAR system and cameras. In fact, due to the complex nature of the environment (e.g., indoor scenarios with numerous rooms and/or partitions), non-line-of-sight (NLOS) coverage represents a large portion of the target service area. In this case, 6G-based mapping can help provide the most up-to-date knowledge of an environment, based on which high-accuracy localization is still possible even in NLOS scenarios. Specifically, we can map the locations of objects in an environment to provide additional localization information. The accurate mapping information can then be applied to determine the multipath reflection points via ray tracing techniques. For this reason, with 6G-based SLAM, the location of objects can be traced back through both line-of-sight (LOS) and NLOS paths. The amount of multipath information may indirectly increase the probability of detection, thereby improving the localization performance. We believe that when the distance between a sensing device and an object is within the normal indoor range (approximately 10 meters), 6G-based SLAM can accurately locate the object at the centimeter level.

3.2.2   Indoor Imaging and Mapping

6G-based sensing opens up the realm of possibilities in 3D indoor imaging and mapping, which in turn enable various applications such as indoor scene reconstruction, spatial localization, and indoor navigation. Such applications usually require

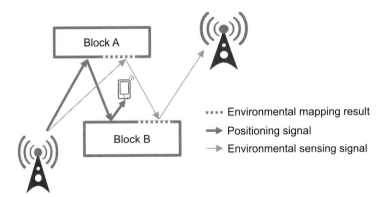

Figure 3.4 Localization in the NLOS scenario.

super-resolution and high accuracy. Owing to the fact that scattered signals bounce multiple times where the LOS surfaces act as mirrors, compensated images of NLOS objects can be reconstructed by applying mirroring.

An imaging system that can sense beyond corners would be highly valuable and applicable in numerous new scenarios. Once the environment is reconstructed, the next step will be the localization and imaging of NLOS targets. The target locations can be detected with good accuracy when prior information regarding the scene's geometry is known. Even if some errors are introduced into the assumed geometry, the obtained locations are still centered on the true tracks. Figure 3.4 illustrates how target localization can be used in the NLOS scenario with environmental reconstruction. For accurate NLOS localization, the environmental reconstruction results (the distances between the walls of blocks A and B) have a 5% margin of error. An error this small in the assumed distances will alter the determined target locations, but the locations will still be centered on the true tracks. In this case, in an indoor corridor where the distance between the walls is assumed to be 2 meters, a 5% error equates to an environmental mapping accuracy of 10 cm [3].

3.2.3 Outdoor Imaging and Mapping

The sensors on mobile vehicles usually have restricted view and limited coverage due to weather, obstacles, and the sensors' power control. That said, nearby stationary base stations may have a greater field of view, longer sensing distance, and higher resolution. As such, the vehicles can achieve higher levels of autonomy by utilizing the maps reconstructed by the base stations to determine their next move. Furthermore, the sensing resolution and accuracy performance significantly improve because of the fusion of imaging results that are shared globally through the network with cloud-based services. To acquire high resolution for target recognition in autonomous driving, 3 cm or less is a reasonable requirement for today's automotive LIDAR systems [4]. The densely distributed base stations in an urban area and ISAC make environmental reconstruction and 3D localization possible, which in turn form the virtual city.

Since the distance between buildings is always larger than one meter, meter-level resolution and accuracy may enable building-level imaging and mapping where a sensing range of hundreds of meters can be suitable for urban area reconstruction. It is worth noting that road mapping provides comprehensive traffic information in real time. 6G-based imaging can be used to precisely reconstruct road traffic, as well as the environment. The road signature extracted from a road map contains millions of individual reflection points, which reflect off crash barriers, road signs, and road courses. The reconstructed map can be used for smart traffic control, such as traffic flow monitoring, queue detection, and accident detection.

3.3   Augmented Human Sensing

Example 3.3 Consider a future medical center powered by many 6G-enabled augmented human sensing devices and equipment. Dr. Kleine wears sensing glasses with ultra-high resolution during surgery. This sensing equipment helps Dr. Kleine see a patient's millimeter-resolution distribution of blood vessels and lymphatic system, thereby reducing the possibility of permanent damage during surgery. In addition, these glasses can penetrate the patient's skin and detect their heartbeats and bleeding in real time, helping Dr. Kleine monitor their conditions and make quick decisions during surgery with the assistance of AI. During surgery, Dr. Kleine extracts tissues from a patient for diagnostic purposes, and sends the pathological specimen to a medical examination center, where examiner Tommi examines it with sensing gloves. The flexible graphene gloves, called "T-RAY," allow Tommi to review the molecular vibration spectrogram in real time and quickly detect the sample's condition by simply touching it slightly. Similarly, pharmacological testing is conducted in this laboratory by gently placing medical powder onto the sensing gloves.

Augmented human sensing aims to provide a safe, high-precision, as well as low-power, sensing and imaging capability that exceeds human abilities. With ISAC serving as the enabling technology, we can use a portable terminal to sense the surrounding environments, instead of relying on large specialized devices. For example, the sensing devices can be 6G-enabled mobile phones, wearables, or medical equipment implanted beneath human skin. With the help of scientific and technological advancement, we can achieve augmented human sensing to facilitate information collection and integrate the maximum number of environmental messages into the 6G network.

In addition to hospitals, other feasible applications of augmented sensing will include the detection of objects inside packages, water pipes behind walls, defects on products, and sink leakages. Moreover, to obtain information on vascular status, organ status, and other vital signs, we will be able to utilize wearable devices above or beneath the skin in higher-frequency bands (mmWave, THz, or optical bands). This type of technology will be capable of acquiring critical information more accurately

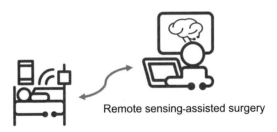
Remote sensing-assisted surgery

Figure 3.5 Remote surgery with an ultra-high resolution imaging monitory system.

and effectively than is possible through human ability alone. In addition, precise location information about mobile devices and associated information about the sensed data will help ensure that patients receive medical care during emergencies.

3.3.1   Seeing Beyond the Eye – Ultra-High Resolution

In the future, sensing techniques will support high-resolution imaging and detection, where with the 6G network they will open the door for numerous applications, such as remote surgery, cancer diagnostics, detection of slits on products, and sink leakage detection. As shown in Figure 3.5, a surgeon can conduct surgery at a different location through the help of an ultra-high-resolution imaging monitory system and remote operation platform system. In addition, intelligent factories will leverage these superior sensing solutions to implement contactless ultra-high-precision detection, tracking, and quality control.

For this type of application, millimeter-level range resolution and ultra-high cross-range resolution are required, which rely on higher bandwidth and increased antenna array aperture, respectively. Using 6G communication technologies, with a high THz frequency and corresponding short wavelength that is less than 1 mm, these augmented human sensing functions can be integrated or installed in portable devices.

3.3.2   Seeing Beyond the Eye – Making the Invisible Visible

While ultra-high-resolution scenarios require higher bandwidth and increased antenna aperture, another application of "seeing beyond the eye" poses different requirements. More precisely, frequencies with a relatively low penetration loss are required to acquire underlying and important information that might be beneath the skin, behind occlusion, or in darkness. In other words, the invisible becomes visible.

Conventional optical imaging technology can only provide LOS imaging at a level similar to the level at which the human eye can observe. However, 6G-based sensing can achieve the NLOS imaging capability, as illustrated in Figure 3.6. For practicability, technology for detecting hidden objects will simply rely on portable devices with a powerful low-latency imaging capability, as opposed to the currently prevalent large and bulky machines. As such, we foresee mobile phones being used to detect pipelines behind walls or perform security scans on packages. Typically, we can

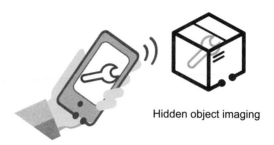

Hidden object imaging

Figure 3.6 Examples of NLOS imaging.

effectively implement this application by utilizing the penetration characteristics of the electromagnetic wave. The penetrability of wireless signals depends on frequencies and transmit power. In light of this, non-ionizing materials such as skin (0.5–4 mm), subcutaneous fat (12–20 mm), suitcases (0.5 cm), and furniture (2 cm) are some of the main examples of where penetration could be applied.

6G sensing can also play an important role in the medical technologies applied to diagnosis, monitoring, and treatments. Nowadays, the number of people living with chronic diseases is rapidly increasing worldwide and burdening healthcare systems. Chronic diseases (asthma, arrhythmia, hypoglycemia, hyperglycemia, chronic pain, etc.) need to be constantly monitored in the long term using wearable devices. For example, the heartbeats of a patient suffering from heart disease need to be monitored 24 hours a day, and the blood sugar level of a patient suffering from hypoglycemia or hyperglycemia needs to be monitored after every meal. Fortunately, 6G sensing technology enables atraumatic medical detection. It not only detects heartbeats, but is also able to obtain a real-time image of the heart blood stream. It therefore goes without saying that this technology helps doctors instantly acquire accurate information on a patient's condition. The atraumatic detection technology will not harm or cause discomfort to patients and it features ultra-high reliability and accuracy, where the apical impulse diameter of 1–2 cm [5] would require a bandwidth ranging from tens to hundreds of GHz. With this type of augmented human sensing, we will be able to repair, replace, and expand our capabilities in the future with the new brain–machine and muscle–machine interfaces.

3.3.3   Seeing Beyond the Eye – Spectrogram Recognition

Spectrogram recognition is based on identifying targets through the spectrogram sensing of their electromagnetic or photonic characteristics. This includes the analysis of absorption, reflectivity, and permittivity parameters, which helps to distinguish the quality of materials. Pollution, calorie calculation, and product quality management have been mentioned as some of the prospective applications of this technology.

These applications depend on the molecular vibration effect, which is related to a material's unique absorption curve. Different reflectivities can be presented under the multi-band THz radiation. For instance, electromagnetic radiation in the THz and

Figure 3.7 Spectrogram recognition for calories.

sub-THz frequency range is strongly absorbed by liquid water, and the transmission or reflection curve clearly varies according to different water contents. This property can be used for the non-destructive testing of industrial products to accurately determine the water content of wood, paper, etc. The distribution and ratio of each component in the mixture can be detected by conducting a transmission or reflection test. To acquire all of the absorption peak in the molecular vibration, a bandwidth of 2 GHz to 8 GHz is required. Also, a frequency jitter of less than ± 10 GHz at 1 THz is required to make spectrogram recognition possible [6].

Spectrogram recognition can be used for food sensing applications. Food content (type and ingredient) can be detected through the transmission and reflection of THz signals. This will help us identify different types of food, calorie content, the presence of contaminated ingredients, etc., as illustrated in Figure 3.7.

In environmental sensing, which includes evaluating air quality and detecting pollution, the PM2.5 transmission spectrum has two distinct absorption bands between 2.5 and 7.5 THz. Correlations between the absorption bands and cross-peaks in the synchronous and asynchronous plots indicate that the metallic oxides' absorption falls within the range of 2.5–7.5 THz. These results [7] verify that the THz spectral analysis of PM2.5 is a promising tool, which can be utilized to understand the composition and mass of pollutants, and the 0.1–7 THz band is preferred for acquiring the absorption bands.

3.4   Gesture and Activity Recognition

As technology advances, the scope of computing has progressed beyond the desktops that we all know. As such, the community for human–computer interfaces has conducted research on a multitude of alternative input strategies. Without doubt, in recent years, there has been tremendous interest in human–computer interaction through the whole body (activities) or hand gestures, amid numerous applications. Device-free gesture and activity recognition using machine learning is the key to promoting human–computer interfaces that allow users to convey commands and conveniently interact with devices through gestures and actions. We can easily integrate this technology into our daily lives and as such, it will considerably alter the way we live. It's safe to say that gesture and activity recognition in the RF spectrum features beneficial properties when being used as a sensing modality for interactive applications and intelligent pervasive systems. Phrased differently, it yields new possibilities when utilizing RF signals as the medium for interpreting gestures and activities, without the need for wearables.

Feature extraction is the main step that gesture and activity recognition involves, and the extracted features are from the time, frequency, and spatial domains, or the corresponding cross-domains. In 6G, the higher-frequency band will enable higher resolution and accuracy, and can also capture finer activities and gestures. On top of that, the detection of Doppler shifts, which are caused by motion activities, is more sensitive in the higher-frequency band. Furthermore, the massive antenna arrays in 6G allow for recognition with significantly improved spatial resolution and accuracy. Unlike video-based sensing, 6G signals are not affected by lighting conditions and can therefore deliver a more robust gesture and activity recognition solution. In addition, NLOS sensing can also be performed owing to the propagation characteristics of the 6G signals. Another important benefit of gesture and activity sensing in 6G is the fact that it does not involve any underlying privacy risks (regarding video surveillance), which can be unsuitable in certain application scenarios (within homes). In a future gesture and activity recognition system that fully utilizes user equipment (e.g., smartphones) and cellular base stations, devices will be collectively used to sense surrounding environments, and this will be possible owing to the densely distributed 6G network. In addition, the abundance and availability of user devices and base stations in the network mean that the sensing range will be extended considerably. The sensing data association and fusion, shared globally through the network with cloud-based services, will significantly improve the overall recognition performance.

3.4.1   Contactless Control – Macro Recognition

Example 3.4 As the co-founder of a flying-car manufacturing startup, Jack works very hard for his business. However, he recently fell ill owing to his long working hours and got admitted into a smart hospital to undergo surgery. The next day, Jack was operated on in a smart operation room, and the surgeon used gestures to control the medical equipment without touching it. After the surgery, Jack had to undergo medical rehabilitation to aid his recovery. Activity recognition ensured that his rehabilitation was effortlessly supervised and his routines were done correctly. At all times during his admittance, Jack's health condition was continuously monitored by detecting if he falls, has any irregular movements, or sneezes and coughs. All this is critical to ensuring his speedy recovery.

There will be advanced gesture and activity recognition features in a smart hospital in the foreseeable future. The medical rehabilitation system in the future smart hospital will enable the automatic supervision of patients. This ensures that their gestures and movements during physiotherapy conform to the standard requirements for rehabilitation exercises. There will be prompt alerts on incorrect movements or gestures, significantly improving patients' rehabilitation. For gesture and movement recognition during physiotherapy, a velocity resolution of 0.01 m/s or better would be required because the movements of recovering patients are slow and gradual. In addition, an alarm alerting the hospital's control center will be raised if a patient falls during an exercise, or if a suspicious person is detected intruding into a restricted area. To enable the fall or intrusion detection functionality, where the displacement of such activities is much larger than gestures, a range resolution of 2 cm, a velocity resolution of 0.05 m/s, and detection probability of over 99.9% are required.

In a smart hospital, infectious diseases are mainly spread through sneezing and coughing. It is therefore imperative that we detect droplets expelled through sneezing and coughing, and the corresponding detection and tracking will be very critical to pinpointing the locations of patients when they sneeze or cough. Mucus and saliva can burst from a person's mouth at nearly 40 m/s and travel as far as eight meters [8]. After a person exhales, droplet-containing gas can remain suspended in the air for several minutes, depending on the size of the droplets. For this purpose, a range resolution of 5 cm, velocity resolution of 0.1 m/s, and detection probability over 99% are required in order to accurately detect and locate sneezes or coughs. Moreover, the sneeze and cough detection and tracking system can detect the corresponding cloud size, as well as its trajectory (which shows how far the clouds can spread). With more information and parameters on human sneezes and coughs, researchers hope to better control and curb the spread of diseases.

3.4.2 Contactless Control – Micro Recognition

Example 3.5 After a long day at his flying-car manufacturing factory, Jack returns home. Upon entering the living room, he simply waves his hands to activate the lights and air conditioner without touching the switch. Then, he would like to watch a TV program and therefore uses hand gestures to control and interact with the smart TV. After dinner, he wants to practice the melody he learned yesterday on the virtual piano.

The future smart home will be equipped with an advanced hand gesture capturing and recognition system. The system will allow us to track a hand's 3D position, rotation, and gesture. As a result, by simply waving our hands, we can turn the lights on or off anywhere in the house without having to hit a switch. Besides lights, we can also control the smart TV through hand gestures using 6G signals. The main interaction with the smart TV system is done through the human face and natural hand gestures. Put differently, a facial recognition system is used to identify the user, and the hand gesture recognition system controls the actual TV (channels and volume). Since the hand gestures are captured by 6G signals, the user does not need to be in front of the TV. To precisely capture a person's gestures, a range resolution of 1 cm, velocity resolution of 0.05 m/s, detection probability over 99%, and coverage range of up to eight meters (the size of a large living room) are required. In addition, to separate the gestures of multiple people in front of the TV, a cross-range resolution of 5 cm or less is ideal.

To take it a step further, more complicated functionalities could be realized by an advanced hand gesture capturing and recognition system. For example, we could play a virtual piano in the air through the sensing function in the 6G network, providing a completely immersive experience anywhere, anytime. Finger kinematics is very sophisticated, and there are studies to show that the average width of the index finger is approximately 1.6–2 cm for most adults [9]. The finger velocity during a pianist's performance at a slow tempo has a maximum value of ± 2 m/s [10]. Therefore, in order to realize the virtual piano concept, the range and cross-range resolutions should be lower than 0.5 cm with a velocity resolution of 0.01 m/s. In addition, a probability of recognition above 99% would ensure that no interruptions occur as the virtual pianist performs. Without doubt, this futuristic concept would open up the realm of possibilities for many more innovative applications related to high-accuracy finger motion detection and tracking.

References

[1] 3GPP, "Study on positioning use cases," 3rd Generation Partnership Project (3GPP), Technical Report (TR) 22.872, 09 2018, version 16.1.0. [Online]. Available: https://portal.3gpp .org/desktopmodules/Specifications/SpecificationDetails.aspx?specificationId=3280

[2] S. Kumar, A. Majumder, S. Dutta, R. Raja, S. Jotawar, A. Kumar, M. Soni, V. Raju, O. Kundu, E. H. L. Behera *et al.*, "Design and development of an automated robotic pick & stow system for an e-commerce warehouse," *arXiv preprint arXiv:1703.02340*, 2017.

[3] T. Johansson, Å. Andersson, M. Gustafsson, and S. Nilsson, "Positioning of moving non-line-of-sight targets behind a corner," in *Proc. 2016 European Radar Conference (EuRAD)*. IEEE, 2016, pp. 181–184.

[4] J. AmTechs Corporation, Tokyo, "VLP 16 puck," 2020, accessed Sept. 2020. [Online]. Available: https://www.amtechs.co.jp/product/VLP-16-Puck.pdf

[5] A. Chandrasekhar, "To evaluate apical impulse." [Online]. Available: http://www .meddean.luc.edu/lumen/meded/medicine/pulmonar/pd/pstep36.htm

[6] J. Advantest Inc., Tokyo, "Terahertz spectroscopic system TAS7400 product specification," 2019, accessed Dec. 27, 2019. [Online]. Available: https://www .advantest.com/documents/11348/146157/spec_TAS7400_EN.pdf

[7] H. Zhan, Q. Li, K. Zhao, L. Zhang, Z. Zhang, C. Zhang, and L. Xiao, "Evaluating PM2. 5 at a construction site using terahertz radiation," *IEEE Transactions on Terahertz Science and Technology*, vol. 5, no. 6, pp. 1028–1034, 2015.

[8] "See how a sneeze can launch germs much farther than 6 feet," 2020, accessed Sept. 2020. [Online]. Available: https://www.nationalgeographic.com/science/2020/04/ coronavirus-covid-sneeze-fluid-dynamics-in-photos/

[9] "Finger-friendly design: Ideal mobile touchscreen target sizes," 2020, accessed Sept. 2020. [Online]. Available: https://www.smashingmagazine.com/2012/02/finger-friendly-design-ideal-mobile-touchscreen-target-sizes/

[10] S. D. Bella and C. Palmer, "Rate effects on timing, key velocity, and finger kinematics in piano performance," *PloS One*, vol. 6, no. 6, p. e20518, 2011.

4 Full-Capability Industry 4.0 and Beyond

Industry 4.0 refers to the Fourth Industrial Revolution, which automates traditional manufacturing and industrial practices using modern smart technology [1]. Compared with Industry 3.0, as shown in Figure 4.1, Industry 4.0 aims for industrial production with significantly improved flexibility, versatility, usability, and efficiency based on advanced cyber-physical systems. Building on that, Industry 4.0+ serves as a vision for future smart manufacturing that further leverages the fast development in wireless communications and automation technologies, as well as real-time AI and machine learning technologies.

Each generation of the manufacturing industry takes decades to develop and evolve, as shown in Figure 4.1. This is very different from the wireless communications industry, where each generation includes a shorter lifecycle. For Industry 4.0, new technologies like wireless communications and IoT are playing major roles, leveraging the ICT domain's advancements.

Even though wireless technologies are being used in today's plants, e.g., Wi-Fi, LTE, Bluetooth, and Zigbee, they have not become the main methods of achieving connectivity for key industrial use cases in factory automation. This is mainly due to technological performance limitations. That said, 5G has large potential due to its strong focus on machine type communication and ultra-low latency, and high reliability communication.

The applications of wireless technology can be classified into five categories: factory automation, process automation, human–machine interface and production IT, logistics and warehousing, as well as monitoring and predictive maintenance [2]. The typical use cases that fall into these categories are: motion control, control-to-control, mobile control panel, mobile robot, massive wireless sensor network, remote access and maintenance, AR, closed-loop process control, process monitoring, and plant asset management [2]. One use case may apply to one or more applications.

Even though some of the use cases could be realized in 5G, certain very challenging aspects (due to the peculiarities of vertical industries) still need to be considered or handled by the next-generation system. For instance, some use cases may have extreme requirements or be positioned in a harsh environment, in terms of ultra-high network performance, challenging propagation environments with potentially high interference, brownfield installations that may require seamless integration, specific safety and security concerns. The latter part of this section will introduce some representative use cases that inspire the new 6G system design, which aims to provide

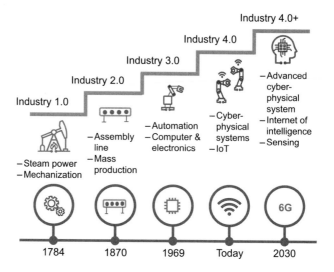

Figure 4.1 Industrial revolution roadmap.

better support for vertical industries. On the other hand, the new features of 6G may also open up active discussion and innovation on Industry 4.0 and beyond.

4.1   Factory of the Future

Example 4.1 Mark works as an engineer for an automotive manufacturing factory. His responsibilities involve supervising the car production in the plant, which has built with 6G technology. Instead of standing by the assembly line in a noisy and sometimes dangerous environment, Mark is now working remotely in a bright control room that is much safer and more environmentally friendly. He can now focus on high-value supervision and management through various types of XR and teleoperation equipment with super-fast, ultra-low latency and ultra-reliable 6G connections.

One afternoon, Mark receives new orders from the Internet. He double checks the customer requirements and then triggers the input of configurations into the intelligent management system on a virtual control tablet. As the plant has no fixed production line, the intelligent system decides the best way to reorganize the production line by itself on the basis of all the gathered information as well as its own experience and knowledge. The production line is thus automatically reorganized to manufacture highly customized cars as demanded.

The assembly line concept is still being used in manufacturing factories almost 100 years after being invented by Henry Ford, owing to its inherent high efficiency. However, this efficiency is specifically suited for mass production, which compromises production flexibility and the possibility of customization. As a result, it cannot meet the demands of mass individualization in future manufacturing [3].

The factory of the future, instead, aims for complete revolution to ultimately realize "lights-out" manufacturing with full automation and flexibility. Put differently, this refers to manufacturing without humans on site, meaning that plants can switch all lights off. In addition to looking and operating differently, the factory of the future also innovates differently.

- **Wireless as a first step towards flexibility:** The main step needed to achieve the flexibility required for such customization involves freeing machines from interconnected cables. In other words, this entails individual and flexible production modules moving around to form an assembly line instantly and collaborating with other robots, AGVs, as well as drones through information exchange over ultra-high-performance radio links. In this way, the focus of the assembly line concept can truly transform from conventional "mass production" to future-proof "customized production."

- **Fast spread of intelligence among robots:** Individual information from all objects in a plant will be collected and formulated as physical representations in the digital virtual world, which will be used to design, simulate, and optimize the production process. Moreover, AI will be widely and deeply adopted by the manufacturing industry. For instance, machines and robots will be able to accumulate manufacturing experience and knowledge, which can then be shared with other machines and robots in the same production line, cross-production lines, or cross-plants globally through advanced mobile communications technology. In this way, one machine or robot can leverage the intelligence of all the other machines or robots, which in turn leads to an optimized and rapidly evolving manufacturing process. Furthermore, in these scenarios, a few robots are required to collaborate on a task, and as such, their commutations will not be based on simple connectivity. This is a typical "connected intelligence" scenario.

- **Real-time sensing for proactive maintenance:** The entire production environment will be monitored and controlled via ubiquitous sensors through the ultra-dense connectivity and intelligent RF sensing system, as discussed in Chapter 3. All this aims at providing proactive maintenance of the entire production environment and process, guaranteeing both production safety and efficiency. Even a minor production defect can be detected in real time and corrected instantly.

- **Low carbon footprint and environmentally friendly factory:** Since there are no people in most areas of the factory, there is no need to keep the lights on or maintain other facilities for human employees. For instance, all the monitoring can be conducted through wireless sensor networks or RF sensing functions integrated in the 6G communications system; therefore, there is no need to keep the lights on for surveillance cameras either. In this case, a much lower OPEX and carbon footprint can be achieved.

The factory of the future features a wide range of use cases and will require enabling technologies from both the OT and ICT sides. We elaborate further on a number of challenging use cases in the following sections.

4.2   Motion Control

Example 4.2 In the factory where Mark works, all the machines and robots are connected wirelessly. Just like "Lego" building blocks, they can be easily moved around and self-constructed with other machines and robots according to product requirements.

The motion control system is the brain of a machine and a vital component of automation process. It is widely used in numerous industry sectors, including the manufacturing, automotive, and medical industries. It is responsible for controlling, moving, and/or rotating the machine's parts in a well-defined manner (e.g., in printing machines, machine tools, or packaging machines). In a closed-loop control process, a motion controller sends commands to one or several actuators in a strictly cyclic and deterministic manner. After receiving the set points from the controller, the actuators perform corresponding actions on one or several processes. For instance, they can rotate a robot's arm or move a certain component on a machine. Meanwhile, sensors are used to determine the current state of the processes (e.g., the current position and/or rotation of one or multiple components) and send the actual values back to the motion controller [2].

Motion control is one of the most challenging use cases in the automation field; it requires ultra-high reliability and a low-latency, deterministic communication capability. For instance, the E2E latency (across the radio access, core, and transport networks) required to support motion control applications like machine tools or packaging machines could even be on the microsecond level, and require reliability higher than 99.9999%. This type of mechanism has been employed in modern manufacturing for decades. Today, wired technologies like industrial Ethernet are extensively used for motion control systems [2, 4].

In order to change the paradigm from the fixed production line to the flexible production line, which could be assembled like Lego building blocks, the first and most fundamental step is replacing wired communication by wireless communications links. That said, achieving the same level of performance as industrial Ethernet sounds easier said than done. Even though use cases regarding motion control have been discussed in 5G, the majority of them can only be realized in 6G, owing to performance limitations. As such, ultra-low latency and ultra-high reliability communication with deterministic features will definitely be a challenge for the design of the entire mobile communications system.

4.3   Collaborative Robots in a Group

Example 4.3 In the car factory, the major work is done by robots instead of humans. During production, the raw materials, spares, and accessories of a car are transported from the warehouse to the production line by robots, AGVs, or drones. Large parts are usually carried by multiple robots via collaboration. 6G network-enabled real-time localization and synchronization help maintain the accuracy level needed for complicated collaborative work, e.g., collaborative carrying.

Assembly work is done by a group of collaborative robots without any human intervention. Local and joint machine learning is a key technique that enables the robots to deal better with different situations. If one robot experiences and solves an unknown situation, the related knowledge and experience can be rapidly shared with other robots in the same plant or cross-plants through the 6G network.

Owing to advanced robotic technology, industrial robots have found their way into the industrial world of production. They can perform welding, painting, and soldering, among numerous other tasks. As such, they will continue to play a vital role in the future manufacturing industry. One use case is collaborative carrying [2], where large or heavy parts are carried from one place to another by multiple robots. These robots need to collaborate to ensure safety and working efficiency, and this is achieved through a cyber-physical control application that controls and coordinates their movements. The communication between the robots requires ultra-high reliability and service availability, as well as strict synchronization. It is also worth noting that control commands are given and feedback exchange is conducted through periodic deterministic communication.

The collaborative workforce can be used to carry rigid or fragile parts that require very precise coordination among the robots, or used to carry more flexible or elastic parts that allow some freedom in the coordinated movements. For both cases, networks that perform better in terms of synchronization, latency, and localization accuracy can help improve the collaboration efficiency. In general, the E2E latency needed to support robots can be approximately 1 ms, and the reliability is higher than 99.9999%.

In the factory of the future, robots will work in complete harmony with each other and machines. This collaboration involves more than just predefined programs. Put differently, it is based on robots sensing the environment, and the intelligent inference of obtained information and knowledge based on AI.

4.4   From Intelligent Cobots to Cyborgs

Example 4.4 A warning alarm suddenly goes off, and Mark walks into the production area. The potential location of the problem is displayed over his smart glasses, which instantly guide him to the site. Information on the possible cause of the problem is displayed on his tablet. The intelligent management system helps Mark analyze and determine the optimal maintenance solution.

Once the solution is determined, the assistant robot brings all the tools needed and halts the production line. It interacts with Mark smoothly and provides suggestions if it has a different opinion on certain actions.

Mark's colleague, Bernd, is blind owing to a congenital disease. He implants a device that can receive external information and content, helping him reconstruct his eyesight. Because of the technology, his disability did not stop him from pursing his dream of becoming an automotive engineer.

An artificial word "cobot" (a portmanteau of "collaboration" and "robot") has appeared in the manufacturing industry. It is a concept used for advanced Industry 4.0 and beyond. Being different from the use case mentioned above, which describes the situation when robots are working with each other, cobot refers to robots that can collaborate closely and interactively with people, like partners. In contrast to conventional industrial robots, cobots are not separated from humans by conventional protective mechanisms such as plexiglass walls or separated areas. Cobots carry out tasks that are too difficult or dangerous for people.

Intelligence, precision, persistence, and trustworthiness are the keys to enabling efficient human–machine interaction in everyday life or in physically demanding jobs within dangerous environments. Cobot can improve the working environment and quality for people in industrial and service sectors. Unlike conventional industrial robots, cobots are not restricted to fixed tasks or areas. They can be trained, and learn through observation and inference from their human mentors or electronic co-workers. To achieve this, the underlying infrastructure needed to support the cobots' mobility and connectivity should feature industry-grade performance.

Equipped with sophisticated sensing technology and high-performance communication technology, the cobots can react to and interact with humans. This capability offers a wide range of opportunities in industrial sectors, especially considering the fact that the current global trend is towards an aging population. Older employees can work longer if cobots support them in physically demanding activities. Meanwhile, cobots could easily expand their professional field of application from vertical sectors to customer-oriented services in retail or catering businesses. These cobots can enhance our capabilities in both professional and leisure capacities through mechanical but intelligent methods.

If we view cobots as the embodiment of intelligent cybernetics representing the top-level fusion of AI, ICT, and OT, then cyborgs are the next step. The cyborg concept was laid out in 1960, and originally defined as follows: "the cyborg deliberately incorporates exogenous components extending the self-regulatory control function of the organism in order to adapt it to new environments" [5]. In short, cyborgs are cybernetic organisms, or, phrased differently, they are human bodies enhanced with machines. With the recent advanced developments in medical, mechanical, and human–machine interface fields such as neuro-control, cyborgs will turn from science fiction to reality in the near future. This will bring unprecedented benefits, especially to physically challenged people. For example, these people will be able to see and lift things, and as such, will be on an equal footing with everyone else in work environments. Even people without physical challenges will be able to protect themselves in dangerous professional environments, while enhancing their knowledge and environmental awareness. To conclude, neuroscience and robotics will develop in tandem with mobile communications technology, and humans themselves could serve as new terminals, with 6G being the key to interconnection.

References

[1] Wikipedia, "Fourth industrial revolution," 2020, accessed Sept. 21, 2020. [Online]. Available: https://en.wikipedia.org/wiki/Fourth_Industrial_Revolution

[2] 3GPP, "Service requirements for cyber-physical control applications in vertical domains," 3rd Generation Partnership Project (3GPP), Technical Specification (TS) 22.104, 12 2019, version 17.2.0. [Online]. Available: https://portal.3gpp.org/desktopmodules/Specifications/SpecificationDetails.aspx?specificationId=3528

[3] Y. Koren, *The global manufacturing revolution: product–process–business integration and reconfigurable systems*. John Wiley & Sons, 2010.

[4] 5GACIA 5G alliance for connected industries and automation, "White paper, 5G for connected industries & automation (Second Edition)," 2018, accessed Sept. 20, 2020. [Online]. Available: https://www.5g-acia.org

[5] M. E. Clynes, "Cyborgs and space," *Astronautics*, vol. 14, pp. 74–75, 1960.

5   Smart City and Smart Life

Smart city and smart life refer to a very broad scope of use cases, which will transform all aspects of urban life. These use cases aim at providing efficient municipal governance, high-quality public services, and sustainable economic development. This transformation has been under way since the 4G era, beginning from IoT support in mobile communication, and evolving to 5G as mMTC. The next phase involves an evolution towards 6G, which may generate more stringent requirements for mobile communication systems.

5.1   Smart Transportation

Example 5.1 Summer holidays are coming, and Mark, his wife Jennifer, and three kids are on a road trip. It will take them 12 hours to get to their destination, and, as such, long-distance driving would have been both tiring and dangerous, but not this time. Their new car is equipped with the Level 5 autonomous driving capability, which makes the journey fun, relaxing, and enjoyable. The intelligent vehicle completely takes over driving and Mark and Jennifer can now enjoy the beautiful scenery while also playing games with their family.

Even though swarms of people travel with their cars during the summer holidays, Mark and Jennifer do not have to deal with road congestion. This is because "flying cars" turn the "road" from 2D to 3D, as shown in Figure 5.1, considerably improving its capacity. An additional benefit is the fact that they don't have to learn how to fly the car. With the Level 5 autonomous driving capability, all cars in the 3D traffic system are well managed and travel to their destinations via their respective optimal routes.

Cellular V2X communication trends began from 4G Long Term Evolution-Vehicle and transformed the entire transportation ecosystem. This not only improves the driving experience, but also delivers more safety and higher resource efficiency. Cellular V2X has a very broad usage scope, which includes: safety, vehicle operations management, convenience, autonomous driving, platooning, traffic efficiency, environmental friendliness, as well as society and community [1]. Among them, autonomous driving is the most challenging usage and will bring most innovations.

Figure 5.1 Flying cars with Level 5 autonomous driving.

The Society of Automotive Engineers (SAE International) uses different levels to define the automation capabilities of vehicles, among which, Level 0 refers to fully manual vehicles and Level 5 refers to fully autonomous vehicles [2]. Sensing, connectivity, and communications are the fundamental factors propelling the development of autonomous driving vehicles.

Among the various sub-use cases of autonomous driving, teleoperation could be the first to arrive on the market. Some of the areas it could be applied to include industrial mining, quarry, construction, and agriculture. With autonomous driving, all heavy machines (e.g., cranes and drilling machines) in perilous areas could be operated remotely, representing the new workforce of driving. The existence of Level 5 autonomous driving vehicles or drones in consumers' daily lives will redefine our understanding of car travel. Without rush-hour traffic and the need to plan accordingly, Level 5 autonomous driving will impact our lives just like the transformation from horse power to steam power did. For Level 5 autonomous driving, the most challenging factors are unknown environmental situations. In light of this, sensing and AI services provided by mobile communication systems will assist vehicles in obtaining accurate information for decision-making. Driving efficiency (e.g., vehicle speed and traffic density) goes hand in hand with the performance of communication systems. As such, needless to say, service-level requirements demand ultra-low latency (e.g., millisecond level or lower) and high reliability (e.g., 99.9999% or higher) communications as well as precise localization (for instance, to determine the distance between vehicles). Security and safety will also be crucial to the formulation of service-level requirements. To put this into perspective, the consequences of being hacked by malicious attackers could be unimaginable.

From an architectural perspective, some practical issues should also be considered. One such issue involves upgrading the edge computing platform, which involves multiple operators and vendors to eliminate performance bottlenecks, especially for high-speed terminals. Other than high-performance wireless links, system intelligence is also a must. It does not only include adaptive decisions based on the duration of situations, but also predicts hazardous situations and provides actions or suggestions accordingly.

Vehicles feature a very complicated terminal format – a mobile intelligent entity without extensive energy constraints – and are therefore vastly different from conventional mobile devices such as smartphones. They generate high amounts of data, and collect data from nearby peers; therefore, they can act as moving edges of a fixed infrastructure, which will bring interesting but challenging questions regarding system design.

5.2   Smart Building

Example 5.2 Mark and Jennifer arrive at their hotel, where parking has already been reserved and check-in completed. A cobot waits for them, helps them carry the luggage, and guides the family to the room. Over the next few days, the cobot also serves as a guide for the family during city tours tailored to their preferences and hobbies.

The building industry itself is a self-contained ecosystem with diversified use cases. Like other business sectors, it is also in the process of digital transformation. Smart building technology does not only involve installing the latest electronic products (e.g., elevators and TVs) in buildings, but also involves managing and controlling the buildings as intelligent entities with seamless information flowing between the electronic products, smart materials, and building control and security systems. It also involves flexibly achieving multi-purpose buildings with greatly extended energy efficiency and intelligence.

The first trend is building system integration and automation, in which a single building may contain up to 10 different subsystems that are interconnected using heterogeneous communication methods. These subsystems could include, for instance, a surveillance camera system, elevator control system, fire alarm system, parking management system, air conditioning system, power supply and distribution system, as well as building automation control system. Given this, the maintenance and operation of such buildings could incur high costs, while also being inefficient. The key to introducing mobile communication technology for the smart building industry is to use it as a common infrastructure to unify different subsystems, which has strict and customized resource isolation and support for native trustworthiness. The smart building industry will also extensively adopt ICT from mobile communications to IoT. For example, in a modern commercial building, the number of deployed sensors (for environmental monitoring, infrastructure monitoring, personnel control, etc.) could reach the level of 10 to the power of 5 or higher. Hence, it is essential that an extensive amount of connectivity support is provided for the smart building industry. On top of that, some sensors used for environmental monitoring (for instance, smoke detectors) usually have very high requirements in terms of energy consumption. A battery should typically be able to last for at least a couple of years, and ideally for over 20 years. Moreover, buildings will also be equipped with smart materials and objects like robots and cobots, which need communication and are interactive.

The second trend is to interconnect buildings across different regions. In the future, buildings will no longer be isolated; instead, they will be interconnected in the same region or across different regions. Mobile communications infrastructure is therefore anticipated to provide the digital foundation for cross-platform and cross-domain trustworthiness.

5.3 Smart Healthcare

Example 5.3 Due to the fact that Lisa, Mark's mother, lives alone and has been diagnosed with several chronic diseases, Mark sets up her home with smart healthcare technology, including a sensing system to monitor her health condition, and equipment for starting three-way holographic telepresence. This ensures that doctors receive vital data in the case of an emergency, and that Mark is immediately informed. Lisa feels as if Mark and her doctor are by her side, and she can receive the best medical advice immediately.

In order to maintain her health, Lisa needs to do aerobic and stretching exercises regularly. She has an "intelligent haptic cloth" that can monitor her movements and provide both audio and haptic feedback, just like the voice and hands of a human coach. With this haptic cloth, Lisa can download different programs for various exercises, and, if necessary, a real human coach wearing a similar cloth can transmit real-time haptic information to help her learn new exercises.

Pervasive and customized health service provisioning without any geographical constraints is the eHealth vision for the next decade. This vision relies strongly on the progress of mobile communication systems, which poses critical requirements on system reliability, availability, security, and privacy handling.

Combined with AI, real-time analysis on the data collected from patients could help in the prediction of serious diseases. In that regard, advanced infrastructure with sensing capabilities will go a long way in helping to control major pandemics, such as COVID-19. Combined with advanced video, holographic and haptic technologies will take immersive professional interaction and operation experience to the next level. (The relevant KPI requirements for 6G were described in Chapter 1.) We can therefore say that remote diagnosis, remote surgery, and dynamic monitoring, as well as holographic medical training will become prevalent in the future, as shown in Figure 5.2. Use cases like remote surgery may require ultra-low latency and

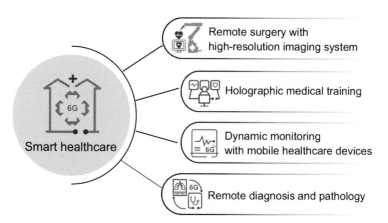

Figure 5.2 Exemplary use cases of smart healthcare.

ultra-high reliability communication over a cross-continent distance (e.g. for a doctor in China to perform cardiac surgery for a patient in Africa). This will be essential for reducing the pressure in aging societies, especially in regions with insufficient medical resource.

5.4   Smart Services Enabled by UAV

Example 5.4 Lisa, Mark's mother, cannot go on the family vacation because she had an accident, which led to her having limited mobility. Mark uses a drone to make XR calls with Lisa, and he also commands the drone to capture the Alps from the sky. This enables Lisa to enjoy a full immersive experience, and she feels as though she traveled with the family.

During the vacation, Mark buys a souvenir for Lisa in the local shop. It is a breath-taking ring, which Lisa adores and can't wait to wear. In order to deliver the ring to Lisa as soon as possible, Mark calls in a courier drone that takes off immediately, like a messenger pigeon in ancient times, as shown in Figure 5.3. When power runs short en route to Lisa, the courier drone flies to the nearest charging station, enabling it to cover very large distances.

UAVs, also referred to as drones, are a type of aircraft without a human pilot on board. They can be remotely controlled via a UAV controller, and feature a range of partly autonomous flight capabilities, which will be fully autonomous in the future. UAVs come in a wide variety of sizes and weights, and can be used in numerous different business sectors. For instance, as shown in Figure 5.4, they have commercial and industrial applications, such as process automation for unmanned inspection, first response for smart and safe city, environmental monitoring, and smart logistics as mentioned in Example 5.4 and illustrated in Figure 5.3.

Figure 5.3 Package delivery by UAVs.

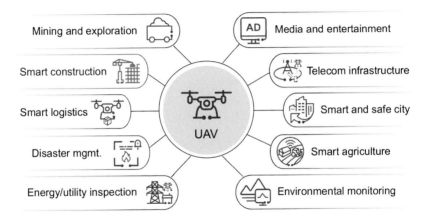

Figure 5.4 UAV application areas and use cases.

Drones can also be used to push the boundaries of the above industries with new potential use cases. According to Keystone, the UAV technology market will reach US$41–$114 billion globally over the next five years [3].

Regardless of the type of industry sectors, the main feature provided by UAVs is air mobility. As such, they serve as a platform that can naturally be used to carry various sensors and high definition cameras. For this reason, UAVs will play an essential role in enhancing the sensing capability of 6G. UAVs are not only a terminal function, but can also be used as a relay or access point to form a temporary network and extend mobile communication coverage, especially in special scenarios (for instance, mountain accidents or natural disasters). Put differently, the flexibility and agility of UAVs could make them serve as good alternatives in critical situations without fixed infrastructure support.

UAVs are an existing 5G use case, and they will evolve with 6G in terms of higher KPI requirements, some of which include ultra-high data rates (e.g., 360 degree video streaming with higher than 10 Gbit/s throughput), ultra-high reliability and ultra-low latency (for seamless control or autonomous flying). UAVs pose high-accuracy positioning requirements, and a case in point is the fact that centimeter-level localization accuracy may be needed for 8K video live broadcast, laser mapping/HD patrol, and periodic still photos. Other than KPI requirements, industrial-grade trustworthiness will be a must-have for UAVs. This is applicable in scenarios where, for instance, UAVs become a source of private user data. Within this context, new research for mobile communication systems will also focus on guaranteeing data sovereignty.

However, battery capacity limits the capability of UAVs to operate for extensive periods of time. As such, green communication will definitely be beneficial in terms of providing better support for UAV communication.

References

[1] 5GAA, "White paper, C-V2X use cases, methodology, examples and service level requirements," 2019, accessed Sept. 2020. [Online]. Available: https://5gaa.org/wp-content/uploads/2019/07/5GAA_191906_WP_CV2X_UCs_v1-3-1.pdf

[2] SAE On-Road Automated Vehicle Standards Committee *et al.*, "Taxonomy and definitions for terms related to driving automation systems for on-road motor vehicles," SAE International: Warrendale, PA, USA, 2018.

[3] HUAWEI Technologies (UK) Co., Ltd., "White paper: Connected drones. A new perspective on the digital economy," 2017, accessed Sept. 2020. [Online]. Available: https://www-file.huawei.com/-/media/corporate/pdf/x-lab/connected_drones_a_new_perspective_on_the_digital_economy_en.pdf?la

6 Global Coverage for Mobile Services

The next-generation wireless network will need to provide global coverage for mobile services so that people and things can access the Internet anywhere and at any time. In 6G, terrestrial networks can integrate non-terrestrial access nodes, such as satellites, drones, HAPs, and flying cars. This will form a global integrated communication network that offers seamless coverage spanning land, sea, and sky, as shown in Figure 6.1. Furthermore, it will allow users to connect to the 6G network regardless of whether they are on foot, in a vehicle, or aboard an aircraft, even if some infrastructure is faulty or damaged, ensuring an uninterrupted service experience.

In 6G, terrestrial and non-terrestrial networks will be fully integrated at the system level, achieving convergence of services, air interfaces, networks, and terminals. By organically integrating these two access media into one converged, multi-layer heterogeneous network spanning the globe, 6G will provide users with a consistent service experience. The capability to provide global coverage of mobile services will be an important aspect for the 6G network.

The integrated 6G network will expand service capabilities and provide various types of services. For example, by integrating terrestrial and non-terrestrial networks, 6G will enable broadband and IoT services in regions and areas that lack terrestrial network coverage, including remote farms and even aboard ships and aircrafts. Furthermore, 6G will facilitate the expansion of new applications, such as high-precision satellite–ground positioning for vehicles and high-precision real-time imaging for agriculture.

6.1 Broadband Wireless Access for the Unconnected

The integrated 6G network will provide 3D coverage, eliminating all coverage gaps worldwide. For example, it will allow people in remote and unpopulated areas, aboard types of moving transportation such as ships and aircrafts, and at sea on oil-drilling platforms to connect to the Internet via high-speed wireless links. People will be able to connect to the 6G network via terrestrial and non-terrestrial media using the same mobile phone or handheld device anywhere and at any time. Furthermore, the integrated 6G network will be resilient to natural disasters, meaning that it can ensure connectivity for first responders and disaster relief.

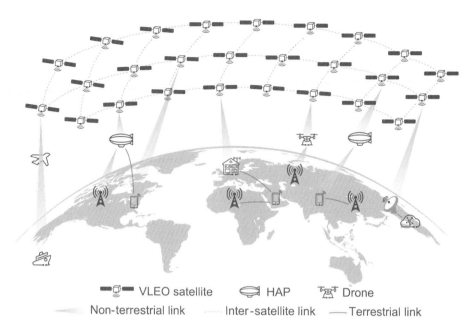

VLEO satellite   HAP   Drone

Non-terrestrial link   Inter-satellite link   Terrestrial link

Figure 6.1 Global integrated 6G coverage.

6.1.1 Mobile Broadband for the Unconnected

Example 6.1 Carl owns a farm far from the city center. The cellular coverage at the farm is weak, with service outages occurring sometimes. A telecommunications operator tells him that it is now offering 6G services with high-quality broadband access regardless of his location. Carl orders the new 6G service and updates his mobile phone to a 6G-capable one that integrates terrestrial and non-terrestrial access. In downtown, the 6G-capable mobile phone can connect to the terrestrial network, just like the old phone. At the farm, Carl can now connect to both the terrestrial and non-terrestrial networks transparently, improving his broadband experience. Regardless of his location, Carl can now obtain broadband connectivity.

Today, an estimated 3.7 billion people still do not have basic Internet access [1]. Most of them live in rural and remote areas, where the terrestrial and non-terrestrial networks can be integrated to improve the user experience of broadband services, as shown in Figure 6.2. For maritime applications, the integration of broadband satellites, ship–earth stations, and terrestrial ground stations provides higher bandwidth, enabling high-speed, real-time, and low-cost communications.

In addition to fixed and moving relay applications, the direct connection between non-terrestrial stations and mobile phones is an attractive prospect. For many years, such a connection was extremely expensive and offered a low transmission rate.

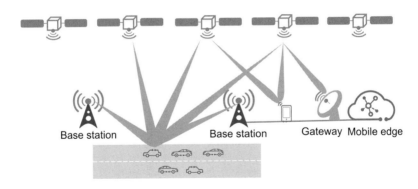

Figure 6.2 Integrated service for pedestrians and cars.

It also required people to carry two different phones: one to access the satellite network and another to access the cellular network. In the future, 6G will make it possible to integrate all types of access services into one mobile phone, ensuring seamless switchover between services.

To provide a consistent, high-quality service experience in rural areas, the integrated 6G network should offer an experienced data rate similar to that offered via 5G rural macro base stations; specifically, 50 Mbit/s for downloading and 25 Mbit/s for uploading per user [2].

6.1.2 Broadband Connection on the Move

Example 6.2 Celia is a penguin researcher and had been stationed in Antarctica for the last six months. Having completed her assignment, she is preparing to go home, which necessitates taking an icebreaker to Buenos Aires, and then taking a plane. On the icebreaker, she uploads the videos of penguins to social media via the icebreaker's broadband satellite service. After arriving in Buenos Aires, she boards the plane home and, using the plane's 6G communication system, makes a video call with her family mid-flight.

Enabling people to access Internet services at any time via the same device regardless of their location, be it at home or aboard a plane, is one of the ultimate goals for wireless networks. Types of moving transportation include cars, trains, aircrafts, and ships, as shown in Figure 6.3. Taking aircraft as an example, more than 4 billion people travelled via aircraft in 2019 [3]. Most of them had no Internet connection during their flights; for those that did, the connection was slow and expensive. In the future, integrated 6G systems should provide MBB connections for all aircraft passengers. The data rate experienced by each user should exceed 15 Mbit/s and 7.5 Mbit/s for downloading and uploading, respectively [2], assuming a 20% activity factor and 400

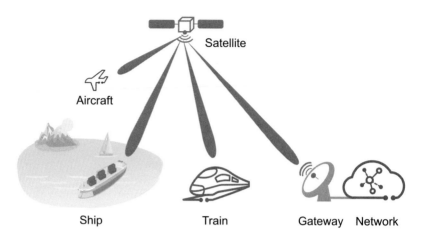

Figure 6.3 Broadband connection on the move.

users per plane. In this case, the overall experienced data rate for one plane should be no less than 1.2 Gbit/s and 600 Mbit/s for downloading and uploading, respectively.

6.1.3   First Responder Communication and Disaster Relief

Example 6.3 Andrea works for a national emergency management center, where she is responsible for providing disaster alerts and organizing emergency rescues if disasters occur. During the tornado season, Andrea broadcasts tornado information to fishermen and ships via satellite networks. Typically, if an earthquake strikes, part of the cellular network is damaged, meaning that Andrea must use non-terrestrial networks (including satellites, HAPs, and drones) to communicate with the onsite rescue teams. 6G enhances emergency rescue services by integrating the undamaged terrestrial and non-terrestrial networks. It also enables mesh networks to be constructed so that members of rescue teams can communicate with each other and the emergency control center.

A reliable, uninterrupted emergency communication system is critical in disaster management scenarios to provide disaster prediction, warning, emergency response, and emergency communication. In some cases, natural disasters may disrupt terrestrial networks; however, 6G integrates terrestrial and non-terrestrial networks to ensure non-stop services, supporting the management of public emergencies. The emergency management communication system involves multiple systems, such as wired and wireless voice, data network, and video systems. With the help of the emergency management and dispatch software, the command center, associated units, expert teams, and onsite rescue personnel can be quickly contacted.

6.2 Wide-Ranging IoT Services Extended to Unconnected Locations

---

Example 6.4 John is a marine biologist studying the impact of the marine environment on fish. He has dropped many buoys to collect information about the ocean. These buoys are powered by solar panels and can connect to the 6G network. John can collect information from the buoys, including seawater pollution, seawater temperature, wave height, and wind speed, as well as their precise location, through the 6G network. He can download the updated information from his office computer to track aquatic changes.

Currently, IoT communication is based on terrestrial network coverage. However, this approach cannot guarantee uninterrupted connectivity in many scenarios. For example, collecting information from buoys in the ocean or reporting information from containers during ocean transportation might experience interruptions. In the future, IoT devices should be able to connect and report information anywhere and at any time. Wide-ranging IoT services will be extended to cover unconnected locations such as remote areas, oceans, and more.

The cellular service is limited in remote areas, which may lack sufficient resources for massive IoT devices. In many unpopulated areas, such as Antarctica and deserts, people deploy IoT devices to collect information. However, these areas are not covered by terrestrial networks, meaning that information upload and collection are restricted. 6G, by integrating terrestrial and non-terrestrial networks, will provide access for IoT devices in such areas, making it easy to collect information such as the status of penguins or polar bears, and monitor crops at remote farms.

In some maritime scenarios, people need to deploy buoys in the ocean to measure wave heights, water temperature, and wind speed. Such information can help mariners avoid areas of strong winds and waves or take necessary precautions. In addition, reporting information about containers during long-distance ocean transportation is important. If people can obtain information about each container in real time, they can check the temperature, humidity, and location of containers during the entire transportation process.

6.3 High-Precision Positioning and Navigation

Example 6.5 After landing at the airport, Celia decides to rent a self-driving car. The car, using precise positioning and prediction, stops in front of her when she reaches the terminal door. Celia gets into the car, which drives her home with the help of precise positioning and navigation while she surfs the Internet. Celia lives a long way away from downtown, so when the car loses terrestrial coverage, it dynamically switches to

6G's non-terrestrial network coverage for the navigation and communication services. After driving for about one hour, the car arrives safely at Celia's home where her family await her.

Current satellite navigation systems, such as the Beidou and global positioning system, are used extensively and achieve an accuracy of about 10 meters. A global navigation satellite system (GNSS) and LEO constellation integrated positioning technology can implement high-precision positioning and navigation, achieving an accuracy of about 10 cm in outdoor scenarios. Using integrated terrestrial and non-terrestrial networks, high-precision positioning and navigation can be realized around the world, regardless of whether it is used in city centers or remote areas. This will facilitate the deployment of many new services, such as high-precision user location services and navigation services for self-driving cars, precise agricultural applications, and mechanical construction.

High-precision navigation can effectively improve the efficiency of agricultural applications, for example, by enhancing the positioning accuracy of agricultural vehicles. Enhanced positioning and navigation technologies will improve the quality of operations and provide a strong technical guarantee for operations such as precise land preparation, sowing, cultivation, fertilization, plant protection, harvesting, spraying, mechanical picking, and more.

In the future, most cars will be able to connect to the 6G network. The terrestrial network can provide high-quality V2X services for vehicles in the city, while the non-terrestrial network can provide accurate (within a few dozen centimeters) positioning and navigation services for vehicles in remote and unpopulated areas, with help from LEO mega constellations, as shown in Figure 6.2. Furthermore, drones and UAVs will be used extensively in the future, when the integrated 6G network should provide higher accuracy navigation services.

6.4   Real-Time Earth Observation and Protection

Example 6.6 On a trip to visit a national park, Celia and her family witness an accident where a fuel tanker has overturned and caught on fire. The 6G-assisted earth observation system detects this situation and sends the data back to the control center in real time. The control center analyzes the composition of fuel and finds that the tanker is at risk of explosion. The control center immediately informs Celia of the risk via the car's navigation system, warning her to keep a safe distance from the accident area. Celia switches to a new route accordingly and arrives safely at the national park. As well as notifying Celia, the earth observation system also sends alarms to traffic police and firefighters, helping to avoid a potentially severe explosion.

One typical use case of the current satellite system is performing remote sensing and earth observation via imaging techniques. Such techniques include optical imaging, which obtains image data using visible light cameras and partial infrared band sensors, and RF imaging, which obtains image data by recording and analyzing radio waves reflected from earth with a synthetic aperture radar. However, owing to the limited communication capability, it takes time before people can use these images. In the future, the integrated 6G network will enable earth observation and communication to be performed simultaneously within a single system.

References

[1] ITU Broadband Commission for Sustainable Development, "State of broadband report 2019," 2020. [Online]. Available: https://www.itu.int/dms_pub/itu-s/opb/pol/S-POL-BROADBAND.20-2019-PDF-E.pdf

[2] 3GPP, "Service requirements for the 5G system," 3rd Generation Partnership Project (3GPP), Technical Specification (TS) 22.261, 07 2020, version 17.3.0. [Online]. Available: https://portal.3gpp.org/desktopmodules/Specifications/SpecificationDetails.aspx?specificationId=3107

[3] E. Mazareanu, "Air transportation – statistics & facts," 2020. [Online]. Available: https://www.statista.com/topics/1707/air-transportation

7 Connected Machine Learning and Networked AI

Since 2012, breakthroughs in the technologies associated with DNN-based machine learning (ML) have given rise to AI technologies, driven by advances in super-scalar computing power and the modern Internet making big data accessible. The advanced technology revolution is focusing on AI, which has reached – and in some cases surpassed – human ability. For example, AI outperforms humans in sophisticated gaming such as chess and Go, and even in tasks such as image and voice recognition.

A recent study [1] predicts that AI will be able to perform a wide range of tasks during the lifetime of 6G (i.e., from 2030 to 2040). The use cases for 6G to deliver AI services are not easy to predict, but the space for innovation will be huge due to the combined power of 6G and AI with ML. This means that AI has the potential to have a significant impact on society and to become the most important service and application that 6G will enable and deliver.

From the mathematical perspective, we can understand the ML techniques in AI as solving a series of complicated optimization algorithms. With the evolution of technologies (especially DNN), the enhancement in computing capabilities, and the availability of massive real-time data in mobile networks, AI with ML is expected to solve higher-dimensional optimization problems that cannot be solved with traditional algorithms.

As was discussed in Chapter 1, 6G is a key enabler for making AI available to every person, anywhere and at any time. Consequently, 6G is not simply a connecting pipe for AI services and applications; rather, it will be designed from the ground up to optimize AI in a more efficient way. Advanced mobile transmission technologies, as well as the computing power offered by network nodes, will be leveraged to augment large-scale intelligence with widespread distributed learning. This will make it possible for society and numerous industries to benefit from the capabilities offered by AI. Knowledge is acquired from massive data sensed in a target environment, and then used to guide intelligent actions and help stimulate and improve productivity for the target applications. With the help of AI and 6G, we therefore believe that a worldwide revolution of cognition will happen in the near future.

As shown in Figure 7.1, AI will play an important role in both enhancing communications systems and supporting smart use cases. The following sections focus on two categories of basic use cases, namely, AI-enhanced 6G services and operations, and 6G-enabled AI services.

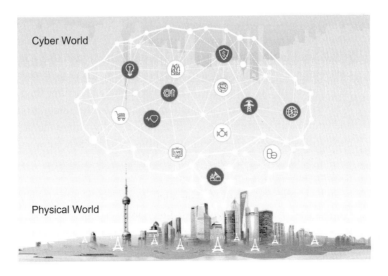

Figure 7.1 Native AI in communications systems and smart use cases.

7.1   AI-Enhanced 6G Services and Operations

Service provider networks – especially wireless networks – currently in operation require a large workforce to manage, maintain, and operate them. For more than 250 operators worldwide, their expenditure on network operations, administration, and maintenance (OA&M) is a major financial burden. It is hoped that one day we will no longer need personnel to perform operations and maintenance – this is the ultimate zero-touch networking.

7.1.1   AI-Enhanced 6G Network Performance

Example 7.1 Tom is responsible for operations in a smart factory that employs few human personnel. AI in 6G helps the network to provide extremely reliable and prompt communication among Industry 4.0 robots. During his lunchbreak, Tom plays a football game virtually with friends. AI in 6G ensures that the network latency is optimal, allowing Tom to enjoy an immersive 360° VR experience without any discomfort, such as motion sickness. When he feels thirsty, Tom gives verbal instructions to a robot, which understands the instructions thanks to AI in 6G empowering the robot with semantic sensing capabilities. This allows the robot to deliver a drink to Tom just like a human waiter. After work, Tom drives to another city to meet some friends. AI in 6G enables the network to support smart transportation with Level 5 autonomous driving, which ensures that his trip is both safe and relaxing. During his trip, AI dynamically adapts the provision of his 6G network coverage according to his location and requirements of services.

By 2030, smart devices around the world will include a growing number of personal and household devices, various citywide sensors, unmanned vehicles, smart robots, and more. These smart devices will require wireless connectivity to facilitate the complete coordination and cooperation of unattended tasks, calling for higher throughput, higher reliability, and lower latency and jitter than the communications systems available today can support. Thanks to the integration of communications and computing platforms, we will be able to fully utilize the perception and learning capabilities of networks. Communications systems will automate resource configurations intelligently, provide highly intelligent and customized services for users, and achieve near-optimal performance.

In Example 7.1, AI in 6G can adapt to the environmental changes and implement the optimal solution in potential emergencies, helping to ensure that the network meets the stringent reliability and latency requirements of Industry 4.0 for information exchange among robots and the central system. For immersive 360° VR, as discussed in Section 4.1, compression and local rendering help to partially mitigate the strict throughput and E2E latency requirements on the network. Furthermore, AI functions as an optimization engine, optimizing both communications and computing resources intelligently. The AI capabilities of 6G networks in these areas will directly affect the quality of service.

AI used in 6G networks can empower robots with semantic capabilities, enabling them to understand the context of semantic instructions. This will help realize a more human-like effect for sensing and localization services. Furthermore, AI can enable intelligent information "source coding," allowing robots to communicate at the semantic level rather than at today's bit level.

AI can also be applied to synthesis network slices in order to achieve heterogeneous networks, such as integrated non-terrestrial and terrestrial networks. AI in 6G will help to coordinate complex multi-layer heterogeneous networks, and eventually provide users with optimal coverage.

7.1.2   AI-Enhanced Network Operations

Example 7.2 Esther works for an operator as a network planning engineer. She has been tasked to prepare for a concert that is expected to attract many people. To accommodate the increase in traffic, Esther plans to add several UAV or low-orbit satellite nodes to provide additional network capacity for the concert. Similar tasks in the past have needed her to configure many parameters and perform lengthy tests. But now, using 6G with native AI capabilities, she can simply drag the nodes to the desired location via the intelligent network planning system's control tablet, following which the subsequent work will be done automatically.

As large numbers of UAVs, HAPSs, and VLEOs become part of the 6G infrastructure, operating such a multi-layered system can be an extremely complex task. AI will be a vital tool for tackling issues such as smart heterogeneous access, energy consumption, optimization of tens of thousands of parameters, and enhancement of system performance. For instance, the AI agent will learn and adapt dynamically to optimize the usage of spectrum, power, antennas, and energy. This will help ensure that different communication nodes, such as satellites, drones, and ground stations, can cooperate effectively.

Moreover, enhanced intelligence will allow some 5G use cases to be further extended in 6G. The key is to leverage AI to fully automate network operations, with the ultimate goal of reducing and eventually eliminating human intervention in network OA&M tasks and realizing zero-touch operations.

Another promising example is the automatic creation of specialized network slices according to service characteristics. This meets operators' network tailoring needs, helping them transform to real factories for network slices.

7.2   6G-Enabled AI Services

In many areas, AI has achieved human-level intelligence. In terms of computing, however, it is not as efficient as the human brain; there is still a significant gap to be closed. The computing speed of the current computing platform doubles every 18 months, adhering to Moore's law [2]. However, the demand for AI computing is increasing much faster, about 40–100 times faster than Moore's law, over the past 8–10 years [3].

A recent study of AI's computing limit suggests that using ML for huge tasks will be expensive and not environmentally friendly [4]. Fully leveraging the global learning capabilities through a highly efficient network like 6G will eliminate the duplication of ML computations and minimize the transfer of big data necessary for ML across the wireless network. This will be the driving factor for 6G design. 6G, with advanced communication and distributed computing capabilities, is expected to help close the gap between AI computing and the human-level benchmark. Distributed and collaborative ML enabled by low-latency and high-capacity communications are required to fully maximize the computing efficiency. We provide two examples in the following sections to show the native support of different AI services in 6G systems.

7.2.1   6G for Collaborative Intelligence and Real-Time Control

Example 7.3 Tom is working in a building when an earthquake occurs, trapping him inside. A fleet of mobile robots is sent out to locate him, and ideally help him to escape. These robots are empowered with collaborative AI capabilities with a 6G network. Individually, they search every possible route that may lead to Tom's position. Each robot can see, hear, and sense its surroundings for the signs of life, fire, poisonous gases, and more. The local computing capability partially trains its AI model on data collected from the surrounding areas. At the same time, locally trained models together with some raw sensing data (if needed) from the entire fleet of robots are transmitted to the network. Network nodes leverage their global information and enhanced computing capabilities to further refine the global AI model. This global model is then distributed to each robot to enhance local inference, thereby improving and synchronizing the intelligence of all robots. The robots eventually find Tom, thanks to their collaborative efforts, and instruct him to reach a safe location, determining the best route there by processing the global information from all robots. Despite their tiny size, the robots work together using their combined strength of cooperative carrying to clear some larger obstacles along the route.

In this case, some specific requirements for the 6G network can be summarized from three aspects:

- **Low latency and massive capacity uplink access:** In collaborative AI, each robot transmits its locally trained AI model (and potentially some of the raw data) to the network, contributing to the global intelligence. The 6G network needs to support low latency and massive uplink connections, each of which is required to provide bandwidth reaching several Gbit/s.
- **Deep-edge node intelligence:** Tom may have concerns about data privacy (e.g., on-scene pictures and voice). Owing to the critical nature of the situation, it is essential to ensure low-latency communication among network nodes and robots. To address Tom's concerns while also ensuring low latency, instead of sending all information to the cloud a local edge node is used. This edge node is physically close to the scene, and provides dedicated and prompt services to some degree.
- **Distributed ML:** Transmitting all raw data to the central cloud imposes a significant burden on network bandwidth, and processing this large amount of data imposes a burden on the central node's computing capabilities. By organizing computing tasks in a distributed manner, collaborative AI and ML can help alleviate these burdens.
- **ML data compression for air interfaces:** Transmitting the large amount of data required for ML and DNN training over wireless links is costly and inefficient. Studying emerging information theories, such as information bottleneck, is crucial for learning how to compress the large volume of data required for training, since the compression philosophy is very different from the existing visual/audio source coding which has been optimized for human perception.

7.2.2   6G for Large-Scale Intelligence

Example 7.4 Forests are a complicated ecosystem. Since the 1930s, it has taken scientists studying forests 60 years to understand the raw data and realize that naturally occurring forest fires help keep forests healthy. New policies were subsequently adopted in the 1990s to stop interference in natural forest fires. However, forests replanted by humans are more vulnerable to natural fires because they are 2–3 times denser than natural ones on average. Without human prevention, a natural forest fire would spread too widely to be controlled. This was identified as the main cause of frequent epidemic forest fires since 2015, when most of the trees replanted after the 1990s had grown into mature forests. It has taken us another 25 years to realize that replanted forests are too dense.

Given the long period needed to accumulate data, gain knowledge from the data, and then determine the optimal policy, the policy is likely to be obsolete before it can even be implemented.

The 6G-enabled cyber world would address these issues. With three-dimensional full coverage and sensing capabilities of the 6G network, different sensors can be widely deployed in the forests. High-density forestry data, such as temperature, wind, soil, moisture, treetop heights, and so on, are continuously collected. Then, AI can be used to identify the correlation and periodicity, recognize patterns, perform simulations, and compare the data with other forests and historical data. Different policies can be evaluated in the cyber world, allowing policies to be developed more efficiently and correctly, and ensuring that damage to physical nature is minimized.

In this case, wireless network coverage of the forest is a fundamental requirement. However, because there is only limited human activity in the forest, it is not economically viable to cover the forest with only a terrestrial mobile communication system. On the other hand, satellite services are currently too expensive to implement sensing and data collection on a large enough scale for replication in the cyber world. The power consumption and potentially short battery life of satellite terminal devices also prevent large-scale deployment in the forest. This requires the new capabilities of 6G.

6G will provide three-dimensional full coverage by integrating non-terrestrial and terrestrial networks. With the development of VLEOs, UAVs, and HAPs, together with pervasive intelligence and sensing capabilities, replication of the physical world (i.e., the forest) in the cyber world will facilitate the development of optimal policies for environmental sustainability.

In this case, large amounts of data are collected from forests all over the world. These data are fed into distributed local deep-edge nodes to perform ML, and the trained neural network parameters are exchanged among all deep-edge nodes for collaborative AI. Abundant data will significantly improve the intelligence level regarding forest issues. The prediction results of different development policies will be more accurate and efficient, without requiring extended periods of data collection.

References

[1] K. Grace, J. Salvatier, A. Dafoe, B. Zhang, and O. Evans, "When will AI exceed human performance? Evidence from AI experts," *Journal of Artificial Intelligence Research*, vol. 62, pp. 729–754, 2018. *arXiv preprint arXiv: 1705.08807*, 2018.

[2] Wikipedia, "Moore's law," accessed Sept. 2020. [Online]. Available: https://en.wikipedia.org/wiki/Moore%27s_law

[3] R. Perrault, Y. Shoham, E. Brynjolfsson, J. Clark, J. Etchemendy, B. Grosz, T. Lyons, J. Manyika, S. Mishra, and J. C. Niebles, "The AI index 2019 annual report," AI Index Steering Committee, Human-Centered AI Institute, Stanford University, Stanford, CA, 2019.

[4] N. C. Thompson, K. Greenewald, K. Lee, and G. F. Manso, "The computational limits of deep learning," *arXiv preprint arXiv: 2007.05558*, 2020.

Summary of Part II

6G will bring a wealth of new possibilities to various aspects of wireless technologies. The six categories of use cases that were discussed in this part cover the future usage scenarios of 6G that we currently foresee. However, from the perspective of all possible 6G applications, these 6G use cases are merely examples of what the future holds.

The numerical analysis on the performance requirements in these typical use cases leads to the summary of target KPIs introduced in Chapter 1, especially from the RAN aspect. Among the six categories, the most stringent requirements determine the specific KPI values. For example, in terms of the data rate, 3D holographic communication poses the highest requirement. And in terms of transmission latency, teleoperation in highly dynamic environments and industry motion control have posed the highest requirements. The industry use cases are the most demanding for deterministic communication, where the requirements on jitter and reliability are extremely stringent. In future smart cities and smart life, IoT use cases are the main source of requirements on connection density and sensor battery life. More importantly, new use cases such as sensing and AI bring new dimensions in evaluating performance metrics such as sensing resolution and probability. For some of these new dimensions, further research may be needed into their associated measures, such as the flexibility and scalability to support native AI services and the level of trustworthiness of the network.

Part III

Theoretical Foundations

Introduction to Part III

Wireless networks are rapidly evolving into extremely complex systems, with an exponentially increasing number of devices demanding even more connections under increasingly strict requirements. 6G networks in particular will be characterized by:

- **Unprecedented network size and density:** In 2020, wireless communications represented 78% of global IP traffic, an increase of more than 10 billion connected devices compared with 2015. Looking ahead, network infrastructure will become fully integrated with the environment through transmitters embedded in walls, data caching, and wireless sensors located all around us. The communications environment will evolve into an intelligent wireless fabric, where objects like buildings, walls, cars, and road signs will be equipped with intelligent surfaces capable of amplifying incoming electromagnetic signals, performing computations, and storing data.
- **Unprecedented level of stochasticity:** With unpredictable network dynamics and connectivity demands, network resources will be opportunistically deployed on the basis of the evolution of traffic conditions and user mobility, while the network infrastructure itself will also become mobile. This opens up new possibilities to process and store data, which can now also be done at the network edge through emerging techniques such as mobile edge computing and edge AI.
- **Unprecedented heterogeneity in terms of services types and QoS requirements:** Future wireless networks will need to simultaneously provide a communication rate of 1 Tbit/s for MBB services, air interface latency of 1 ms for URLLC, localization accuracy of 10 cm for automotive applications, the ability to handle 100 million terminals per km^2 in massive IoT scenarios, coverage enhanced by 10 dB with respect to 5G, and energy efficiency 10 times higher than that of current networks.

In order to keep pace with rapidly evolving wireless technologies, simply increasing the performance of transmission technology is not a viable strategy. Realizing the 6G vision requires a paradigm shift in the way wireless networks are designed. As such, the scientific knowledge we have today is simply not sufficient, and new fundamental knowledge needs to be acquired to understand how the different parts of such a complex system can work together to maximize network performance in terms of throughput, energy efficiency, bit error rate (BER), localization accuracy, and processing/communication latency.

These tools are a combination of theoretical mathematical and physics techniques that model and analyze the interactions among a very large number of nodes and the energy consumption caused by the computations required to process such a huge amount of information. Such a foundation will enable the design and development of the 6G AI network in the future.

8 Theoretical Foundations for Native AI and Machine Learning

8.1 Fundamental AI Theory

Native AI technology has been developing at a rapid pace, particularly since neural networks techniques were invented and empowered by high-performance GPUs. In the first part of this section, we will start with common definitions in machine learning and then describe in more detail the fundamentals of AI theory.

8.1.1 Definitions

Artificial Neural Network (ANN)

An ANN is composed of interconnected elementary processing units called neurons, which are typically organized into some consecutive layers. The input layer receives input data, which are then processed by one or more hidden layers, and finally by the output layer, which provides an output for the ANN. If an ANN has only one hidden layer, it is known as shallow, whereas if more than one hidden layer is present, it is referred to as deep. Deep ANNs are usually preferred, as they have been found to require fewer neurons than shallow networks when executing tasks [1]. In deep learning, each neuron performs three operations:

- computes an affine combination of inputs;
- computes an activation function (typically non-linear), whose input is the result of the previously computed affine combination (for example, rectified linear units, sigmoid functions, and hyperbolic tangents);
- forwards the result to the following neurons in the next layer.

Though the operations performed by each neuron are quite simple, by combining several neurons it is possible to execute very complex tasks. It has been shown that ANNs are universal function approximators, whereby their input–output relationships can reproduce any functions, provided that the weights and biases of the affine combinations are properly tuned [2]. Unfortunately, this result is not constructive, as it does not tell users how to tune the weights and biases to accomplish a given task. Instead, this is achieved by training the ANN, which involves feeding a dataset with desired input–output pairs to the ANN, and then using well-established training algorithms and stochastic gradient descent (SGD) [3, 4] to extrapolate suitable weight and bias configurations.

Deep Neural Network (DNN)

DNNs form a subset of ANNs. A DNN consists of several layers, each of which contains a number of neurons. The neurons at one layer are connected to those at the next layer, and each connection has a trainable weight. Each neuron combines all of its inputs into one single output. This combination function is non-linear, similar to a sigmoid or rectified linear unit (ReLU). Current DNNs have several notable limitations that should be carefully considered:

- **Generalization:** In practice, channels vary with time. If a DNN includes a time-varying fading channel, the DNN's generalization capability will be affected.
- **Complexity:** The minimum number of neurons used for the encoder and decoder determines their complexity.
- **Training data size:** DNN performance depends on the training dataset; deciding on and reducing the size of the training dataset is currently an area of intense research.
- **Training period:** DNN training periods are long and variable; this depends on the training data, which presents a challenge for its application in real-time systems.

Convolutional Neural Network (CNN)

Among all DNN types, CNN is the most popular one; it implements 2D convolutional filtering and max-pooling in addition to full connectivity among layers. Through convolutional and max-pooling layers, a high-dimensional input can be reduced to a lower-dimensional representation, avoiding possible optimization issues caused by dimensionality. The lower-dimensional representation is referred to as a latent layer in DNN terms and contains the most essential features, textures, or semantics of the corresponding high-dimensional input, depending on the specific learning issue.

Through such layer-to-layer architecture, Lagrange optimization for SGD can be realized by a chain of differentiable functions. Using the chain rule, the CNN performs backward propagation over an epoch of data to tune all neurons to its training targets, purposes, or objectives. Typical training objectives include mean square error, maximize likelihood, and minimum classification error. In some cases, more than one training objective is used to direct the training (optimization) process.

Recurrent Neural Network (RNN)

In human languages (semantic), a sentence can be considered as a Markov chain, whereby each word is more or less dependent on those preceding it. Similarly, a radio channel can be also simplified into a Markov chain. However, while traditional DNNs are unable to take previous events into account, RNNs utilize loops to address this issue, and allow information to persist. The concept of an RNN makes a lot of sense for the semantic communication to be discussed in Chapter 10. Long short-term memory (LSTM) is an example RNN for natural language processing (NLP) [5]. It is also proposed for chartering wireless channels [6].

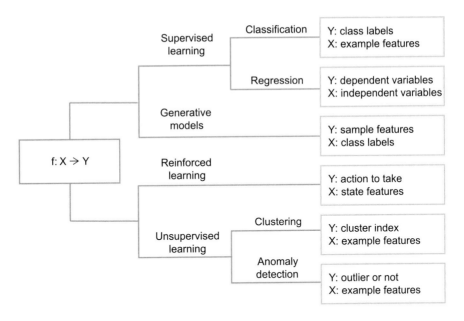

Figure 8.1 Unifying machine learning as a function mapping.

8.1.2 Machine Learning Taxonomy

Mathematically, AI can loosely boil down to the question of finding an unknown function mapping $f : X \to Y$, where X is input feature space representing data points and Y is a label space representing the knowledge outputs. From the differences in the experience available, the objective function, and the specific learning algorithms, as shown in Figure 8.1, the AI tasks can be classified into three categories: supervised learning, unsupervised learning, reinforcement learning. As an improvement, self-supervised (semi-supervised) learning is being developed to reduce the cost of manual labeling. Furthermore, generative models are designed to model data directly, or establish conditional probability distributions between variables.

Supervised Learning

In a supervised learning model, the training dataset is labeled to provide a direct answer for each input, thereby training the DNN to improve its accuracy. Supervised learning generates a predictor from labeled data and expects good generalization on unseen data. It is the most common learning method associated with classification, regression, and ranking problems. While traditional options such as support vector machine and kernel methods are still widely used, supervised learning is the most effective option for big data, especially computer vision and NLP tasks. The success of supervised learning is due to the powerful feature extraction capabilities of DNN, which is known as representation learning. The data structure extractor is a key component of supervised learning; for example, convolution is extremely important for image data. However, different data structures require their own "convolution" equivalents.

Unsupervised Learning

In an unsupervised learning model, the training dataset is unlabeled, and the DNN must learn how to extract features, patterns, semantics, or textures by itself. Consequently, unsupervised learning makes predictions for unseen data based on experience with unlabeled data. With this model, it is difficult to precisely evaluate the generalization capability of a learner. The learner will learn the intrinsic structures and representations of unlabeled data and make predictions on the basis of the structures and representations. The most common tasks associated with unsupervised learning are clustering and dimensionality reduction. Auto encoder is a widely used unsupervised learning tool.

Self-supervised (or semi-supervised) learning is a potentially powerful representation learning method, which automatically generates some kind of supervised signal to solve a particular task (for example, to learn data representations or automatically label a dataset).

Reinforcement Learning (RL)

As one of the most important research directions of ML, RL has attracted much attention over the last 20 years. The agent in an RL process periodically observes environment states, makes decisions, obtains results, and adjusts its policy to achieve optimal performance. One issue which has hindered RL from being widely used in practice is that it takes a significant length of time for the agent to explore all possible states before converging to the best policy.

Deep Reinforcement Learning (DRL)

The rapid development of deep learning [1] in recent years has created new possibilities in the development of RL because the use of deep learning, especially DNNs, can dramatically speed up the training and inference of RL. The new technology of deep reinforcement learning (DRL) embraces the advantage of DNN algorithms and dedicated hardware, such as GPU and NPU. DRL also provides much faster learning speeds and better performance. Moreover, when a group of autonomous and interacting entities exist in a common environment, each entity can be equipped with a DRL agent, helping with cooperation, competition, and coordinating with others by making proper decisions to achieve global objectives.

Most of the decision-making problems under uncertain and stochastic environments can be modeled by a so-called Markov decision process (MDP) [7], which is usually solved by dynamic programming [8]. However, with the scale of the system increasing, the computational complexity rapidly becomes unmanageable. Moreover, precise modeling of the system is sometimes impractical. As a result, DRL provides an alternative solution to overcome the challenge. In Figure 8.2, we provide the main differences between RL and DRL.

The above AI or ML categories need to be implemented by using a specific neural network structure, such as ANN, DNN or CNN, etc. In the following, we will discuss first the information theoretical principles of one neural network structure known as DNN and then describe several implementation classes of DNN.

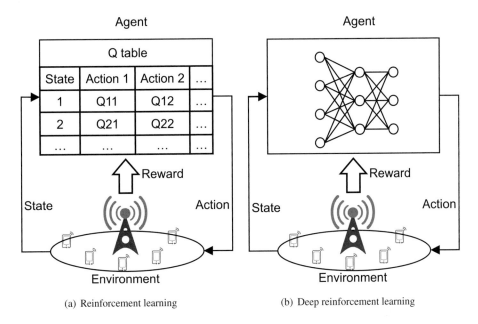

(a) Reinforcement learning   (b) Deep reinforcement learning

Figure 8.2 RL and DRL frameworks in the context of wireless cellular networks.

8.1.3 Information Theoretic Principle of DNN

Many theories are used to interpret the DNN with a focus on deep learning technology, such as optimization theory, expectation maximization algorithms, topology theory, graph theory, semantic theory, dynamics system theory, and approximation theory. Choosing the right theory to interpret the DNN depends on the actual problem. Information theory and information bottleneck theory in particular are among the most suitable theories in the context of wireless communication, as they originate from the same information theory perspective as wireless systems. The following focuses on an interpretation from this perspective.

One metric for the effectiveness of communications is information bottleneck theory [9]. This theory lies at the frontier between machine learning and prediction, statistics, and information theory.

Shannon defined information as uncertainty from a probabilistic view, which provides the building blocks for deep learning theory.

First, the mutual information $I(X;Y)$ reflects the degree of probabilistic interaction between two random variables: X and Y. If X is subject to any changes, Y will vary to a certain extent; and vice versa: $I(X;Y) = I(Y;X)$. The DNN can be modeled as a flow of mutual information from the input to the output.

Second, an adversary relationship over mutual information $I(X;Y) = I(P_X; P_{Y|X})$ includes a concave function $I(X;Y)$ of varying P_X and fixed $P_{Y|X}$, and a convex function $I(X;Y)$ of varying $P_{Y|X}$ and fixed P_X. This results in source coding for rate distortion to minimize $I(X;Y)$ and channel coding for matching to maximize $I(X;Y)$.

Shannon argued that a reliable communication system should include both. Similarly, the DNN also accommodates the two adversary steps. While rate distortion dominates training in some epochs, and matching dominates in others, both are indispensable for training. At the end of training, an equilibrium is reached.

From an information theory perspective, information bottleneck theory is proposed in [9] to explain what happens on a DNN by using a flow of mutual information, adversary rate distortion, and matching.

To train a DNN, we expect the output layer Y to be as similar to the input layer X as possible, and the latent layer Z to be as different from the input layer X as possible. In terms of information theory, this is outlined as maximizing $I(X;Y)$ while minimizing $I(X;Z)$. Rate distortion happens from X to Z, and matching happens from Z to Y. Assume that an information bottleneck γ is applied at the latent layer Z: if the entropy of Z is less than γ, then all Z's information can pass by; otherwise, only the most essential γ part of Z can pass by.

The mutual information between X and Z is defined as follows:

$$I(X;Z) = \sum_x \sum_z P_{Z,X} \cdot \log \frac{P_{Z,X}}{P_Z \cdot P_X} = \sum_x \sum_z P_X \cdot P_{Z|X} \cdot \log \frac{P_{Z|X}}{P_Z}$$

The cross entropy in Y is:

$$H_{cross}(Z,Y) = -\sum_z P_Z \cdot \log P_{Y|Z} = E_{x \sim P_X}\left[-\sum_z P_{Z|X} \cdot \log P_{Y|Z} \right]$$

Given the information bottleneck γ, the final loss function of the DNN is:

$$f_{loss} = E_{x \sim P_X}\left[-\sum_z P_{Z|X} \cdot \log P_{Y|Z} \right]$$
$$+ \beta \cdot \sum_x \sum_z P_X \cdot P_{Z|X} \cdot \max\left(\log \frac{P_{Z|X}}{P_Z} - \gamma, 0 \right)$$

Since it is mathematically difficult to minimize

$$\sum_x \sum_z P_X \cdot P_{Z|X} \cdot \max\left(\log \frac{P_{Z|X}}{P_Z} - \gamma, 0 \right)$$

an alternative is to minimize its upper bound $E_{x \sim P_X}[D(P_{Z|X}|Q_Z)]$, where $D(P_{Z|X}|Q_Z)$ is the Kullback–Leibler divergence of $P_{Z|X}$ over Q_Z. Finally, the loss function becomes:

$$f_{loss} = E_{x \sim P_X}\left[-\sum_z P_{Z|X} \cdot \log P_{Y|Z} \right] + \beta \cdot E_{x \sim P_X}[D(P_{Z|X}|Q_Z)]$$

The loss function defines the information bottleneck [10], and it reveals the underlying relationship among the training target, neural network architecture, and input data distribution.

Several learning techniques can use information bottleneck theory to optimize their architecture. Variational inference for example constructs a lower bound on the information bottleneck objective and is trained with the collected data without prior knowledge about its distribution [11]. This framework can be extended into cases with various distortions required by high-level tasks, or cases with multi-view or multi-task problems [12].

For the neural network architecture and training method, it is easiest to split a neural network model pre-trained for high-level tasks into two parts: one used at the transmitter side and one used at the receiver side. One can then insert neural layers between the two parts. The whole model is trained and then fine-tuned with the channels considered [13].

8.1.4 DNN Implementations

Auto Encoder (AE)
If the rate distortion ($E_{x \sim P_X}[D(P_{Z|X}|Q_Z)]$) is implemented implicitly by a DNN, it ends up with an AE [14].

The distortion rate $|Z|/|X|$ fits the data sample X, and the target on Y is not a priori known. When $Y = X$, the AE becomes a learning-based source for a fixed-length block code with the lowest distortion rate $|Z|/|X|$, which is lower-bounded by $H(X)$ in general. If the dimension of X is not sufficiently large (small block), the AE may help to find a source code with a distortion rate $|Z|/|X|$ lower than $H(X)$.

Variational Auto Encoder (VAE)
VAE [15] introduces a prior distribution (Q_Z) into the latent layer, and this distribution is a perceptual metric on top of the rate and distortion. A generic VAE is illustrated in Figure 8.3. Different prior distributions (Q_Z) will result in different DNN coefficients. If Q_Z is a uniform multi-dimensional Gaussian distribution, the projection on each dimension is a score along that dimension. In theory, Q_Z can be any other distribution,

$$\Omega_Y(X; \theta)$$

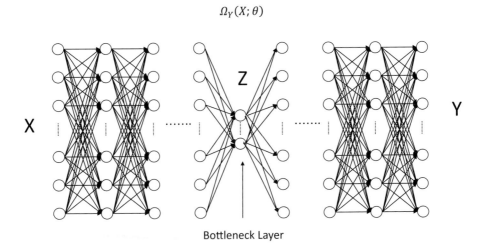

Bottleneck Layer

Figure 8.3 The information bottleneck of VAE such that $|X| > |Z|$ and $|Y| > |Z|$, i.e. latent layer Z is bottleneck.

and therefore non-uniform sub-channels of a wireless system can also be modeled as Q_Z, enabling VAEs to be customized to a particular prior known distribution.

Unlike AEs, which only maximize $I(X; Y)$, VAEs request two adversary training targets to minimize $I(X; Z)$ and maximize $I(X; g_Y(Z; \theta_g))$ simultaneously. At the end of training, the remaining $I(X; g_Y(Z; \theta_g))$ is the minimum mutual information (Z) to be kept at the bottleneck layer. In Chapter 23 of the book, we will explain how VAEs help to generate a joint source coding scheme.

Generative Adversarial Network (GAN)

GAN is a big leap forward from VAEs [16]. In a typical Wasserstein GAN, prior distributions (Q_Z) are real instead of Gaussian images, and the metric is no longer the mutual information but the Wasserstein distance instead. Though GAN is one of the best generative DNNs currently available, VAE is still preferred in wireless communication systems owing to its Gaussian latent layer – linear operations can be performed over Gaussian distributions.

8.2   Distributed AI Theory

To date, deep learning has been employed only in conventional centralized architectures, whereas its use in distributed architectures has seldom been explored. This is a critical issue as far as 6G objectives are concerned, as distributed architectures are essential to guaranteeing the scalability and flexibility of a communication system. However, applying deep learning in a distributed architecture will pose several fundamental problems. For example, each wireless node will have its own ANN, which will be trained using its own dataset acquired from local measurement and experience. As a result, different wireless nodes need to learn how to behave on the basis of datasets that might differ in both quantity (different nodes might have different measurement and storage capabilities) and quality (different nodes might experience different data perturbations owing to measurement sensor variations). This could potentially lead to instability and, in the worst cases, cause the wireless network to crash. In addition, in a distributed architecture, each node will try to optimize its own performance rather than the system-wide utility, thereby treating other nodes as possible adversaries. This might cause undesirable or dangerous situations in which a wireless node can learn to cheat in order to maximize its own performance. As such, it is imperative to develop a solid knowledge of all these issues, so as to ensure correct behavior in ANN-based wireless networks. Finally, the majority of deep learning results do not consider distributed scalable solutions that account for the large dimensions and unpredictable temporal evolution in 6G wireless scenarios. As a result, it becomes very challenging to build suitable training sets, as common supervised training approaches assume that ANNs will operate in similar conditions to those experienced during the training phase. Instead, sudden traffic and/or topology changes might render available training sets useless. Consequently, new ANN training methodologies capable of anticipating, or at least adjusting to, sudden changes in the wireless network are urgently required.

The following explains two such methodologies: federated learning and multi-agent reinforcement learning (MARL).

Federated Learning

Federated learning is a machine learning setting where a central model is trained in a decentralized manner [17, 18] by using the data on nodes. The specific process is: in each round, each node independently computes an update to the central model using its stored data, and then transmits that update to a central server. The central server aggregates the updates sent from different nodes and then computes an improved global model.

Modern mobile devices retain a wealth of data in their memory, and the data can be used to train models. While such data could greatly improve models, it is often composed of sensitive information and therefore privately secured. Owners are reluctant to share such data with the central server for training purposes. To work around this issue, instead the server could send the central model to nodes for training using local data. After generating an updated model, the nodes send it to the central server. In this way, the central server will receive a large amount of model updates from many nodes, which it then aggregates into an improved global model. Then another round of the learning process starts: the central server sends the improved model to new nodes, and the new nodes train it further using their private data. After many rounds, the model parameters can converge, resulting in a well-trained model.

The principal scope of federated learning is to decouple model training from direct access to raw training data, enabling the use of vast amounts of data stored on nodes. Below is a general description of the federated learning process that is mostly based on [17]. Many other extensions also exist, such as [19, 20].

Assume that the model to be trained is a neural network, and $f_i(w)$ is the loss function of the prediction on the example (x_i, y_i) with parameter w. Also assume that there are K users over which data is partitioned, P_k is a set of indexes of data points on client k, and $n_k = |P_k|$. Then the objective function to be minimized can be expressed as:

$$f(w) = \sum_{k=1}^{K} \frac{n_k}{n} F_k(w)$$

where

$$F_k(w) = \frac{1}{n_k} \sum_{i \in P_k} f_i(w)$$

Let C be the fraction of the nodes that will be chosen in each round in order to compute the updates and gradients of the loss function. If $C = 1$, data on all nodes are used to compute a full-batch (non-stochastic) gradient descent.

A typical implementation of federated learning chooses t random devices in every round. The central server sends parameter w_t of the network to every node; each user k then computes $g_k = \nabla F_k(w_t)$, which is the average gradient on its local data in the current model w_t, and then updates parameter $w_t^k \leftarrow (w_t^k - ag_k)$ (where a is the learning rate). This step is iterated E times on the users' devices in order to ensure the most reliable w_t^k is sent back to the central server. The central server then aggregates these w_t^k and applies the update $w(t+1) \leftarrow \sum_{k=1}^{K} \frac{n_k}{n} w_t^k$. The amount of computation is controlled by three key parameters: C, the fraction of nodes involved in computation in each round; E, the number of training passes that each client makes over its local dataset in each round; and B, the local mini-batch size used for client updates. If $B = \infty$, a full local dataset is treated as a single mini-batch. For a user with n_k local examples, $u_k = E \frac{n_k}{B}$ local updates will be made per round.

Federated learning provides a number of advantages, including the following.

- **A large amount of real data can be exploited for model training:** Since there are a lot of nodes, and each node contains a certain amount of data, aggregating all the data increases the possibility of optimal model training.
- **Training is performed on nodes in a decentralized manner,** mitigating the computation burden on the central server.
- **The privacy concern [20] is respected,** as users upload only model parameter updates rather than their own data.

On the other hand, some disadvantages also exist:

- **The central server should still be trusted to an extent,** as the process of inspecting the model updates sent by a user can sometimes result in the retrieval of some user information.
- **Communication cost is a primary concern,** as communication resources must be consumed twice per user: once to send parameters to the user, and again to return updates to the server. The randomness of wireless communication makes such updates rather challenging.
- **Additional concerns** relate to the fact that the acquired training data are not taken from independent and identically distributed sources, as each user has their own particular dataset which cannot represent the population distribution. Individual users also possess different volumes of local training data, and these variations may impact the convergence of the learning algorithm.
- **Decentralized training is more difficult than centralized training** in terms of duration and tuning. The values of hyperparameters and the architecture of the neural network should be carefully chosen in advance.
- **The performance of federated learning** is lower than that offered by a centralized method.

Multi-Agent Reinforcement Learning (MARL)

In reinforcement learning [21], the decision maker (i.e., the agent) aims to achieve optimal behavior in the presence of uncertainties by interacting with the environment, which is usually modeled as an MDP. With the advancement of deep learning [22], reinforcement learning has been shown to achieve outstanding performance in fields such as board games [23, 24], autonomous driving [25], and robotics [26]. Publication

[27] provides an overview of the recent achievements in deep reinforcement learning. In most applications, the agent learns how to react to an uncertain environment or a single adversary that responds to its actions. However, in many important applications, a great number of agents interact with one another, some of which might be devices implementing the same or different strategies for deciding their own actions. MARL has the potential to systematically analyze environments with strategically interacting agents. The study of MARL combines reinforcement learning and MDP techniques with game theory, particularly the branch of game theory known as mean field game [28]. It seeks to provide algorithms conducive to learning stable, best-response policies in a multi-agent system with more than a few agents, which remains a significant challenge in most cases as the complexity grows with the number of agents [29].

The difficulty in extending classical learning algorithms (such as reinforcement Q-learning, adaptive heuristic critic, or regret-minimizing multi-armed bandit) to multi-user or multi-player cases stems from the balance that must be maintained between exploiting the information gained during learning and adequately defining the search space. For example, in cases of very large numbers of interacting devices, mean field MARL can be highly effective.

There are various examples of MARL applications in wireless networks. For instance, the problem of fully decentralized MARL is considered in [30], where agents are located at the nodes of a time-varying communication network. The authors propose two decentralized actor-critic algorithms with function approximation, which are applicable to large-scale MARL where both the number of states and the number of agents are extremely large. Under the decentralized structure, the actor step is performed individually by each agent with no need to infer the policies of others. Meanwhile, for the critical step, a consensus update over the network is adopted. Both algorithms are fully incremental and can be implemented online. Convergence analysis is also available for these algorithms when the value functions are approximated within the class of linear functions.

8.3 Dynamic Bayesian Network Theory

Timely and cost-effective signal and information processing in RANs requires the exploitation of the spatio-temporal dependencies among network variables. These dependencies occur in multiple domains, such as:

- **User activation domains,** which include MTC/IoT devices reporting/reacting to correlated events.
- **Channel domains,** with certain users experiencing similar propagation characteristics.
- **Data content domains,** with users in a social structure sharing or demanding similar content.

These dependencies open up opportunities to harness and exploit performance gains in wireless networks. As a result, learning the dependencies between network variables is essential for the provisioning of advanced information and signal processing algorithms for improved radio access.

Dynamic Bayesian inference is a natural framework used to study the dependencies between variables over time, including interference and traffic patterns, channel structure, and user activity/data detection. The learned time-varying structure may then be used for dynamical clustering in cell-free massive MIMO, for example, yielding improved radio access solutions. Many such problems may be cast in terms of estimating a sparse time-varying inverse covariance matrix, which reveals a dynamic network of interdependencies between the entities. However, compared with time-invariant inference, time-variant network learning is computationally expensive, as more parameters exist and additional coupling is required. Furthermore, standard methods do not scale well for larger networks, where low complexity and approximate inference methods are preferred. Ideally, these methods should also appropriately account for the learning overhead, by trading off channel resources and performance. In addition, the learning algorithms should operate with incomplete, non-linear measurements; for example, when data are fused using capacity-limited links, as in the case of distributed MIMO.

To provide scalable solutions, we will build on approximate inference methods (such as message passing), and distributed (multi-processor) versions therein. Message passing and approximate message passing have recently demonstrated scalable performance in generalized inverse linear problems, and will serve as the starting point for the methods applied here. The existing solutions for static time-invariant scenarios will be extended to account for time dynamics. The representative theoretical tools include the following.

Bayesian Network (BN)

A BN is a type of graphical model that is represented as a directed acyclic graph. Nodes are graphical representations of objects and events that exist in the real world, and are usually referred to as variables or states. For each variable in the directed acyclic graph, there is a probability distribution function, whose dimensions and definition depend on the edges leading into the variable.

Bayesian networks can be defined as a special case in a more general class, known as a graphical model, wherein nodes represent random variables and a lack of arcs represents conditional independence assumptions between variables. Taking the graph as a whole, the conditional probabilities, BN structure, and joint probability distribution can be used to determine the marginal probability or likelihood of each node. This procedure is known as marginalization. The power of belief calculation in the BN takes effect whenever one of these marginal probabilities changes. The effect of the observation is propagated throughout the network, and in every propagation step, the probabilities of different nodes are updated. According to [31], in simple networks the marginal probability or likelihood of each state can be calculated from the knowledge of joint distributions, according to Bayes' theorem.

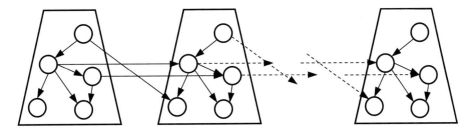

Figure 8.4 Time slice representing a snapshot of an evolving temporal process.

Dynamic Bayesian Network (DBN)

Most events in our everyday lives are not detected at a particular point in time, and can be described through multiple states of observations that yield a judgment of one complete final event. The DBN model describes a system that is dynamically changing or evolving over time. This model will enable users to monitor and update the system continuously, and even predict system behavior. DBN is usually defined as a special case of singly connected BN specifically aimed at time series modeling, as stated in [32]. All the nodes, edges, and probabilities that form a static interpretation of the system are identical to those in the BN. The variables here can be denoted as DBN states, as they include a temporal dimension, and the state of the system described by the DBN satisfies the Markovian condition, as shown in Figure 8.4.

Hidden Markov Model (HMM)

An HMM is a stochastic finite automaton, where each state generates (emits) an observation. We will use X_t to denote the hidden state and Y_t to denote the observation. If there are K possible states, then $X_t \in \{1, K\}$. The model parameters include the initial state distribution, transition model, and observation model. The transition model is usually characterized by a conditional multinomial distribution $A(i; j) = P(X_t = j | X_{t-1} = i)$, where A is a stochastic matrix. The stochastic matrix A is often sparse, and its structure is usually depicted graphically, where nodes represent states, and arrows represent allowable transitions (i.e., transitions with non-zero probability).

Shifting focus to the technical characteristics of DBNs, these can be summarized in three categories.

- **Inference:** As only a subset of states can be observed at each time slice, we need to calculate all the unknown states in the DBN. This is achieved through a procedure known as inference, whereby different types of DBNs require different types of estimations and calculations based on their specific structures. In certain circumstances, it may be more appropriate not to estimate the conditional probability density function, and instead estimate the sufficiency statistics of the probability density function. If there is a strong similarity between the DBNs and singly connected BNs, an efficient forward–backward algorithm can be employed. Smoothing and prediction algorithms are also used to calculate inference in DBNs.

- **Sequence decoding:** Another problem relating to DBNs is finding the most likely sequence of hidden variables, given the observations. Since DBN nodes can have more than one state, it is necessary to determine the sequence of hidden states with the highest probabilities. This procedure is usually denoted as sequence decoding, and can be achieved using the dynamic programming Viterbi algorithm. It should be noted that if we intend to decode the sequence of hidden states using the Viterbi algorithm, it is necessary to possess a complete set of observations. A variation of the algorithm, known as a truncated Viterbi algorithm can be used after a fixed number of observations has been received.
- **Learning:** The representation of real-world problems in a DBN structure often requires the introduction of several nodes for which conditional probabilities cannot be exactly determined. Even expert knowledge cannot offer solutions for some conditional relationships in particular domains. In such circumstances, it becomes necessary to learn these specific probability distributions. This learning process is complex and based on the expectation maximization or general expectation maximization (GEM) algorithm.

Table 8.1 summarizes the most effective methods to use depending on whether the structure is known. Each method also presents unique advantages and disadvantages, as listed in Table 8.2.

Many potential applications exist for DBNs in 6G. For example, the learned dependencies between network variables may be exploited for potential performance gains in various wireless services. In particular, the learned time-varying structure can be used for dynamic user or network clustering (typically cell-free massive MIMO) in the different domains (channel, activity, and content), with a direct effect on radio access applications. A prominent example is the time-varying correlation structure of a

Table 8.1 Methods for different structure states.

Structure state	Observability	Method
Known	Full	Simple statistics
Known	Partial	Dynamic programming
Unknown	Full	Maximum likelihood

Table 8.2 Methods for different structure states

Method	Advantages	Disadvantages
Simple statistics	Simple calculation, computationally cheap	Full structure and observability required, unlikely to happen
Dynamic programming	Known algorithm: Viterbi	Can get stuck at local optimal
Maximum likelihood	EM/GEM algorithm, good behavior	Can get stuck at local optimal

wireless channel, which – if learned – can be used to enhance the RAN performance. Huge performance gains are expected in relation to massive MIMO, as the knowledge of the channel's structure (specifically, the MIMO channel) would provide an opportunity to dynamically cluster users in the RAN on the basis of their channel conditions. The channel structure and user clustering facilitate the provisioning of channel estimation schemes with lower overhead, by decreasing both the amount of channel resources for training and the amount of feedback associated with the channel estimates. The Bayesian framework will be extremely effective in this scenario, as it delivers beliefs relating to the quality of the involved estimates. These beliefs can be assessed both locally at the user side and network edge, as well as globally at the central server and macro base station, to facilitate decisions regarding channel estimation and data decoding. In the context of massive MIMO, distributed MIMO, or cell-free networks, this approach will result in novel channel estimation and feedback schemes. To be specific, dynamic user clustering will affect the design and length of the pilot sequences used to train (both in the uplink and in the downlink) a cellular system where base stations are equipped with a large number of antennas. This will effectively decrease the training overhead, freeing up channel resources for data transmission. In particular, the CSI feedback design can be optimized on the basis of the user clustering and soft information (beliefs) output by the Bayesian inference procedure. New proposals based on quantized explicit feedback can now be investigated.

In addition, the Bayesian framework integrates user activity detection by including a user activation statistics model (which can also be learned). It extends over the non-adaptive and adaptive probabilistic group testing methods to incorporate dependencies among users and the probability distributions they follow. This is of particular relevance in massive MTC/IoT scenarios, where the channel estimation and user activity detection issues should be jointly addressed in order to reduce both the signaling overhead and the latency associated with the random access procedure. In particular, in massive connectivity applications featuring two-phase transmission (initial access followed by data transmission), the Bayesian framework assesses the optimal amount of resources dedicated to the initial access (first phase) versus that dedicated to data transmission (second phase), with immediate implications on resource allocation for the data transmission phase. In general, it is expected that in overloaded systems where the number of active users far exceeds the number of antennas at a base station, Bayesian inference-assisted user scheduling can significantly improve the overall spectrum efficiency. This will affect the designs of both scheduled schemes and grant-free (GF) schemes that target MTC/IoT applications with small-packet transmissions, and integrate initial access and data transmission.

The Bayesian framework may also be applied to localization and trajectory prediction. In addition, the distributed prediction of channel states and user locations is seen as a key ingredient to the design of robust wireless communication systems. For instance, by predicting users' (short-term) locations or the evolution of their angular power spectrum (i.e., the function showing the average energy per angular unit in an antenna array), schedulers gain useful information on how to precode training symbols so that CSI acquisition can be performed for multiple users with the same pilot.

DBNs may affect the directions of many technologies, including user activity detection, channel estimation, and mobility tracking.

References

[1] I. Goodfellow, Y. Bengio, A. Courville, and Y. Bengio, *Deep learning*, Vol. 1. MIT Press, Cambridge, 2016.

[2] K. Hornik, M. Stinchcombe, H. White *et al.*, "Multilayer feedforward networks are universal approximators." *Neural Networks*, vol. 2, no. 5, pp. 359–366, 1989.

[3] J. Duchi, E. Hazan, and Y. Singer, "Adaptive subgradient methods for online learning and stochastic optimization," *Journal of Machine Learning Research*, vol. 12, no. 7, 2011.

[4] D. P. Kingma and J. Ba, "Adam: A method for stochastic optimization," *arXiv preprint arXiv:1412.6980*, 2014.

[5] S. Fernández, A. Graves, and J. Schmidhuber, "An application of recurrent neural networks to discriminative keyword spotting," in *Proc. International Conference on Artificial Neural Networks*. Springer, 2007, pp. 220–229.

[6] D. Madhubabu and A. Thakre, "Long-short term memory based channel prediction for siso system," in *Proc. 2019 International Conference on Communication and Electronics Systems (ICCES)*. IEEE, 2019, pp. 1–5.

[7] M. L. Puterman, *Markov decision processes: Discrete stochastic dynamic programming*. John Wiley & Sons, 2014.

[8] D. P. Bertsekas, *Dynamic programming and optimal control*, Vol. 1, no. 2. Athena Scientific, Belmont, MA, 1995.

[9] N. Tishby, F. C. Pereira, and W. Bialek, "The information bottleneck method," *arXiv preprint physics/0004057*, 2000.

[10] R. Shwartz-Ziv and N. Tishby, "Opening the black box of deep neural networks via information," *arXiv preprint arXiv:1703.00810*, 2017.

[11] A. A. Alemi, I. Fischer, J. V. Dillon, and K. Murphy, "Deep variational information bottleneck," in *Proc. International Conference on Learning Representations*, 2016.

[12] I. Estella-Aguerri and A. Zaidi, "Distributed variational representation learning," *IEEE Transactions on Pattern Analysis and Machine Intelligence*, to be published.

[13] A. E. Eshratifar, A. Esmaili, and M. Pedram, "Bottlenet: A deep learning architecture for intelligent mobile cloud computing services," in *Proc. 2019 IEEE/ACM International Symposium on Low Power Electronics and Design (ISLPED)*. IEEE, 2019, pp. 1–6.

[14] G. E. Hinton and R. S. Zemel, "Autoencoders, minimum description length and helmholtz free energy," in *Proc. Conference on Advances in Neural Information Processing Systems*, 1994, pp. 3–10.

[15] D. P. Kingma and M. Welling, "An introduction to variational autoencoders," *Foundations and Trends in Machine Learning*, vol. 12, no. 4, 2019.

[16] I. Goodfellow, J. Pouget-Abadie, M. Mirza, B. Xu, D. Warde-Farley, S. Ozair, A. Courville, and Y. Bengio, "Generative adversarial nets," in *Proc. Conference on Advances in Neural Information Processing Systems*, 2014, pp. 2672–2680.

[17] B. McMahan, E. Moore, D. Ramage, S. Hampson, and B. A. Y. Arcas, "Communication-efficient learning of deep networks from decentralized data," in *Proc. Conference on Artificial Intelligence and Statistics*. PMLR, 2017, pp. 1273–1282.

[18] J. Konečný, H. B. McMahan, F. X. Yu, P. Richtárik, A. T. Suresh, and D. Bacon, "Federated learning: Strategies for improving communication efficiency," in *Proc. NIPS Workshop on Private Multi-Party Machine Learning*, 2016.

[19] B. Hitaj, G. Ateniese, and F. Perez-Cruz, "Deep models under the gan: Information leakage from collaborative deep learning," in *Proc. the 2017 ACM SIGSAC Conference on Computer and Communications Security*, 2017, pp. 603–618.

[20] V. Smith, C.-K. Chiang, M. Sanjabi, and A. S. Talwalkar, "Federated multi-task learning," in *Proc. Conference on Advances in Neural Information Processing Systems*, 2017, pp. 4424–4434.

[21] R. S. Sutton and A. G. Barto, *Reinforcement learning: An introduction*. MIT Press, 2018.

[22] I. Goodfellow, Y. Bengio, A. Courville, and Y. Bengio, *Deep learning*, Vol. 1 MIT Press, 2016.

[23] D. Silver, A. Huang, C. J. Maddison, A. Guez, L. Sifre, G. Van Den Driessche, J. Schrittwieser, I. Antonoglou, V. Panneershelvam, M. Lanctot *et al.*, "Mastering the game of Go with deep neural networks and tree search," *Nature*, vol. 529, no. 7587, pp. 484–489, 2016.

[24] D. Silver, J. Schrittwieser, K. Simonyan, I. Antonoglou, A. Huang, A. Guez, T. Hubert, L. Baker, M. Lai, A. Bolton *et al.*, "Mastering the game of Go without human knowledge," *Nature*, vol. 550, no. 7676, pp. 354–359, 2017.

[25] S. Shalev-Shwartz, S. Shammah, and A. Shashua, "Safe, multi-agent, reinforcement learning for autonomous driving," *arXiv preprint arXiv:1610.03295*, 2016.

[26] J. Kober, J. A. Bagnell, and J. Peters, "Reinforcement learning in robotics: A survey," *International Journal of Robotics Research*, vol. 32, no. 11, pp. 1238–1274, 2013.

[27] Y. Li, "Deep reinforcement learning: An overview," *arXiv preprint arXiv:1701.07274*, 2017.

[28] B. Jovanovic and R. W. Rosenthal, "Anonymous sequential games," *Journal of Mathematical Economics*, vol. 17, no. 1, pp. 77–87, 1988.

[29] Y. Shoham and K. Leyton-Brown, *Multiagent systems: Algorithmic, game-theoretic, and logical foundations*. Cambridge University Press, 2008.

[30] H. Tembine, R. Tempone, and P. Vilanova, "Mean-field learning: A survey," *arXiv preprint arXiv:1210.4657*, 2012.

[31] R. Sterritt, A. H. Marshall, C. M. Shapcott, and S. I. McClean, "Exploring dynamic Bayesian belief networks for intelligent fault management systems," in *Proc. 2000 Conference on Systems, Man and Cybernetics*, vol. 5. IEEE, 2000, pp. 3646–3652.

[32] K. P. Murphy, "Dynamic Bayesian networks: Representation, inference and learning," Ph.D. thesis, 2002.

9 Theoretical Foundations for Massive Capacity and Connectivity

9.1 Electromagnetic Information Theory

This line of research is concerned with the formulation of fundamental wireless communication and antenna engineering problems, which occur at the crossroads of the well-established fields of Maxwell's electromagnetic theory and Shannon's information theory. The interdisciplinary field constituted by such wave and information theoretic problems and their solutions can be descriptively termed electromagnetic information theory (EIT).

Work in this area, which combines wave physics with information theory, has a long history, dating back to the origins of information theory [1] and, in particular, to the pioneering work on light and information [2, 3]. The concept of degrees of freedom (DoF) was introduced to represent the number of effective dimensions for communication. Rigorous methodologies to determine the number of DoF using wave theory were established during the 1960s, and further progress in this field has been intermittent since then.

Channel capacity is defined as the maximum mutual information,

$$C = \max_{p(x)} \{I(x, y)\}$$

where x and y are the Tx and Rx vectors, respectively. In order to consider the impact of the laws of electromagnetism on channel capacity, the spatial capacity S is defined in [4] as the maximum mutual information between the Tx and Rx vectors, as well as the channel information (assuming that perfect channel state information (CSI) is known at Rx):

$$S = \max_{p(x), \mathbf{E}} \{I(x, \{y, \mathbf{G}(\mathbf{E})\})\}$$

$$\text{Constraints:} \quad \langle x^+ x \rangle \le P_T, \quad \nabla^2 \mathbf{E} - \frac{1}{c^2} \frac{\partial^2 \mathbf{E}}{\partial t^2} = 0, \quad \mathbf{E} = \mathbf{E_0}, \quad \forall \{\mathbf{r}, t\} \in B$$

where $\mathbf{E}$ is the electric field used to transmit data, B is the boundary condition (depending on the scattering environment), and $\mathbf{G}$ is the channel matrix. The number of DoF of the electromagnetic field is essential to understanding the physical limitations of wireless (radiating) communication systems.

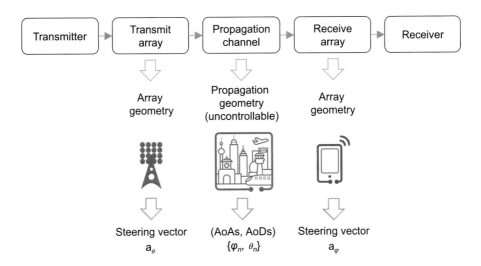

Figure 9.1 Schematic diagram of a generalized wireless channel.

As shown in Figure 9.1, a generalized wireless channel consists of a Tx array, a propagation channel (the environment), and an Rx array. The Tx and Rx arrays impact at least the radiation pattern and non-linear coupling effect of the channel. The complex generalized wireless channel $\mathbf{G}(\mathbf{E})$ is usually characterized as a convolution of array steering vectors and multipath physical propagation channels:

$$\mathbf{G}(\mathbf{E}) = \mathbf{H} = f(\{\varphi_n, \theta_n\}) = \sum_n h_n \mathbf{a}_{\theta_n} \mathbf{a}_{\varphi_n}^H$$

where n is the number of multipaths, φ_n and θ_n are the angle of arrival (AoA) and the angle of departure (AoD) of the physical propagation channels respectively, $\mathbf{a}_\theta$ and $\mathbf{a}_\varphi$ are the steering vectors of the transmit and receive arrays, and h_n is the impact response of the ith multipath physical channel.

Following the success of MIMO communications in the 2000s, interest in EIT has been growing steadily, with the final objective of understanding the MIMO channel independently of the antenna configuration [5–8]. In any case, the optimal number of antennas and the MIMO capacity are limited for a given aperture size, regardless of the scenario (for example, rich or sparse scattering) [4]. Nowadays, massive MIMO is a mature technology whose key ingredients are included in the 5G NR standard. The advantages of massive MIMO in terms of spectrum efficiency and energy efficiency are well understood and recognized, and the natural follow-up questions are: "Is it feasible to approach the limits of infinite antennas?" and "What can we expect next?" [9]. The answers can be found in EIT, as EIT provides a general framework for wireless network design and performance analysis. Initial research in this direction has been carried out in [10–12].

In order to increase the number of DoF of the generalized wireless channel $\mathbf{G}(\mathbf{E})$, potential technical research directions can be divided into three categories: (1) the physical propagation channel, (2) the antenna array pattern and coupling effect, (3) the electromagnetic physical characteristics.

Physical Propagation Channel

From an uncontrollable to a controllable physical propagation environment.

Representative disciplines include reconfigurable intelligent surfaces (RIS) [13] and intelligent reflecting surfaces (IRS) [14], and the smart environment concept has been attracting increasing amounts of attention as a result of these investigations. Its main advantage is the possibility of actively constructing a full-rank channel matrix. Taking into consideration a series of constraints which include hardware resolution, feedback overhead, and terminal mobility, the theoretical capacity limits of smart environments are unknown. EIT has the potential to play a key role in the related research.

From an irreversible to a reversible behavior of the electromagnetic propagation path.

Time reversal (TR) [15] is a representative discipline in this area; it depends on the temporal symmetry of the electromagnetic wave equation's solution (also known as the principle of optical path reversibility). Using the TR principle, EIT is expected to play a key role in obtaining super-resolution of the spatial and temporal dimensions in ultra-wideband (UWB) systems.

From electromagnetic far field to near field, and even surface waves.

Representative disciplines include large intelligent surface (LIS) [16] and surface wave communication (SWC) [17]. LIS must take into account the spatial non-stationarity of spherical waves and the receiver equalization algorithm, while SWC takes advantage of the long-neglected Zenneck waves, which are tightly bound near the medium's surface and cannot spread to their surroundings. Both LIS and SWC require EIT's mathematical tool to guide their antenna design and micro-nanostructure machining.

Antenna Array Patterns and Coupling Effect

From invariant to variant (physical or virtual) geometric topologies in antenna arrays.

In addition to conventional reconfigurable antennas, the latest area of interest is information metasurfaces [18]. Based on the flexible control of metamaterials, the possibility of electromagnetic modulation in precise spatial dimensions can be realized using index modulation [19]. Although it is not thought that pure index modulation can improve the NDF of channels, it is generally agreed that during EIT development, there are many other technical paths that can be pursued based on metamaterials.

From half-wavelength to ultra-compact arrays.

Super-directivity [20] is a representative discipline in this area. Unlike a traditional antenna array, which minimizes mutual coupling among M elements to maintain the beamforming gain at M through a maximum-ratio transmission, a super-directive antenna array creates strong mutual coupling by spacing antennas less than half a wavelength apart, resulting in a beamforming gain greater than M. Mathematically speaking, the subspace of the impedance matrix comprised eigenvectors with small eigenvalues, representing a portion of the propagation vectors. Physically speaking, a spectrum of plane waves has a great number of super-wave components, and the waves are evanescent in the terminal direction. The evanescent waves decay exponentially. EIT can solve the strong mutual coupling from the perspectives of Maxwell electromagnetic theory and Kirchhoff circuit theory [20, 21].

From small to extremely large aperture arrays (ELAAs).

Representative works here include large intelligent surfaces [16] and ELAAs [22], the latter of which causes the spatial non-stationary field properties along arrays. To reduce the computational burden of the ELAA–MIMO scheme, it is recommended to divide the signal processing of the whole array into smaller disjoint sub-arrays [22]. Electromagnetic information theory must provide a new analysis framework to better address the spatial non-stationary field properties that arise for various reasons.

Electromagnetic Physical Characteristics

From single- (or dual-) to triple- (or multi-) polarization.

Although Tse *et al.* analyzed the DoF gain in triple-polarized elements early in 2011 [23], it has existed as a theoretical result for a long time, but the use of three-dimensional (3D) triple-polarized antenna arrays has not spread for a number of reasons. Owing to the throughput increase in quadrifilar helix antenna (QHA)-based MIMO, 3D antenna arrays and related multi-polarized channel measurements are now attracting increased attention [24]. In the 6G era, multi-polarized channel measurements and modeling are matters that concern researchers, and it is possible to identify some previously neglected channel characteristics in order to increase the DoF of channels. EIT will play an important role in generalized channel modeling.

From linear (energy) to angular momentum.

Orbital angular momentum [25] is a representative area of work in this area. Further research has shown that orbital angular momentum is unlikely to become a new independent DoF, and is instead only a special case of MIMO [26]. However, in some quasi-paraxial and near-field environments, orbital angular momentum can complement MIMO to achieve some beneficial effects. Whether new electromagnetic characteristics can be found for wireless communication under the guidance of EIT remains unanswered.

Traditional analysis of the DoF of electromagnetic waves in space is based on whether the free space is boundaryless (or has an infinitely large boundary). The complex wavefront of an electromagnetic wave is first decomposed into multipole expansion terms with standard spherical harmonics, and then a wireless channel response matrix is obtained through linear superposition of the Green's function propagated by the electromagnetic wave. Finally, an independent orthogonal basis is determined in the eigenspace [27, 28]. In the 6G era, new trends in antenna arrays and channel environments are driving us to discover more powerful electromagnetic analysis tools to cope with the complexity of the electromagnetic environment caused by super-directive antenna arrays, ELAAs, and smart metasurfaces, in order to acquire a more accurate understanding of channels.

Computational electromagnetics (CEM).

This is a new cross-field discipline capable of solving complex electromagnetic theory and engineering problems [29]. It permeates every field of electromagnetics, and is interrelated with electromagnetic field theory and engineering. For electromagnetic field engineering, CEM is used to solve the modeling, simulation, optimization, and design problems of electromagnetic fields with increasingly complicated actual scenarios. For electromagnetic field theory, CEM can provide methods, means, and results for complex numerical and analytical calculations, and investigate the laws and mathematical equations of electromagnetic fields. Common CEM methods include the finite element method [30], the finite-difference time-domain method [31], transmission line matrices [32], the method of moments [33], and various calculation-acceleration methods. It is necessary to pay attention to the development of CEM theory and computer-aided design simulation tools to improve our understanding and application of the Maxwell Green's function and the wireless channel response matrix in complex 6G electromagnetic environments. As we will discuss further in Chapter 13, CEM is required in 6G channel modeling for new spectrum bands and usage scenarios.

As 6G continues to develop, many wireless technologies are beginning to emerge which may affect the development of electromagnetic information theory. These include: LIS, RIS, digital controllable scatterer, orbital angular momentum, holographic MIMO, time reversal, surface wave communications, ultra-massive MIMO, and ELAA.

9.2   Large-Scale Communication Theory

In many modern engineering domains, systems tend to be organized in increasingly large networks (such as ultra-massive MIMO and dense networks). In addition, system dynamics become increasingly vital as fast processing in rapidly changing environments is now demanded. As a common denominator, such networks now face the challenge of having to simultaneously cope with a growing system size (more nodes in the networks, and more actors interacting with them), an increased level of stochasticity (random changes in the topology, and random environment evolution), and a need for decentralized and self-organized processing (network nodes must be autonomous and capable of dynamically adapting to the changing environment). These three aspects constitute the basic ingredients required to model, analyze, and optimize the performance of future networks. Traditional methods used to tackle such problems in large complex systems involve heavy simulation (Monte Carlo method) or heuristic tools, such as neural networks or particle swarms. However, these come with a number of serious drawbacks, such as not being mathematically stable and not lending themselves well to simple interpretation, whereby system improvements and optimizations are either based on trial-and-error or simply impossible. As a result, these methods are often seen as backup solutions in cases where mathematical models do not possess solid foundations.

Consequently, the development of novel theoretical tools that allow for a systematic, reliable, and interpretable analysis of large random networks is required. These tools must be generic enough to encompass a wide scope of network models while also enabling easy specialization for practical networks. As will be presented in further detail below, various recent mathematical tools have already been identified as potential candidates to address part of the challenges inherent to these large systems. However, most of these tools are still in the early stages of mathematical development, or have just begun to make their way into engineering research. Of these new tools, the following three have emerged as the most promising:

- **Random matrix theory:** Random matrix theory, and more specifically large-dimensional random matrix theory [34], technically constitute the study of large-dimensional Hermitian matrices with random entries. This field experienced a surge of interest with regard to wireless communications in the early 2000s during the analysis of the emerging multi-antenna and spreading code technologies of that era, whose mathematical models were based on random matrix communication channels. While scientific advancement on this subject continued from 2000 to 2010 (based on the mathematical results available by 2000), few researchers developed the mathematical bases of random matrix theory much further as was needed in order to tackle challenging problems. More recently, however, new theoretical tools related to wireless communication network analysis have emerged in the context of massive MIMO and ultra-massive MIMO.

  Random matrix theory aims to study the spectral properties (eigenvalues and eigenspaces) of matrices with random entries whose sizes are assumed to be very large. Similarly to how random variables or finite-size random vectors exhibit deterministic limits as a large sample set of these variables is drawn (mostly based on the law of large numbers or the central limit theorem), the spectral measures of some random matrix models exhibit deterministic behaviors as both row and column dimensions grow towards infinity. Having long remained the exclusive domain of mathematicians, it became evident in the early 2000s that random matrices may adequately model wireless communication channels, which are stochastic and rapidly changing in nature [35]. Following several years of exploration, it became clear to wireless communication researchers that most random matrix models proposed by mathematicians were either inadequate or too simple to effectively study most realistic communication channels. A second wave of research was then initiated around 2010 to further investigate the mathematical tool of random matrix theory for the specific needs of wireless communication engineers. Recent advances in this field have delivered important new results in the modeling, analysis, and performance optimization of numerous wireless channels, such as MIMO point-to-point (P2P) Rician fading channels [36], MIMO multipath channels [37], MIMO multiple access (MA) channels [38], linearly precoded broadcast channels [39], communication channels with unitary precoders [40], multi-hop channels [41], and single-cell/multi-cell networks [42, 43]. While these results are sometimes based on very complicated system models, they have been delivered in a relatively simplified format and lend themselves well to optimization. The essential ingredient introduced by random matrix theory is the large-dimensional approximation, which efficiently manages to analyze large complex stochastic systems with deterministic quantities, known as deterministic equivalents [44]. Large-dimensional system components are measured on the basis of the number of antennas in wireless devices, number of users in a single cell, or number of cells in a given network. Random matrix theory is therefore acknowledged as an efficient tool for analyzing many local wireless communication systems, where "local" is understood as "single-scale" (either the number of antennas or the number of users is large). Thanks to further generalized tools, such as iterative deterministic equivalents [40], random matrix theory is now expected to become an adequate tool for modeling complex telecommunication systems, in which many large-dimensional components are taken into account, making it possible to envision the joint analysis at the macroscopic and microscopic scales. In addition, random matrix theory tools have also made an impression in the signal processing community, by proposing innovative detection and estimation (statistical inference) schemes, such as MIMO radar technology, for large-array processing [45]. Finally, explorations have been made to design decentralized algorithms for optimizing the performance of multi-cell MIMO networks [46].

- **Decentralized stochastic optimization:** The second tool of interest is a recent framework gathering stochastic approximation and gossip algorithms for decentralized stochastic optimization. The origin of decentralized stochastic optimization can be traced back to the work in [47], in which methods are derived to perform decentralized complex calculus on potentially large clusters of processors. However, the initial work was limited by the fact that inter-processor communications should be inexpensive and independent of the distance between processors, and therefore independent of the underlying network topology. In parallel, decentralized consensus algorithms have been developed to solve a quite different problem: reaching a common decision [48] on an interconnected graph of sensors with low processing and memory capabilities. A classic example of decentralized consensus algorithms is an algorithm for reaching a consensus in a distributed MIMO system. That is, assuming that each access point in a cluster makes an initial measurement, maintains a single value in memory at all times, and can only send this value to its closest neighbor, fast algorithms must be designed to help the cluster ultimately reach a joint decision on the average of the initial measurements.

   Topological considerations here are essential to the feasibility and convergence properties of the algorithms. Decentralized processing in large interconnected networks with limited communication capabilities has emerged [49]. This approach aims to solve a network-wide problem which can be divided into several non-independent sub-problems. Each network node optimizes its own sub-problem and then sends the result to its closest neighbor (deterministically or randomly), thereby a consensus is reached in which each node is eventually aware of the problem's solution. Among the various extensions of these schemes, some algorithms are proposed based on a stochastic approximation [50], which assumes that each network node possesses only partial knowledge of the sub-problem information. In multi-terminal networks, decentralized processing has already been subject to practical investigations [51]. In decentralized MIMO, various problems, such as maximizing the total network throughput, demand a solution in which a central entity has full knowledge of the network. Owing to limited calculus and memory capabilities, as well as the short communication ranges of access points, decentralized processing and gossip algorithms are also essential ingredients for maximizing the performance of dense networks. In particular, such decentralized processing and gossip algorithms are expected to constitute an adequate alternative to game-theoretic and learning approaches, which have long been considered in wireless communications despite having rarely provided compelling arguments for their usage in self-organized networks. It should be noted that game-theoretic considerations, primarily based on a succession of trials and feedback, often require network actors to make inappropriate decisions before an expected Nash equilibrium is reached. If the optimization problems to be solved pose strict inter-node constraints, game-theoretic approaches often lead to algorithms whose outcomes cannot initially satisfy the constraints. This is not the case for decentralized processing and gossip techniques where the exchanged information is not used before being converged.

- **Tensor algebra and low-rank tensor decomposition:** In its simplest incarnation, a tensor can be construed as a D-dimensional data structure. This is akin to the generalization of vectors and matrices to more than two dimensions (in classical linear algebra, vectors are one-dimensional data structures, whereas matrices have two dimensions, namely, rows and columns). The resulting D-way array can be interpreted as either the representation of a multilinear application (as in Newtonian physics, where coordinate-free tensor representations are used to model physical laws) or a data structure, which is naturally indexed by D dimensions (or modes) [52]. In the latter case, the notion of tensor rank plays a critical role. A rank-1 tensor of order D is defined as the outer product of D vectors of appropriate dimensions. This is analogous to a rank-1 matrix, which is defined by the outer product of only two vectors: a column vector and a row vector. The minimum number of rank-1 terms required to reconstruct a given tensor is known as the rank.

The tensor decomposition problem was first identified in the late 1920s [53] and has become deeply rooted in the experimental sciences. Specifically, the decomposition of a given tensor into the sum of rank-1 components (known as canonical polyadic decomposition) has practical significance in numerous applications because it reveals the internal structure of data. In the 1970s, low-rank tensor decomposition was applied to the field of psychometrics, where it was given the name PARAFAC [54]. It was extended to the field of chemometrics in the 1980s [55].

Tensor algebra is underpinned by a profound mathematical theory. Despite the relatively familiar setup provided by the analogy with the matrix case, the properties of the canonical polyadic decomposition depart in major ways from the intuition available in the matrix case [56]. Specifically, high-order ($D > 3$) tensors can have high rank even for moderate tensor sizes. Unlike for matrices, where the rank is limited by the minimum over the row and column dimensions, the expected rank of a generic tensor scales much faster [57]. Furthermore, for tensors of sub-generic rank (i.e., those that contain data arising from a structured model), the canonical polyadic decomposition is essentially unique under mild conditions. Again, this is in contrast to matrices. For example, if we take into account the QR or singular value decomposition, the orthogonality conditions imposed on the factors that make matrix decomposition unique are merely technical and do not always correspond to a meaningful constraint in the considered problem. These unique properties of low-rank tensor decomposition can be exploited in several ways in wireless communications. For instance, if the low-rank tensor model arises from a specular radio frequency propagation model coupled with an antenna array, tensor decomposition enables us to perform blind source separation and direction estimation with more relaxed assumptions and a larger number of sources compared with the classical multiple signal classification (MUSIC) or estimation of signal parameters via rotational invariance techniques (ESPRIT) approaches [58]. Such relaxed assumptions may include using much shorter data samples based on high-order statistics. The blind source separation capability is also at the core of the non-coherent, non-orthogonal multi-user separation approach proposed in [59] in the context of massive access, which will be discussed in Chapter 24.

Contemporary extensions of tensor theory aim to address increasingly complex setups, such as coupled canonical polyadic decomposition [60] (where coupling arises from the considered application) and the storage-efficient tensor-train decomposition [61] applicable to large-size, high-order tensor problems. Recent theoretical developments also include the study of random tensors, focusing on spiked models in particular. In this context, the goal is to characterize analytically the conditions under which a low-rank informative component can be reliably separated from additive measurement noise through the low-rank tensor approximation, and to predict the achievable accuracy. One approach is to use statistical physics tools [62]. Other attempts at developing a clean-slate spectral theory of random tensors – a notoriously difficult problem – are also emerging [63]. Furthermore, the generalization capability of low-rank tensor decomposition is being investigated in the context of "missing data" formulations [64].

As 6G continues to develop, many technologies that are closely related to large-scale communication theory are beginning to emerge, including dense network, federated learning, cell-free massive MIMO, user-centric communication, edge-based communication, and distributed MIMO.

References

[1] C. E. Shannon, "A mathematical theory of communication," *ACM SIGMOBILE Mobile Computing and Communications Review*, vol. 5, no. 1, pp. 3–55, 2001.

[2] D. Gabor, "CIII. Communication theory and physics," *London, Edinburgh, and Dublin Philosophical Magazine and Journal of Science*, vol. 41, no. 322, pp. 1161–1187, 1950.

[3] D. Gabor, "IV light and information," in *Progress in optics*, Vol. 1. Elsevier, 1961, pp. 109–153.

[4] S. Loyka, "Information theory and electromagnetism: Are they related?" in *Proc. 2004 10th International Symposium on Antenna Technology and Applied Electromagnetics and URSI Conference*. IEEE, 2004, pp. 1–5.

[5] J. Y. Hui, C. Bi, and H. Sun, "Spatial communication capacity based on electromagnetic wave equations," in *Proc. 2001 IEEE International Symposium on Information Theory*. IEEE, 2001, p. 337.

[6] J. W. Wallace and M. A. Jensen, "Intrinsic capacity of the MIMO wireless channel," in *Proc. IEEE 56th Vehicular Technology Conference*, vol. 2. IEEE, 2002, pp. 701–705.

[7] M. A. Jensen and J. W. Wallace, "Capacity of the continuous-space electromagnetic channel," *IEEE Transactions on Antennas and Propagation*, vol. 56, no. 2, pp. 524–531, 2008.

[8] F. K. Gruber and E. A. Marengo, "New aspects of electromagnetic information theory for wireless and antenna systems," *IEEE Transactions on Antennas and Propagation*, vol. 56, no. 11, pp. 3470–3484, 2008.

[9] E. Björnson, L. Sanguinetti, H. Wymeersch, J. Hoydis, and T. L. Marzetta, "Massive MIMO is a reality. What is next? Five promising research directions for antenna arrays," *Digital Signal Processing*, vol. 94, pp. 3–20, 2019.

[10] T. L. Marzetta, "Spatially-stationary propagating random field model for massive MIMO small-scale fading," in *Proc. 2018 IEEE International Symposium on Information Theory (ISIT)*. IEEE, 2018, pp. 391–395.

[11] A. Pizzo, T. L. Marzetta, and L. Sanguinetti, "Spatial characterization of holographic MIMO channels," *arXiv preprint arXiv:1911.04853*, 2019.

[12] A. Pizzo, T. L. Marzetta, and L. Sanguinetti, "Degrees of freedom of holographic MIMO channels," in *Proc. 2020 IEEE 21st International Workshop on Signal Processing Advances in Wireless Communications (SPAWC)*. IEEE, 2020, pp. 1–5.

[13] E. Basar, M. Di Renzo, J. De Rosny, M. Debbah, M.-S. Alouini, and R. Zhang, "Wireless communications through reconfigurable intelligent surfaces," *IEEE Access*, vol. 7, pp. 116753–116773, 2019.

[14] Q. Wu and R. Zhang, "Intelligent reflecting surface enhanced wireless network via joint active and passive beamforming," *IEEE Transactions on Wireless Communications*, vol. 18, no. 11, pp. 5394–5409, 2019.

[15] I. H. Naqvi, G. El Zein, G. Lerosey, J. de Rosny, P. Besnier, A. Tourin, and M. Fink, "Experimental validation of time reversal ultra wide-band communication system for high data rates," *IET Microwaves, Antennas & Propagation*, vol. 4, no. 5, pp. 643–650, 2010.

[16] S. Hu, F. Rusek, and O. Edfors, "Beyond massive MIMO: The potential of data transmission with large intelligent surfaces," *IEEE Transactions on Signal Processing*, vol. 66, no. 10, pp. 2746–2758, 2018.

[17] K-K. Wong, K-F. Tong, Z. Chu, and Y. Zhang, "A vision to smart radio environment: Surface wave communication superhighways," *arXiv preprint arXiv: 2005.14082*, 2020.

[18] H. Wu, G. D. Bai, S. Liu, L. Li, X. Wan, Q. Cheng, and T. J. Cui, "Information theory of metasurfaces," *National Science Review*, vol. 7, no. 3, pp. 561–571, 2020.

[19] J. A. Hodge, K. V. Mishra, and A. I. Zaghloul, "Reconfigurable metasurfaces for index modulation in 5G wireless communications," in *Proc. 2019 International Applied Computational Electromagnetics Society Symposium (ACES)*. IEEE, 2019, pp. 1–2.

[20] T. L. Marzetta, "Super-directive antenna arrays: Fundamentals and new perspectives," in *Proc. 2019 53rd Asilomar Conference on Signals, Systems, and Computers*. IEEE, 2019, pp. 1–4.

[21] M. T. Ivrlač and J. A. Nossek, "The multiport communication theory," *IEEE Circuits and Systems Magazine*, vol. 14, no. 3, pp. 27–44, 2014.

[22] A. Amiri, M. Angjelichinoski, E. De Carvalho, and R. W. Heath, "Extremely large aperture massive MIMO: Low complexity receiver architectures," in *Proc. 2018 IEEE Globecom Workshops*. IEEE, 2018, pp. 1–6.

[23] A. S. Poon and N. David, "Degree-of-freedom gain from using polarimetric antenna elements," *IEEE Transactions on Information Theory*, vol. 57, no. 9, pp. 5695–5709, 2011.

[24] B. Yang, P. Zhang, H. Wang, and W. Hong, "Electromagnetic vector antenna array-based multi-dimensional parameter estimation for radio propagation measurement," *IEEE Wireless Communications Letters*, vol. 8, no. 6, pp. 1608–1611, 2019.

[25] B. Thidé, H. Then, J. Sjöholm, K. Palmer, J. Bergman, T. Carozzi, Y. N. Istomin, N. Ibragimov, and R. Khamitova, "Utilization of photon orbital angular momentum in the low-frequency radio domain," *Physical Review Letters*, vol. 99, no. 8, p. 087701, 2007.

[26] O. Edfors and A. J. Johansson, "Is orbital angular momentum (OAM) based radio communication an unexploited area?" *IEEE Transactions on Antennas and Propagation*, vol. 60, no. 2, pp. 1126–1131, 2011.

[27] M. R. Andrews, P. P. Mitra, and R. DeCarvalho, "Tripling the capacity of wireless communications using electromagnetic polarization," *Nature*, vol. 409, no. 6818, pp. 316–318, 2001.

[28] A. S. Poon and N. David, "Degree-of-freedom gain from using polarimetric antenna elements," *IEEE Transactions on Information Theory*, vol. 57, no. 9, pp. 5695–5709, 2011.

[29] D. B. Davidson, *Computational electromagnetics for RF and microwave engineering*. Cambridge University Press, 2010.

[30] O. C. Zienkiewicz and P. Morice, *The finite element method in engineering science*. McGraw-Hill, London, 1971.

[31] Y. Liu, R. Mittra, T. Su, X. Yang, and W. Yu, *Parallel finite-difference time-domain method*. Artech, 2006.

[32] C. Christopoulos, "The transmission-line modeling (TLM) method in electromagnetics," *Synthesis Lectures on Computational Electromagnetics*, vol. 1, no. 1, pp. 1–132, 2005.

[33] W. C. Gibson, *The method of moments in electromagnetics*. CRC Press, 2014.

[34] Z. Bai and J. W. Silverstein, *Spectral analysis of large dimensional random matrices*. Springer, 2010.

[35] D. N. C. Tse and S. V. Hanly, "Linear multiuser receivers: Effective interference, effective bandwidth and user capacity," *IEEE Transactions on Information Theory*, vol. 45, no. 2, pp. 641–657, 1999.

[36] W. Hachem, P. Loubaton, J. Najim *et al.*, "Deterministic equivalents for certain functionals of large random matrices," *Annals of Applied Probability*, vol. 17, no. 3, pp. 875–930, 2007.

[37] F. Dupuy and P. Loubaton, "On the capacity achieving covariance matrix for frequency selective MIMO channels using the asymptotic approach," *IEEE Transactions on Information Theory*, vol. 57, no. 9, pp. 5737–5753, 2011.

[38] R. Couillet, M. Debbah, and J. W. Silverstein, "A deterministic equivalent for the analysis of correlated MIMO multiple access channels," *IEEE Transactions on Information Theory*, vol. 57, no. 6, pp. 3493–3514, 2011.

[39] S. Wagner, R. Couillet, and M. Debbah, "Large system analysis of linear precoding in MISO broadcast channels with limited feedback," *arXiv preprint arXiv:0906.3682*, 2009.

[40] J. Hoydis, R. Couillet, and M. Debbah, "Deterministic equivalents for the performance analysis of isometric random precoded systems," in *Proc. 2011 IEEE International Conference on Communications (ICC)*. IEEE, 2011, pp. 1–5.

[41] N. Fawaz, K. Zarifi, M. Debbah, and D. Gesbert, "Asymptotic capacity and optimal precoding in MIMO multi-hop relay networks," *IEEE Transactions on Information Theory*, vol. 57, no. 4, pp. 2050–2069, 2011.

[42] J. Hoydis, M. Kobayashi, and M. Debbah, "Optimal channel training in uplink network MIMO systems," *IEEE Transactions on Signal Processing*, vol. 59, no. 6, pp. 2824–2833, 2011.

[43] H. Sifaou, A. Kammoun, L. Sanguinetti, M. Debbah, and M.-S. Alouini, "Max–min SINR in large-scale single-cell MU–MIMO: Asymptotic analysis and low-complexity transceivers," *IEEE Transactions on Signal Processing*, vol. 65, no. 7, pp. 1841–1854, 2016.

[44] R. Couillet and M. Debbah, *Random matrix methods for wireless communications*. Cambridge University Press, 2011.

[45] X. Mestre and M. Á. Lagunas, "Modified subspace algorithms for DoA estimation with large arrays," *IEEE Transactions on Signal Processing*, vol. 56, no. 2, pp. 598–614, 2008.

[46] H. Asgharimoghaddam, A. Tölli, L. Sanguinetti, and M. Debbah, "Decentralizing multi-cell beamforming via deterministic equivalents," *IEEE Transactions on Communications*, vol. 67, no. 3, pp. 1894–1909, 2018.

[47] D. P. Bertsekas and J. N. Tsitsiklis, *Parallel and distributed computation: numerical methods*. Prentice Hall, 1989.

[48] F. Benaych-Georges and R. R. Nadakuditi, "The eigenvalues and eigenvectors of finite, low rank perturbations of large random matrices," *Advances in Mathematics*, vol. 227, no. 1, pp. 494–521, 2011.

[49] S. S. Ram, A. Nedić, and V. V. Veeravalli, "Distributed stochastic subgradient projection algorithms for convex optimization," *Journal of Optimization Theory and Applications*, vol. 147, no. 3, pp. 516–545, 2010.

[50] C. Kubrusly and J. Gravier, "Stochastic approximation algorithms and applications," in *Proc. 1973 IEEE Conference on Decision and Control Including the 12th Symposium on Adaptive Processes*. IEEE, 1973, pp. 763–766.

[51] J. J. P. Bianchi, "Distributed stochastic approximation for constrained and unconstrained optimization," *arXiv preprint arXiv: 1104.2773*, 2011.

[52] P. Comon, "Tensors: A brief introduction," *IEEE Signal Processing Magazine*, vol. 31, no. 3, pp. 44–53, 2014.

[53] F. L. Hitchcock, "The expression of a tensor or a polyadic as a sum of products," *Journal of Mathematics and Physics*, vol. 6, no. 1–4, pp. 164–189, 1927.

[54] R. A. Harshman *et al.*, "Foundations of the PARAFAC procedure: Models and conditions for an 'explanatory' multimodal factor analysis," UCLA Working Papers in Phonetics, vol. 16, pp. 1–84, Dec. 1970.

[55] C. J. Appellof and E. R. Davidson, "Strategies for analyzing data from video fluorometric monitoring of liquid chromatographic effluents," *Analytical Chemistry*, vol. 53, no. 13, pp. 2053–2056, 1981.

[56] T. G. Kolda and B. W. Bader, "Tensor decompositions and applications," *SIAM Review*, vol. 51, no. 3, pp. 455–500, 2009.

[57] L. Chiantini, G. Ottaviani, and N. Vannieuwenhoven, "An algorithm for generic and low-rank specific identifiability of complex tensors," *SIAM Journal on Matrix Analysis and Applications*, vol. 35, no. 4, pp. 1265–1287, 2014.

[58] N. D. Sidiropoulos, R. Bro, and G. B. Giannakis, "Parallel factor analysis in sensor array processing," *IEEE Transactions on Signal Processing*, vol. 48, no. 8, pp. 2377–2388, 2000.

[59] A. Decurninge, I. Land, and M. Guillaud, "Tensor-based modulation for unsourced massive random access," *arXiv preprint arXiv:2006.06797*, 2020.

[60] M. Sørensen, I. Domanov, and L. De Lathauwer, "Coupled canonical polyadic decompositions and multiple shift invariance in array processing," *IEEE Transactions on Signal Processing*, vol. 66, no. 14, pp. 3665–3680, 2018.

[61] I. V. Oseledets, "Tensor-train decomposition," *SIAM Journal on Scientific Computing*, vol. 33, no. 5, pp. 2295–2317, 2011.

[62] T. Lesieur, L. Miolane, M. Lelarge, F. Krzakala, and L. Zdeborová, "Statistical and computational phase transitions in spiked tensor estimation," in *Proc. 2017 IEEE International Symposium on Information Theory (ISIT)*. IEEE, 2017, pp. 511–515.

[63] L. Qi and Z. Luo, *Tensor analysis: spectral theory and special tensors*. SIAM, 2017.

[64] M. Nickel and V. Tresp, "An analysis of tensor models for learning on structured data," in *Proc. Joint European Conference on Machine Learning and Knowledge Discovery in Databases*. Springer, 2013, pp. 272–287.

10 Theoretical Foundations for Future Machine Type Communications

10.1 Semantic Communication Theory

The word "semantics" (meaning) comes from languages (natural or formal) and the concept of compositionality, which states that the meaning of a sentence is determined by three ingredients:

- sentence composition rule (syntax);
- meaning (semantics) of each component;
- context.

Many notions of semantics have been proposed in order to define what a semantic communication could be, but until now, none have been satisfactory. The lack of any commonly accepted mathematical definition of semantics has greatly restricted further development in this area.

However, such a theory of semantic information becomes necessary when we expect that the form of communication involves no human intervention. In traditional information theory, we do not consider the meaning of what we transmit as there will always be a "human brain" to interpret it. Images, videos, text, speech, audio excerpts; all this information is interpreted by our brain. If we hear a dog barking, we understand that "there is a DOG, and it is BARKING" because our brain is able to interpret the sounds. When intelligent machines need to communicate without any human intervention, they send information encoded using their own internal language (like human beings using natural language).

An important new use case of 6G may involve communication between intelligent machines. However, 6G is not expected to arrive for 10 years, by which time machines will have acquired more advanced capabilities, especially in terms of compositionality.

The semantics of natural language is generative. Competent users of a natural language are able to understand and generate an indefinite number of different sentences, even though the vocabulary of the language is finite. By applying a finite number of syntactical rules to a finite vocabulary, we are able to generate and understand an almost infinite number of meaningful sentences. This remarkable ability is what we mean when we say that human natural language is "generative." The semantics of natural language is also compositional, as meaningful expressions are built up from other meaningful expressions. We can analyze the meaning of a sentence such as "Bob is a teacher and a violinist" by noting that the expression contains two component sentences: "Bob is a teacher" and "Bob is a violinist." Similarly, the meaning of a complex sentence like "If John buys the tickets then either Ben or Mary will pay for snacks" is determined in part by the meanings of the component sentences "John buys the tickets," "Ben will pay for snacks," and "Mary will pay for snacks." In linguistics, the "principle of compositionality" states that the meaning of a complex expression is determined by its structure and the meanings of its constituents.

Propositional logic (as well as predicate calculus) was designed to model this kind of compositional semantics, and suggests an obvious relationship between logic and language: logical systems can represent or model important structural features of natural language, such as generativity and compositionality. Of course, we could always question whether natural language really is generative and compositional in the ways suggested here. Many have thought so, but this is an empirical question that can only be settled through empirical investigation. (Natural language may not turn out to be strictly compositional, but there are compelling arguments that natural language is by-and-large compositional.)

What is the present situation regarding a semantic information theory? The following is an excerpt from the 1949 Shannon and Weaver monograph [1]: "Relative to the broad subject of communication, there seem to be problems at three levels. So it seems reasonable to ask, serially:

- **Level A.** How accurately can the symbols of communication be transmitted? (The technical problem).
- **Level B.** How precisely do the transmitted symbols convey the desired meaning? (The semantic problem).
- **Level C.** How effectively does the received meaning affect conduct in the desired way? (The effectiveness problem)."

While many applaud Shannon's and Weaver's attempt to frame information theory as an umbrella covering syntax, semantics, and pragmatics, the overall results must seem disappointing. For those who regard Shannon's equations as technical models of signal transmission, Weaver's extension into questions of meaning and effectiveness must seem distasteful, as Shannon showed little interest in the semantic meaning of a message or its pragmatic effect on the listener. Like the manufacturers of state-of-the-art audio processing, he didn't care whether the channel carried Beethoven or Mozart, or whether the listener preferred the beat of rock, the swing of jazz, or the counterpoint of Bach. Instead, his theory aimed at solving the technical problems of high-fidelity sound transfer. Shannon was somewhat wary about the wholesale application of his mathematical equations to the semantic and pragmatic issues of interpersonal communication. But his hesitation was not shared by Warren Weaver, an executive with the Rockefeller Foundation and the Sloan-Kettering Institute on Cancer Research, and a consultant to a number of private scientific foundations. Shannon's published theory was paired with an interpretive essay by Weaver that presented information theory as "exceedingly general in its scope, fundamental in the problems it treats, and of classic simplicity and power in the results it reaches." The essay suggested that whatever the communication problem, reducing information loss was the solution. Shannon, however, has a technical definition for the word *infor-mation* that doesn't equate *information* with *meaning*. He emphasizes that "the seman-tic aspects of communication are irrelevant to the engineering aspects." For Shannon, information refers to the opportunity to reduce uncertainty. It gives us a chance to reduce entropy.

From Carnap, Bar Hillel *et al.* [2] to modern concepts like Groenendijk and Stokhof's semantics of questions [3] or the information flow of Barwise, Seligman *et al.* [4], it is typical to qualify a deductive inference as being analytical. In informational terms, this characteristic can be formulated as follows: the information contained in the conclusion is a subset of the information contained in the premises. In other words, the information contained in the conclusion is already contained in the premises. In terms of the truth function – a function that accepts truth values as input and produces a unique truth value as output – it is common to use the claim that the conclusion of a valid deduction retains the true value contained in the premises, or that the set of possible words that verifies the set (or conjunction) of all premises is a subset of the set of possible words that verifies the conclusion. Reconciliation between the two views can be achieved from a semantic perspective of information theory. More precisely, both views assume that every proposition of interest is a union of disjoint elementary propositions Z, and that they define the content of proposition P as the union of all Zs such that the negation of Z is implied by proposition P.

Neural networks may also be viewed in this manner. The answers to the questions asked are visible at the output layer of a trained neural network, and the logical content is maximized at this output layer. However, even if the "regular" distortion (which does not take into account the meaning) at the output layer is huge, the output of the network gives us what we require.

- **An example:** Assume that a DNN has been trained to recognize a cat in an image. The logic associated with this DNN is very simple: one type (Cat), no composition-ality, and one proposition:

$$\{x, \top x \in Cat\}$$

  where x is the input image and $\top$ means True. So $\{x, \top x \in Cat\}$ means that the image x represents a cat (the image is of type Cat). Of course, the DNN has been trained on a dataset labeled by humans, and those who labeled the training dataset are the context. Truth is relative. A semantic communication would only transmit one bit corresponding to the statement "it is a cat, True (1) or False (0)." So the DNN can be seen here as a semantic source encoder which encodes one image into one bit.

  Shifting to more complex situations, a neural network may have to answer many questions, for example the language used may contain several propositions, several types, conjunctions, and disjunctions. The neural network will compress data and transform it into propositions requiring far fewer bits. As a suitable case of semantic communication, machine-learning-based joint source and channel coding will be discussed in Chapter 23.

- **Mathematical perspective:** It is evident that semantic information is not easy to acquire, primarily owing to a lack of appropriate mathematical tools. However, some promising directions of research do exist. Without dwelling too much on the details, Baudot and Bennequin, in [5], rederived Shannon's information measures from a categorical/topological viewpoint.

  In order to define Level-A information measures, Shannon used an axiomatic approach. He said: "This theorem, and the assumptions required for its proof, are in no way necessary for the present theory. It is given chiefly to lend a certain plausibility to some of our later definitions. The real justification of these definitions, however, will reside in their implications." The implications showed he was right and, in this case, scalar quantities such as entropy and mutual information are sufficient and successful.

  The extension of this axiomatic approach to the semantic case has not been fruitful up to now, and it seems that the connections between different points of views on entropy – algebraic, probabilistic, combinatorial, or dynamic – are still poorly understood. This was a primary motivation in [5] to introduce a categorical framework as a formalism general enough to integrate different theories, and we believe that it could lead to a theory of semantic information. Using such a framework, Vigneaux-ariztia in [6] found non-Shannon information measures such as Tsallis entropy, quantum entropy, or combinatorial information theory, where probabilities are replaced with frequencies.

  The departing point is the definition of an information structure. It is a pair (S, E), where S is a category [7] representing the variables (morphisms of S are given by conditioning), and E is a covariant functor from S to the category of measurable spaces. Morphisms are then defined between information structures in order to create a category of such structures. In modern topology, spaces are studied though their (co)homology, which encodes their shapes to some extent. Here, the natural notion of space is that offered by Grothendieck, the topos [8], and interesting toposes here are the categories of contravariant functors of sets in category S for a given information structure and the coarsest Grothendieck topology (all presheaves are sheaves). These presheaves and the natural transformations between them form an abelian category, which entails the possibility of defining cohomological functors.

  Grothendieck found that the Shannon entropy is a cohomology of toposes. Furthermore, we may change S to another more general category. (Investigations are currently being carried out in the syntactic category, corresponding to a given language, that depends on the probabilities of propositions.) Also, in the semantic context, information measures may not be scalars, but actually larger objects such as posets and sets.

  One final point relates to the very strong connection that toposes introduce between geometry and logic, as each topos has an internal language.

6G will drive many technologies to evolve and develop in tandem with semantic communication theory, such as joint source channel coding and source compression.

10.2 Super-Resolution Theory

Super-resolution theory refers to the representation of source information in such a way that it can be reconstructed as close to its original format as possible [9]. Events such as spikes and discontinuity points can be super-resolved with infinite precision from just a few low-frequency samples by solving convenient convex programs. The direct use case for super-resolution is to further enhance resolution beyond the physical limitations of the sensor system (e.g., a microscope). Super-resolution can also be used in other cases where it is desirable to extrapolate fine scale details from low-resolution data or resolve sub-pixel details (e.g., sparsity and compress sensing).

In wireless communications, transmitted signals typically reach the receiver through multiple paths owing to reflection from objects (such as buildings). Assume that $h(t)$ is the transmitted signal, $z(t)$ is the channel response, and d_i and t_i are the complex amplitude and the delay in the ith path; then the time-domain received signal is:

$$x(t) = z(t) * h(t) = \sum_{i=1}^{r} d_i h(t - t_i)$$

where r is the number of paths, which we will hereafter refer to as modes. Using a Fourier transform, we can obtain the spectral domain signal as follows:

$$\hat{x}(f) = \hat{z}(f)\hat{h}(f) = \sum_{i=1}^{r} d_i \hat{h}(f)e^{j2\pi f t_i}$$

Because the transmitted signal $\hat{h}(f)$ is a priori known, the observed data can be expressed as:

$$\frac{\hat{x}(f)}{\hat{h}(f)} = \sum_{i=1}^{r} d_i e^{j2\pi f t_i} \quad \forall f : \hat{h}(f) \neq 0$$

Finally, we can acquire the classic parameter estimation model of super-resolution theory, which is a signal mixture of r modes:

$$x[t] = \sum_{i=1}^{r} d_i \psi(t; f_i) \quad t \in \mathbf{Z}$$

where d_i is the amplitude, f_i is the frequency, ψ is the (known) model function (for example, $\psi(t; f_i) = e^{j2\pi f t_i}$), and r is the number of modes or order of the model. Our purpose is to estimate the $2r$ unknown parameters $\{d_i\}$ and $\{f_i\}$. A simple example is provided in Figure 10.1.

The first widely used resolution theory was the Shannon–Nyquist sampling theorem [10], which requests a sampling frequency at least twice as high as the highest frequency of the source information. This is known as the Nyquist sampling

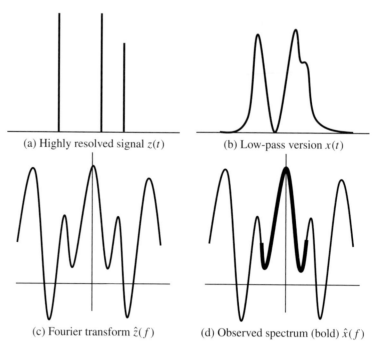

(a) Highly resolved signal $z(t)$ (b) Low-pass version $x(t)$

(c) Fourier transform $\hat{z}(f)$ (d) Observed spectrum (bold) $\hat{x}(f)$

Figure 10.1 Super-resolution that extrapolates the high-end spectrum (fine scale details) from the low-end spectrum (low-resolution data).

frequency and guarantees the similarity and integrity of the reconstructed information. This theorem is based on Fourier transforms which linearly convert source information into a new coordinate system containing an infinite number of cosine wave bases. With a Fourier transform, a time-serial signal or sequence can be represented by a summation of projections (by an inner product) over an infinite number of bases. The Nyquist sampling frequency results in a uniform resolution over all projections.

The Shannon–Nyquist sampling theorem has dominated communication design. It has instructed us on how much bandwidth is required for a reliable transceiver, or for codecs relating to audio, video, images, and analog signals. However, the last four decades have witnessed a tremendous increase in occupied radio bandwidths owing to generations of wireless systems. Bandwidth has become more expensive and carrier frequencies have grown increasingly higher, further driving the study of super-resolution theory.

Tracing back to the source, you may wonder whether it is truly necessary to uniformly represent a signal or information according to its highest frequency component, and this can be further divided into two questions. (i) Is a uniform resolution necessary? (ii) Is the Nyquist sampling frequency necessary?

Uniform resolution or non-uniform resolution: to answer this question, we must consider a source or signal whose receiver or decoder is interested in only some of its frequency components. Video serves as an example of this concept, with human vision paying more attention to the foreground than the background. Accordingly, the foreground consists of higher-frequency components with lower ones relegated to the background. In this scenario, it is logical and more efficient to utilize a higher resolution for the foreground, rather than implementing a uniform resolution.

This consideration gives rise to a second group of widely used theories: principal component analysis) [11], short-time Fourier transform, and wavelets. Principal component analysis uses singular vector decomposition to separate components. However, unlike Fourier transform, principal component analysis adopts a left or right eigenmatrix rather than a Fourier basis as its new coordinate. The JPEG encoder uses principal component analysis to separate and retain principal components, while MIMO transceivers also use principal component analysis to detect principal layers. However, principal component analysis loses some information owing to uniform resolution through Heisenberg's uncertainty principle. For a wideband signal, uniform resolution favors some parts but penalizes others. To solve this problem, Gabor first proposed short-time Fourier transform – and later wavelets – to realize multi-resolution. If we transform a wavelet signal back to the original ambient space (as a decoder), it is likely that the resolution is not actually uniform at all. JPEG-2000 uses wavelets to compress images. Some have even proposed the replacement of OFDM with wavelet transform for wideband signals.

Uniform sampling or non-uniform sampling: we must also consider a source or signal whose high-frequency components are not of great magnitude or critical for its receiver. Audio is a simple example. As the human ear can only perceive audio up to 20 kHz, anything over this threshold makes no sense for humans. As such, it is pointless to sample a high-definition audio signal beyond that range. In this regard, an 8 kHz sampling frequency is sufficient for good voice quality.

This consideration gives rise to the third widely used theory, compressed sensing [12], which is based on the sparsity of natural information. For example, source information x, though non-sparse in its ambient space, can be linearly transformed into a sparse representation s with only K non-zero entries in a new coordinate system. The transformation can be a fast Fourier transform (FFT), a discrete Fourier transform (DFT), or by wavelets, as mentioned above. For this sparse-in-nature information x, an M-by-N random matrix (known as a measurement matrix) defines $M(K < M \ll N)$ random sampling points in order to sample x into an M by 1 vector y. Apparently, y is not uniformly sampled, and its sampling frequency is much lower than twice the highest frequency. At the decoding side, compressed sensing uses the $L1$ norm rather than $L2$ norm to reject outliers, and matching pursuit (MP) or orthogonal matching pursuit (OMP) can be used to approximate the sparse vector [13–16]. Compressed sensing removes the need for a Nyquist sampling frequency to sample sparse-in-nature information or signals. In 6G, sensors can exploit non-uniform random sampling schemes to significantly save radio bandwidth.

Super-resolution for 6G: over the past 40 years we have witnessed how non-uniform resolution is better than uniform resolution and how non-uniform sampling is better than uniform sampling in cases such as source coding or image compression.

Super-resolution theory integrates both non-uniform resolution and non-uniform sampling [9], and can be utilized in many 6G fields.

A simple example is source coding for machines. Along with the greater number of machinery receivers connected in 6G systems, machine-type receivers may function differently from humans. Higher resolution may be applied to backgrounds, and a multiple-resolution source coding strategy can adapt flexibly to different object requirements and save transmission bandwidth.

A final example is low-power non-uniform sampling. High-band 6G systems will pose challenges to dynamic quantization, and the Nyquist sampling frequency will result in massive power consumption. If the source information is sparse in nature, then low-power non-uniform sampling becomes feasible in order to reduce power consumption.

Super-resolution theory may affect the directions of many technologies, including localization algorithms, compressed sensing, and wireless imaging algorithms.

References

[1] C. E. Shannon and W. Weaver, *The mathematical theory of communication*. University of Illinois Press, 1949.

[2] R. Carnap, Y. Bar-Hillel *et al.*, "An outline of a theory of semantic information," Technical Report no. 247, Research Laboratory of Electronics, MIT, 1952.

[3] J. A. G. Groenendijk and M. J. B. Stokhof, "Studies on the semantics of questions and the pragmatics of answers," joint Ph.D. dissertation, University of Amsterdam, 1984.

[4] J. Barwise, J. Seligman *et al.*, *Information flow: The logic of distributed systems*. Cambridge University Press, 1997.

[5] P. Baudot and D. Bennequin, "The homological nature of entropy," *Entropy*, vol. 17, no. 5, pp. 3253–3318, 2015.

[6] J.-P. Vigneaux-ariztia, "Topology of statistical systems: A cohomological approach to information theory," thesis, Max Planck Institute for Mathematics in the Sciences, Max Planck Society, 2019.

[7] S. Awodey, *Category theory*.   Oxford University Press, 2010.

[8] S. MacLane and I. Moerdijk, *Sheaves in geometry and logic: A first introduction to topos theory*.   Springer Science & Business Media, 2012.

[9] E. J. Candès and C. Fernandez-Granda, "Towards a mathematical theory of super-resolution," *Communications on Pure and Applied Mathematics*, vol. 67, no. 6, pp. 906–956, 2014.

[10] H. Nyquist, "Transmission systems for communications," *AT&T Technical Journal*, vol. 2, p. 26, 1959.

[11] I. T. Jolliffe, *Principal component analysis (Second Edition)*. Springer, 2002.

[12] D. L. Donoho, "Compressed sensing," *IEEE Transactions on Information Theory*, vol. 52, no. 4, pp. 1289–1306, 2006.

[13] E. J. Candes, J. K. Romberg, and T. Tao, "Stable signal recovery from incomplete and inaccurate measurements," *Communications on Pure and Applied Mathematics*, vol. 59, no. 8, pp. 1207–1223, 2006.

[14] E. J. Candès, J. Romberg, and T. Tao, "Robust uncertainty principles: Exact signal reconstruction from highly incomplete frequency information," *IEEE Transactions on Information Theory*, vol. 52, no. 2, pp. 489–509, 2006.

[15] E. J. Candes and T. Tao, "Near-optimal signal recovery from random projections: Universal encoding strategies?" *IEEE Transactions on Information Theory*, vol. 52, no. 12, pp. 5406–5425, 2006.

[16] R. G. Baraniuk, "Compressive sensing [lecture notes]," *IEEE Signal Processing Magazine*, vol. 24, no. 4, pp. 118–121, 2007.

11 Theoretical Foundations for Energy-Efficient Systems

11.1 Energy-Efficient Communication and Computation Theory

Designing wireless networks on the basis of energy efficiency metrics is an extremely challenging task, as traditional power allocation schemes lead to continuous transmission using the maximum power available. However, this view has begun to change recently as bit-per-joule energy efficiency, defined as the amount of information that can be reliably transmitted per joule of consumed energy, has emerged as a KPI for 5G networks [1]. In [2], systematic approaches to solve energy efficiency maximization problems are discussed. In this regard, the framework from [3] is employed in [4] and [5] to develop energy-efficient power control algorithms for massive MIMO systems, and in [6] for energy-efficient small-cell networks. However, all available studies consider two main sources of energy consumption, as follows.

- **Transmit energy:** The energy output by each Tx antenna.
- **Static energy:** The energy consumed by all hardware blocks in each transmitter and receiver (e.g., analog-to-digital converter, digital-to-analog converter, analog filtering, battery backup, and cooling vent).

In fact, analysis of the energy consumption required to perform computations for information processing in the digital domain (for example, to compute digital precoders and receivers, or coding operations) is poorly understood. In this context, only a few results are available, which employ fundamental physics/thermodynamics principles in order to understand whether digital computations intrinsically require energy consumption.

Surprisingly, in theory, the answer is negative: no physics principle dictates that some energy must be consumed in order to perform a computation in the digital domain. Indeed, if the computation performed does not destroy any information, then it is a reversible transformation, which, according to the second law of thermodynamics, does not generate any entropy. As a result, reversible computations can in theory be performed without requiring any energy consumption, as first observed in [7] and [8]. Models of machines that implement this reversible logic have appeared in [9] and [10], and operate by storing both the output and the input of each computation, without erasing any bits from the computer's memory. As such, it is always possible to restore the machine to its initial state. With this in mind, why are modern computers based on

Boolean logic, which is generally irreversible (for example, it is not always possible to deduce the input of an "AND" or "OR" operation from its output)?

The first practical consideration is that even if models of reversible logic machines have been proposed, they are only theoretical. In practice, data processing is performed by macroscopic apparatuses, which dissipate macroscopic amounts of energy. The second consideration is more fundamental; even if we could build a reversible logic machine, it would have unfeasible memory requirements, needing to store both the output and input of each operation it executes, without ever erasing any bits from memory. However, since the amount of data that can be stored in a computer is limited by the volume of the memory bank [11], even reversible logic machines would eventually have to overwrite some memory cells, thus consuming energy.

A similar situation arises when detecting or correcting errors over a communication channel. Detecting an error entails no energy dissipation in theory, but correcting it causes information loss (to the environment), thus generating entropy. This leads to an unrealistic situation whereby, in order to avoid generating entropy, a communication system should also store received messages that contain errors. As a result, irreversible operations need to be considered, but very little is known regarding their fundamental limits concerning energy dissipation in communication systems. The cornerstone contribution in this area is [12]; however, this article considers only an isolated computer rather than a complete communication system. Examples of [12] applied to communication systems are scarce.

In [13] and [14], some results are derived with reference to simple P2P communication systems, while the energy consumption required to run generic resource allocation algorithms is analyzed. The above state-of-the-art approach shows that our knowledge of the energy consumed by wireless networks primarily concerns the transmission of information, rather than the digital computations of its processing. This is mainly a result of the data processing in previous and present wireless generations amounting to fairly simple operations, typically executed by a digital signal processor and representing a small fraction of the total network energy consumption. However, this situation will be reversed in native AI 6G wireless networks. ANNs are trained in the digital domain by processing large datasets, and operating on digital data. As a result, unlike present networks, the energy required for information processing computations will be the main source of energy consumption in 6G.

As described in the previous chapters, 6G's energy efficiency requirements are very strict. Consequently, investigating the computing cost of ANNs in terms of energy consumption is extremely important. In addition, the energy aspect should also be studied in connection with the processing latency requirement of 6G, which is also very strict and requires fast (and energy-consuming) computations. Understanding the fundamental energy limits of information processing computations is an essential step towards the successful deployment of 6G wireless networks. As discussed, if an infinite memory space is assumed, no fundamental limit exists for the energy required to perform reversible computations. However, we have also seen that it is physically impossible to run a network based only on reversible operations. Consequently, from an engineering perspective, it is of fundamental importance to understand the best we can achieve by also implementing irreversible operations.

As today's computers are based on an irreversible (Boolean) logic, it is imperative to fill the gap between reversible and irreversible computation in order to understand the fundamental energy limits of information processing. This analysis is made even more challenging by the increasingly complex wireless networks being investigated, which are far more complex than past and present communication networks, thus requiring the development of dedicated theoretical models to estimate the amount of information that must be processed in order to operate the networks. Besides investigating the energy required for information processing, it is also essential to understand how the fundamental energy limits of information processing are connected to the fundamental limits of communication performance. For example, what is the optimal trade-off between communication performance and network energy consumption? This question has been investigated only with reference to the energy used for communication, whereas almost nothing is known about the energy used for computations related to information processing. In addition, all previous studies consider traditional network architectures and topologies, without considering the impact of using deep learning and ANNs.

The thermodynamics of communication and computation theory may help the development of novel algorithms for energy-efficient 6G networks.

11.2   Green AI Theory

In the future, AI will undergo incredible evolution and permeate every corner of society. During this process, AI and communication will converge, with either the communication network serving and providing data pipelines for AI applications or AI improving the data transmission efficiency of the communication network. However, the electric power consumption and carbon emissions resulting from AI's use of high computational resources is unsustainable in terms of both environmental protection and economic benefits. Consequently, Green AI is gaining increasing attention in the AI community [15, 16], and the transformation from Red AI to Green AI will profoundly affect the design principles of next-generation wireless communication networks.

Red AI indicates a model that is trained by a training dataset and evaluated by a test dataset. To develop a red AI model, we usually need to use a set of training data to iteratively adjust the hyperparameters. Correspondingly, Green AI refers to an AI solution where models, algorithms, and hardware yield novel results without increasing the computational cost, and actually reducing it in ideal scenarios [15].

Generally, the computing cost of an AI model is proportional to the product of three key factors: the cost of executing the model on a single sample, the size of the training dataset, and the number of hyperparameter experiments. The number of floating-point operations using different algorithms to implement the same AI function decreased by a factor of 44 from 2012 to 2019, which is equivalent to doubling the algorithm efficiency every 16 months over a period of seven years [17]. In the meantime, due to the rapid expansion of training datasets and hyperparameters and particularly due to the excessive pursuit of improving training precision based on massive computational resources, the electric power consumption of AI increased by a factor of 300,000 from 2012 to 2018, which is equivalent to doubling the resource consumption every 3.4 months [18]. Emma Strubell *et al.* proposed an estimate for AI energy consumption (p_t) and CO_2 emissions (CO_2e) [16]:

$$p_t = \frac{PUE \times t \times (p_c + p_r + gp_g)}{1000} \quad \text{(kWh)}$$

$$CO_2e = CUP \times p_t \quad \text{(pounds)}$$

where t is the total time expected for model training; p_c, p_r, and p_g represent the average power draw (in watts) from all CPU sockets, from all DRAM (main memory) sockets, and from each GPU, respectively, during the training; and g is the number of GPUs used to accelerate training. PUE is the acronym for power usage effectiveness, which accounts for the additional energy required to support the computing infrastructure (primarily cooling). The PUE coefficient is 1.58 from the global average of data centers [19]. CUP is the acronym for carbon dioxide per unit power, which depends on the level of local power industry development. The US Environmental Protection Agency provides an average CO_2 production coefficient of 0.954 pounds/kWh [20].

Considering NLP as an example, the power consumption and carbon emissions corresponding to different AI models and hardware cannot be ignored from an economic perspective [16]. Without any loss of generation, the average electricity cost in the USA was $0.12 per kWh in 2018. In particular, any models sensitive to datasets and hyperparameters will be very expensive, as they may need to be retrained every time a new situation arises.

In recent years, there been a huge number of academic papers on AI in various application fields, including AI used in wireless communication networks. However, little attention has been paid to the ensuing problems of energy consumption and carbon emissions caused by brute-force computing during AI implementation. Consequently, the AI community is attempting to alert researchers and is currently advocating the following: (i) reporting training time and sensitivity to hyperparameters in AI publications, (ii) equitable access to computation resources during AI implementation, and (iii) prioritizing computationally efficient AI hardware and algorithms.

To measure the efficiency, Roy Schwartz *et al.* suggested reporting the amount of work required to generate a result (to train a model and tune hyperparameters) in AI applications [15]. When reporting the amount of work completed by an AI application, certain quantities should be measured in order to allow for a fair comparison between different models. These include the following.

- **Carbon emissions:** A quantity which Green AI aims to minimize.
- **Energy consumption:** Correlated with carbon emissions while being time- and location-agnostic.

- **Elapsed real time:** The total running time for generating an AI result, which is a natural measure for efficiency.
- **Number of parameters:** The number of parameters (learnable or total) used by the AI model, which is another common measure for efficiency.
- **Floating point operations:** A concrete measure for efficiency in the process of generating an AI result.

In [21], the Green AI concept (Once-for-all), which trains a single network and then specializes it for efficient deployment in order to deliver efficient inference across many devices and meet various resource constraints, was proposed. Conventional approaches either manually design or use neural architecture search to find a specialized neural network and train it from scratch for each case, which is computationally prohibitive and generates a large amount of CO_2 emissions. Conversely, the Once-for-all network supports diverse architectural settings by decoupling training and search, and it is possible to quickly acquire a specialized sub-network without additional training.

If we shift our focus to the design of next-generation wireless communication networks, traditional AI optimization algorithms (such as federated learning) usually consider the bandwidth or latency of wireless links as a weight for distributed multiprocessor data exchange, without considering energy limit or power cost differences between different devices in various regions. This incomprehensive consideration of AI energy limits or power costs may lead to a large deviation between wireless network design and actual AI deployment in the future. As a result, we advocate the need to attach equal importance to Green AI and Green Communications. At the start of an architecture design, the impact of AI models, algorithms, and hardware on energy consumption should be fully considered in order to deliver economic benefits to customers with appropriate system OPEX.

In 6G, energy efficiency computing architectures have become necessary and must ensure consideration of the energy limits of AI. This is even more important as such considerations are equally urgent in the current 4G and 5G networks.

References

[1] S. Buzzi, I. Chih-Lin, T. E. Klein, H. V. Poor, C. Yang, and A. Zappone, "A survey of energy-efficient techniques for 5G networks and challenges ahead," *IEEE Journal on Selected Areas in Communications*, vol. 34, no. 4, pp. 697–709, 2016.

[2] A. Zappone and E. Jorswieck, "Energy efficiency in wireless networks via fractional programming theory," *Foundations and Trends in Communications and Information Theory*, vol. 11, no. 3–4, pp. 185–396, 2015.

[3] A. Zappone, L. Sanguinetti, G. Bacci, E. Jorswieck, and M. Debbah, "Energy-efficient power control: A look at 5G wireless technologies," *IEEE Transactions on Signal Processing*, vol. 64, no. 7, pp. 1668–1683, 2015.

[4] A. Zappone, E. Björnson, L. Sanguinetti, and E. Jorswieck, "Globally optimal energy-efficient power control and receiver design in wireless networks," *IEEE Transactions on Signal Processing*, vol. 65, no. 11, pp. 2844–2859, 2017.

[5] M. Di Renzo, A. Zappone, T. T. Lam, and M. Debbah, "System-level modeling and optimization of the energy efficiency in cellular networks: A stochastic geometry framework," *IEEE Transactions on Wireless Communications*, vol. 17, no. 4, pp. 2539–2556, 2018.

[6] C. H. Bennett and R. Landauer, "The fundamental physical limits of computation," *Scientific American*, vol. 253, no. 1, pp. 48–57, 1985.

[7] D. Deutsch, "Is there a fundamental bound on the rate at which information can be processed?" *Physical Review Letters*, vol. 48, no. 4, p. 286, 1982.

[8] Y. Lecerf, "Machines de Turing reversibles-recursive insolubilite en n ∈ N de l'equation u = $\theta\hat{}$ nu, ou θ est un isomorphisme de codes," *Comptes Rendus hebdomadaires des seances de l'academie des sciences*, vol. 257, pp. 2597–2600, 1963.

[9] C. H. Bennett, "Logical reversibility of computation," *IBM Journal of Research and Development*, vol. 17, no. 6, pp. 525–532, 1973.

[10] S. Lloyd, V. Giovannetti, and L. Maccone, "Physical limits to communication," *Physical Review Letters*, vol. 93, no. 10, p. 100501, 2004.

[11] R. Landauer, "Irreversibility and heat generation in the computing process," *IBM Journal of Research and Development*, vol. 5, no. 3, pp. 183–191, 1961.

[12] J. Izydorczyk and L. Cionaka, "A practical low limit on energy spent on processing of one bit of data," in *Proc. 2008 IEEE 8th International Conference on Computer and Information Technology Workshops*. IEEE, 2008, pp. 509–514.

[13] B. Perabathini, V. S. Varma, M. Debbah, M. Kountouris, and A. Conte, "Physical limits of point-to-point communication systems," in *Proc. 2014 12th International Symposium on Modeling and Optimization in Mobile, Ad Hoc, and Wireless Networks (WiOpt)*. IEEE, 2014, pp. 604–610.

[14] E. D. Demaine, J. Lynch, G. J. Mirano, and N. Tyagi, "Energy-efficient algorithms," in *Proc. 2016 ACM Conference on Innovations in Theoretical Computer Science*, 2016, pp. 321–332.

[15] R. Schwartz, J. Dodge, N. A. Smith, and O. Etzioni, "Green AI," *arXiv preprint arXiv:1907.10597*, 2019.

[16] E. Strubell, A. Ganesh, and A. McCallum, "Energy and policy considerations for deep learning in NLP," *arXiv preprint arXiv:1906.02243*, 2019.

[17] D. Hernandez and T. B. Brown, "Measuring the algorithmic efficiency of neural networks," *arXiv preprint arXiv:2005.04305*, 2020.

[18] D. Amodei and D. Hernandez, "AI and compute," *Heruntergeladen von https://blog.openai.com/aiand-compute*, 2018.

[19] R. Ascierto, "Uptime institute global data center survey," Technical Report, Uptime Institute, 2018.

[20] US Environmental Protection Agency, "Emissions and generation resource integrated database (eGRID2007, version 1.1)," 2008.

[21] H. Cai, C. Gan, T. Wang, Z. Zhang, and S. Han, "Once-for-all: Train one network and specialize it for efficient deployment," *arXiv preprint arXiv:1908.09791*, 2019.

Summary of Part III

Wireless networks are rapidly evolving towards a scenario in which the communication infrastructure will merge with the environment. If 5G networks are moving in the direction of dense deployments of base stations and antenna arrays, 6G will take this approach to the next level, equipping objects in the communication scene, such as buildings, walls, cars, and road signs, with intelligent surfaces capable of amplifying electromagnetic signals, performing computations, and storing data. Driven by the exponential increase in connected devices demanding higher-performance services, the communication environment will be transformed into an intelligent wireless fabric, characterized by unprecedented levels of density, stochasticity, and heterogeneity, as well as very large dimensions. In such a challenging scenario, traditional network architectures will fail, and a paradigm shift is required. The goal of Part III has been to lay a theoretical foundation for communications in the 6G intelligent wireless fabric, establishing the roots of theoretical performance limits as well as the design of practical algorithms to approach these limits. We envisage a network infrastructure composed of multiple distributed segments, each endowed with AI capabilities and acting as subsystems capable of independent decision-making. This vision can be realized through an interplay of theoretical tools from multiple disciplines originating from fundamental mathematics and physics. It is, however, important to note that some knowledge gaps remain, and these must be addressed in order to create a unified framework effectively. These knowledge gaps can be listed as follows.

Knowledge gap 1:
In general, machine learning for wireless networks has addressed only centralized network architectures, wherein a single ANN is configured by the central controller to dictate the policy in the entire network. However, in a distributed network, each node will train its own ANN the basis of different algorithms and local training datasets. If not properly understood and controlled, this may cause system instability, possibly leading to system failure and/or triggering cyber attacks.

Knowledge gap 2:
Traditional learning methods assume that the conditions in which the ANNs will operate are similar to those experienced during the training phase. However, this is difficult to ensure in large and stochastic networks owing to their rapidly time-varying and heterogeneous nature.

Knowledge gap 3:
None of the mathematical/physics frameworks described above has ever been employed in distributed wireless networks. Available methodologies based on random matrix theory and computational physics do not apply to AI-based distributed wireless networks, in which each node learns how to behave by independently processing its own dataset without any explicit centralized control.

Knowledge gap 4:
Each of the mathematical/physics frameworks described above is able to capture a specific aspect of complex wireless networks, but, in order to fully characterize large and stochastic native AI networks, all these frameworks must be used in tandem. At present, no result is available on the joint use of so many sophisticated frameworks for wireless networks.

Knowledge gap 5:
Computing requires energy, and ANNs are trained and operated by numerically processing large datasets. Consequently, computations are expected to be the main obstacle impeding the development of green and sustainable networks. Fundamental limits to the energy consumption of information processing are only available for P2P systems, while nothing is known about multi-user wireless networks.

Knowledge gap 6:
The trade-off between the fundamental performance and the energy limits of computation and communication must be investigated in order to reach the best trade-off between achieving optimal communications and ensuring the sustainability of the wireless ecosystem.

Knowledge gap 7:
The use of sophisticated theoretical tools might lead to complicated solutions that are difficult to implement in practical systems. For example, it may be very complicated to solve the different equations arising from EIT or distributed stochastic optimization analysis. A computational framework capable of implementing theoretical solutions is currently lacking.

These knowledge gaps represent just some of the obstacles that will need to be overcome if we are to achieve a unified fundamental theory of 6G networks. From a theoretical perspective, we are approaching the end of an era that has guided ICT for the last century. Many of the remarkable engineering breakthroughs in communication (the famous "G" era) and computing (the famous "Moore's" era) were based on considerably old fundamentals, with the Nyquist sampling theorem dating back to 1924, Shannon's law to 1948, and Von Neumann architecture to 1946. Today, as we approach the limits of the fundamentals, we find ourselves lacking the guidance required to develop new engineering solutions, and there is a need to push forward the fundamentals shared in Part III in order to kick-start a new era of engineering.

Part IV

New Elements

Introduction to Part IV

6G is expected to incorporate new capabilities and provide novel services utilizing new wireless technologies. In addition to the developments made in wireless transmission technologies, the 6G system will encompass many new elements such as new spectrum, new channels, new materials, new antennas, new computing technologies, and new terminal devices. In this part, we start by introducing the idea of a new spectrum and a potential candidate spectrum, and then discuss the opportunities and challenges of using the THz band for both communication and sensing. We also describe new channel modeling and measurement, along with the higher-frequency bands, for scenarios such as ELAA, NTN, and mmWave/THz sensing. Given that the higher-frequency bands for communication and sensing require new materials and antennas, we explore the applications of silicon photonics, heterogeneous III-V materials, reconfigurable materials, photonic crystals, photovoltaics materials, and plasmonic materials in THz communication and sensing. Furthermore, we discuss several new types of antennas for THz frequency bands. This is of particular importance, as, owing to considerable transmission loss, THz antennas differ from conventional ones, which connect to RF systems through coaxial cables or microstrip lines. And as Moore's law is expected to flatten, this part examines new computing technologies such as brain-inspired computing and quantum computing. Finally, we predict the trends of future terminal devices and discuss their novel capabilities.

12   New Spectrum

Suitable spectrum is the main consideration for every generation of wireless system. First, more spectrum is always desirable in order to support higher data rates and network capacity. Second, a globally unified spectrum will allow for greater economies of scale in terms of both infrastructure and user equipment. It is therefore essential to identify and harmonize the spectrum around the world – something that the wireless industry has done successfully via the ITU-R and world radiocommunication conference (WRC). Third, as radio technologies continue to evolve, multi-band radio technologies will enable us to further utilize incumbent and new bands. Fourth, global roaming and technology standardization are critical for achieving service and application success on a global scale. And fifth, unified spectrum allocation and regulatory rules around the world are critical – this is a unique challenge facing the use of spectrum.

In wireless communication infrastructure, spectrum is a vital component. The use of radio spectrum typically takes two generations to mature and, as the technologies used in a wireless generation evolve, the use of spectrum continues to expand to higher-frequency bands. For example, sub-1 GHz bands were used in the first generation (1G), while in 2G and 3G, the spectrum used for international mobile telecommunications (IMT) systems expanded to cover sub-3 GHz. In 5G, C-band is the most widely used band worldwide for initial deployments, and mmWave frequency bands up to 52.6 GHz are supported in 3GPP Release 15/16 (2020). Future releases of 5G are expected to expand this support to approximately 100 GHz. Looking ahead to 6G, expanding the radio spectrum to even higher-frequency bands such as THz is an important consideration, as the pursuit for higher data rates and new services continues. With higher-frequency bands, the next generation of wireless system will open up new possibilities for both better communication services and new services beyond communication.

In this chapter, we discuss the potential spectrum for 6G, starting with the allocation of 5G spectrum. We then describe 6G spectrum requirements and identify candidate frequency bands according to the demands that new 6G services will impose. In particular, we look at the necessity of expanding to higher spectrum and the importance of integrated multi-layer spectrum usage with all low bands, mid-bands, mmWave bands, and THz bands.

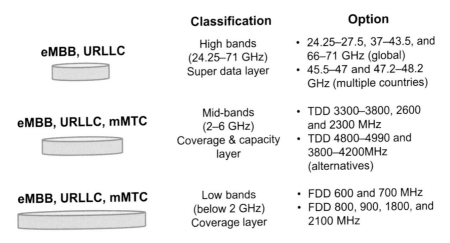

	Classification	Option
eMBB, URLLC	High bands (24.25–71 GHz) Super data layer	• 24.25–27.5, 37–43.5, and 66–71 GHz (global) • 45.5–47 and 47.2–48.2 GHz (multiple countries)
eMBB, URLLC, mMTC	Mid-bands (2–6 GHz) Coverage & capacity layer	• TDD 3300–3800, 2600 and 2300 MHz • TDD 4800–4990 and 3800–4200MHz (alternatives)
eMBB, URLLC, mMTC	Low bands (below 2 GHz) Coverage layer	• FDD 600 and 700 MHz • FDD 800, 900, 1800, and 2100 MHz

Figure 12.1 Multi-layer frequency-band framework and specific frequency-band options for 5G usage scenarios.

12.1   Global Spectrum Allocation of 5G up to 2020

The ITU-R envisions that 5G should include three representative usage scenarios (i.e., eMBB, mMTC, and URLLC) and provide about 20 Gbit/s data rate, millisecond-level latency, and ultra-dense connectivity, representing significant improvements over the previous generation of IMT systems [1].

To this end, 5G employs a multi-layer frequency-band approach, as shown in Figure 12.1. In this approach, the mobile network can access "high bands" (mmWave band from 24.25 to 71 GHz), "mid-bands" (2 to 6 GHz), and "low bands" (below 2 GHz) on demand, supporting the diversified usage scenarios.

Among these frequency bands, the mid-bands (2 to 6 GHz) play the most crucial role for 5G application, offering the best trade-off between coverage and capacity for cost-effective implementation while also supporting most usage scenarios of 5G in wide-area coverage. As of January 2020, many countries have started to allocate the mid-bands for the first wave of 5G commercial deployment [2].

The key mid-bands are the time division duplex (TDD) bands at 3300–3800, 2600, and 2300 MHz. The availability of these bands is increasing worldwide, as network operators consider them the primary frequency bands for large-scale commercial deployment of 5G. Where these bands are not available in the short term or insufficient to satisfy demand, some countries are using the alternative TDD bands at 4800–4990 MHz (e.g., Russia, China, Brazil, and South Africa) and 3800–4200 MHz (e.g., Japan, USA, UK, and South Korea).

Low bands are typically used in combination with mid-bands in order to provide wider and deeper indoor coverage. Some network operators leverage these low bands for independent rapid and cost-effective deployment if enhanced capacity is not the primary concern. The resources in this frequency range mainly comprise two parts:

(a) new frequency division duplex (FDD) bands (e.g., 600 and 700 MHz) allocated at WRC in 2015 (WRC-15) [3], and (b) existing 2G/3G/4G FDD bands (e.g., 800, 900, 1800, and 2100 MHz), which will be shared with 5G via spectrum reframing or dynamic spectrum sharing (DSS).

The mid-bands can be combined with several low bands to achieve wider bandwidth networking or supplemented with low bands to enhance high-speed uplinks and achieve wider coverage. Such innovative uses of bands are called supplementary uplinks (SUL).

High bands, specifically referring to the mmWave frequency bands, are included in the scope of the IMT spectrum for the first time in 5G. These bands are mainly deployed to support the additional capacity and high data rates required by some applications at specific locations; for example, they might be deployed for urban hotspots or fixed wireless access. However, due to the high radio wave propagation loss and limitations of current radio technologies, the use of mmWave bands for seamless, wide-area coverage and mobility applications still faces a number of technological challenges. At WRC-19, the higher bands with a total bandwidth of 14.75 GHz (e.g., the sum of 24.25–27.5, 37–43.5, and 66–71 GHz) and 2.5 GHz (e.g., the sum of 45.5–47 and 47.2–48.2 GHz) mmWave frequency bands were identified for IMT as global harmonization bands and regional harmonization bands, respectively [4]. By employing carrier aggregation technologies, the combination of mmWave and mid- or low bands may help us achieve both wider coverage and higher data rates.

5G spectrum allocation is being fast-tracked around the world to accelerate 5G deployment. By 2018, only a few pioneering countries, such as China, South Korea, and the USA, had completed or announced plans for 5G spectrum allocation, but by January 2020, this number had grown to more than 60 [2]. Most countries and regions are expected to release their own 5G spectrum by 2021 or 2022 at the latest. This means that most of the spectrum we mentioned earlier will be available for use worldwide from 2023 to 2025.

12.2   6G Spectrum Requirements

The ITU-R agenda already includes discussions about the potentially available spectrum for 6G [5].

In general, spectrum allocation is strongly related to usage scenarios, use cases, and network KPIs. As such, many 5G aspects (as discussed in Part II, such as eMBB, mMTC, and URLLC) will continue to evolve into 6G. Furthermore, spectrum allocation emphasizes the continuity of radio policies and regulations, since the lifecycle of allocated spectrum spans decades. This means that the concept of using a multi-layer frequency band framework in 5G will continue to apply for 6G.

Additionally, as new bandwidth-hungry applications such as holographic communication, along with new services and functions such as high-resolution sensing, continue to emerge it is expected that 6G will employ much wider bandwidths – higher than even the mmWave band, and higher frequency up to the THz band or even the

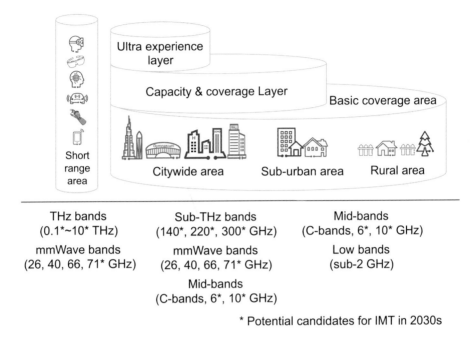

Ultra experience layer

Capacity & coverage Layer

Basic coverage area

Short range area

Citywide area

Sub-urban area

Rural area

THz bands (0.1*~10* THz)	Sub-THz bands (140*, 220*, 300* GHz)	Mid-bands (C-bands, 6*, 10* GHz)
mmWave bands (26, 40, 66, 71* GHz)	mmWave bands (26, 40, 66, 71* GHz)	Low bands (sub-2 GHz)
	Mid-bands (C-bands, 6*, 10* GHz)	

* Potential candidates for IMT in 2030s

Figure 12.2 Potential spectrum requirements and opportunities of 6G.

visible-light spectrum. For detailed use cases of 6G (see Part II), we believe that the spectrum requirements imposed by new technologies, services, and functions in 6G need to be carefully considered.

As part of this, additional mid-bands are needed to ensure that both capacity and coverage can be continuously improved in a cost-effective manner for 6G deployment. Furthermore, the innovative use and the radio technology breakthrough for macro-cellular deployment with mmWave bands will be the key for delivering new 6G services and applications, such as high-resolution positioning and sensing, and for meeting higher capacity requirements in the 2030s. The THz spectrum should also be included in the scope of 6G, as it has great potential in a wide variety of communication and sensing applications. Figure 12.2 shows the potential spectrum requirements and opportunities of 6G.

12.3   Mid-Bands Remain the Most Cost-Effective Way for Wide Coverage

Compared with earlier IMT systems, the requirements of 5G are more diversified. For example, it is necessary to ensure both the user-experienced data rate (100 Mbit/s) and the area traffic density (10 Mbit/(s m^2)). Network operators must therefore consider the cost-effectiveness of a wide network coverage while supporting most of the target application scenarios. Mid-bands, from 3 to 5 GHz, play a crucial role in 5G and are expected to be vital in 6G as well.

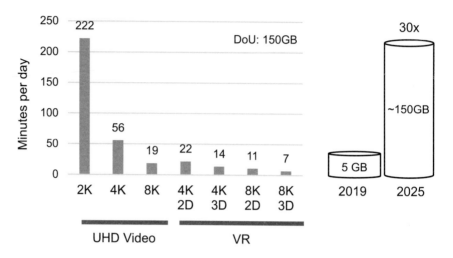

Figure 12.3 Monthly wireless DoU prediction by 2025. Data source is from [6].

Drivers for More Mid-Band Spectrum

In 5G, the importance of the mid-bands (3 to 5 GHz) is fully recognized and many candidate frequency bands are provided, such as 3300–3800 MHz, 2.6 GHz, and 2.3 GHz. However, owing to the differences in how spectrum regulations and allocations are implemented worldwide, frequency bands less than 500 MHz can be deployed globally in the mid-bands. Consequently, in the case of multi-operator co-existence, a single operator can obtain a maximum of only 100 MHz continuous mid-band spectrum on average. This may be sufficient for the initial phase of 5G, but towards the 2030s it will be far from enough, as traffic is expected to grow dozens or even hundreds of times. According to the report [6], the monthly data of usage (DoU) per user in some advanced markets will exceed 150 GB by 2025, driven by the popularity of ultra-high definition (UHD) videos and AR/VR applications. The DoU is expected to increase 30-fold from 2019 to 2025 (see Figure 12.3), and continue to grow into the 2030s.

This means that additional mid-band spectrum will be required in order to support the sustained growth of traffic for 6G towards 2030. Historically, each new generation of IMT system offers at least 10 times better performance than the previous generation. From this, we determine that each generation requires about five times more operating bandwidth than the previous generation. For example, less than 1 MHz was needed in 2G, but the required spectrum for IMT systems increased to 5 MHz in 3G and about 20 MHz in 4G. In 5G, this increases further to 100 MHz, adhering to the "five times more" rule. Following this trend, it is natural to assume that each network operator will need a mid-band spectrum of 500 MHz in 6G in order to ensure sustainable business operations. In the case of multi-operator co-existence, this means that at least 1 to 1.5 GHz additional mid-band spectrum is the optimal target in order to strike the best trade-off between coverage and capacity in the 2030s.

Table 12.1 Comparison of path loss gaps between the candidate mid-bands and 3.5 GHz in urban macro- and micro-cell scenarios.

	6 GHz		10 GHz	
Scenarios	LOS gap (dB)	NLOS gap	LOS gap	NLOS gap
Macro cell (radius = 500 m)	4.7	5.0	9.1	9.7
Micro cell (radius = 100 m)	4.7	5.0	9.1	9.7

Table 12.2 Mobile service spectrum allocation of radio regulations in the range of 6–11 GHz.

		ITU-R regions			WRC-23 agenda item
Bands (GHz)		**1**	**2**	**3**	**for IMT identification**
6G-band	5.825–7.25	×	×	×	6.425–7.025 (R1)
	7.25–8.5	×	×	×	7.025–7.125 (Global)
10G-band	10–10.45	×		×	10–10.5 (R2)
	10.5–10.68	×	×	×	
	10.7–11.7	×	×	×	
	11.7–12.2	×		×	
	12.2–12.5	×	×	×	
	12.5–12.75		×	×	
	12.75–13.25	×	×	×	

Potential Candidates

The 6 GHz (i.e., 5925–7125 MHz) and 10 GHz (i.e., 10–13.25 GHz) bands are compelling candidates within the mid-band range.

From the perspective of spectrum characteristics, these frequency bands have the potential for supporting large contiguous blocks (in theory, at least 1 GHz available bandwidth). Although the propagation attenuation of signals will increase slightly with the higher-frequency bands, such increases are acceptable, especially when we factor in advanced large-scale array technologies (e.g., massive MIMO) along with the wider use of intermediate and radio frequency (IF/RF) components.

Table 12.1 illustrates the increases in signal attenuation compared with the 5G C-band (i.e., 3.5 GHz). We can see that the increases in path loss are approximately 5 and 9 dB for the 6 and 10 GHz bands, respectively, compared with the 3.5 GHz band in the urban micro-cell scenario (where the cell radius is 100 meters), regardless of whether we are considering LOS or NLOS. The same is true in the urban macro-cell scenario (where the cell radius is 500 meters).

From the perspective of radio regulatory policies, these bands have already been allocated to mobile services on a primary or co-primary basis in the Radio Regulations [7], as shown in Table 12.2. Indeed, the ITU-R has launched a new agenda item – WRC-23 agenda item 1.2 – to conduct a feasibility study on some of the bands mentioned earlier (e.g., 6425–7125 MHz and 10–10.5 GHz) for future IMT identification [8].

Consideration should also be given to other spectrum possibilities besides 6 GHz and 10 GHz in the mid-bands, as the study of 3–11 GHz bands for 6G application is important in 6G.

12.4   Millimeter Wave Bands Become Mature in the 6G Era

Compared with lower-frequency bands, mmWave bands have more severe radio propagation characteristics. For example, mmWave bands have larger path loss (reducing the transmission range), deeper penetration loss (reducing the signal strength for indoor services), and sparse clusters (reducing the channel rank and, consequently, lowering the multiplexing gain of multi-user MIMO transmission). The practical implementation of mmWave communications therefore faces more challenges in wide area coverage, owing to coverage and mobility limitations brought about by the blocking effect of mmWave propagation. It is generally believed that mmWave bands are not cost effective for wide eMBB coverage – rather, they are better for supplementing the capacity in limited application scenarios, such as urban hotspots, with extreme traffic density and ultra-high data rates.

However, the development and deployment of mmWave bands for 5G will drive the mmWave technology and associated ecosystem to become more mature. In terms of the transmission range and power consumption of base stations, and the battery life of user equipment, the commercialization and widespread use of mmWave technologies will eventually enable a macro-cell type of deployment in the 6G era.

New Drivers for mmWave Spectrum

In addition to enabling ultra-high data rates for many 6G applications, mmWave bands will play a key role in integrated sensing and communication for 6G.

From the perspective of communications, as discussed in Chapter 1, 6G will be the first mobile communication system that achieves peak data rates in the Tbit/s range. However, the increased amount of spectrum available in lower-frequency bands (such as mid-bands) will not be sufficient to fulfill this goal. Considering the current spectrum assigned to mobile in Europe, as shown in Figure 12.4, we can see that the existing bandwidth of the legacy 5G bands (i.e., low and mid-bands) is about 1 GHz. Even with the additional 1.5 GHz mid-band spectrum, there is only 2.5 GHz of spectrum available for use by 2030. Despite this, if the mmWave bands identified in WRC-19 are gradually used for IMT-2030 (in Figure 12.4, we assume that about half the total 14.75 GHz is used), the increment of the available bandwidth is significant and could be seven times greater than in 5G. Consequently, mmWave bands will play an important role in filling such capacity gaps.

To enable sensing (including accurate positioning, imaging, etc.) as a novel capability for 6G, mmWave bands provide the key spectrum to achieve a sensing resolution of mere centimeters. According to the theory of electromagnetic imaging, the three aspects of sensing resolutions – range resolution, angular resolution, and cross-range resolution – can be calculated using the following formulas [9]:

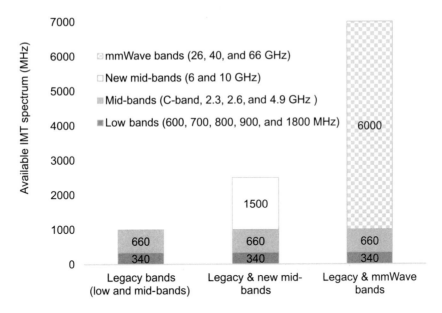

Figure 12.4 Increase in mobile spectrum towards 2030.

$$\text{Range resolution} = \frac{\text{Speed of light}}{2 \times \text{Bandwidth}}$$

$$\text{Angular resolution} = 1.22 \times \frac{\text{Wavelength}}{\text{Antenna aperture}}$$

The cross-range resolution can then be calculated as

$$2d \times \tan(0.5 \times \text{Angular resolution})$$

where d is the sensing distance.

By using the preceding formulas, we can conclude that a contiguous bandwidth of at least 1.5 GHz (i.e., $3 \times 10^8/0.5 \times 10 \times 10^{-2}$) is required to achieve a resolution in the 10-centimeter range. For an even finer resolution, an even larger bandwidth is required, which mid-bands are unable to offer.

In order to achieve cross-range resolution in the 10-centimeter range, the angular resolutions need to be 0.01 degree for a sensing distance of 10 meters. Using mmWave bands around 60 GHz, the antenna aperture needed to achieve such angular resolutions is 0.6 meters. Conversely, using mid-bands such as 6 GHz results in the required antenna apertures increasing to 6 meters. This is difficult to implement in practice.

New Technologies for mmWave Spectrum

In the 6G era, the technical challenges identified in 5G, such as mobility and coverage, as well as hardware implementation issues, will be addressed by more advanced techniques. For instance, channel modeling with higher accuracy will be needed in 6G in order to facilitate the optimal design of radio technologies (Chapter 13 provides further details). In addition, enhancements in beamforming – particularly sensing-assisted beamforming – are needed to improve the performance of mmWave transmission (Chapter 25 provides further details). It is promising to see that the utilization of the mmWave spectrum will be greatly improved as the industries and technologies related to materials, RF components, and signal processing become mature. In this way, we will be able to achieve ultra-high data rates and highly accurate sensing resolutions.

More Candidate Bands for mmWave Spectrum

At WRC-19, 14.75 GHz (24.25–27.5, 37–43.5, and 66–71 GHz) and 2.5 GHz (45.5–47 and 47.2–48.2 GHz) mmWave frequency bands were identified for IMT as global harmonization bands and regional harmonization bands, respectively. However, it should be noted that these bands are not contiguous; rather, they are scattered into several pieces, ranging from 24 to 71 GHz. Jointly utilizing multiple mmWave bands that span tens of GHz (e.g., 26 and 39 GHz, 39 and 66 GHz) will be essential but extremely challenging if we are to achieve data rates in the Tbit/s range or high-resolution sensing in the centimeter range.

Consequently, it may be necessary to identify new mmWave bands supporting large contiguous blocks in future WRC meetings. For instance, E-bands (71–76 GHz and 81–86 GHz) are prime candidates because there will be 10 GHz of contiguous spectrum when the 66–71 GHz band (which has already been identified) is jointly considered. Currently, the 71–76 GHz E-band is allocated to mobile services on a co-primary basis with fixed services or backhaul links. As the integrated access and backhaul (IAB) technologies continue to advance, we expect that the spectrum-sharing issues between these two services will be suitably resolved.

12.5   THz Bands Open New Possibilities for Sensing and Communication

The THz band is the next spectrum opportunity for 6G wireless communications. Compared with lower-frequency bands, the THz band has obvious advantages for very high data rate communication and ultra-high resolution sensing.

One of the most notable features of the THz band is its potential to provide ultra-wide bandwidth and sufficient spectrum resource for the forthcoming 6G. At WRC-19, a total of 137 GHz of bandwidth in the THz frequency region, ranging from 275 to 450 GHz, was allocated to mobile and fixed services (i.e., 275–296, 306–313, 318–333, and 356–450 GHz) [4]. This brings the total amount of spectrum allocated to mobile services in the THz range of 100 to 450 GHz to more than 230 GHz with the addition of the frequency allocated at previous WRCs. Table 12.3 shows how the spectrum was allocated. By providing bandwidth in excess of tens or even hundreds of GHz, it will be possible to achieve a wide range of peak data rates and delay-sensitive applications.

Feasibility of THz Communications

Although the signal attenuation in the THz spectrum is more severe than that in the mmWave spectrum, there are still opportunities to find the appropriate bands for

Table 12.3 Mobile spectrum allocated in the THz range of 100–450 GHz.

Frequency band (GHz)	Contiguous bandwidth (GHz)	Frequency band (GHz)	Contiguous bandwidth (GHz)
102–109.5	7.5	252–275	23
141–148.5	7.5	275–296*	21
151.5–164	12.5	306–313*	7
167–174.8	7.8	318–333*	15
191.8–200	8.2	356–450*	94
209–226	17		

Notes: 1. *New frequency bands identified of mobile at WRC-19 agenda item 1.15
2. Fragmented frequency bands with bandwidth <5 GHz are not listed

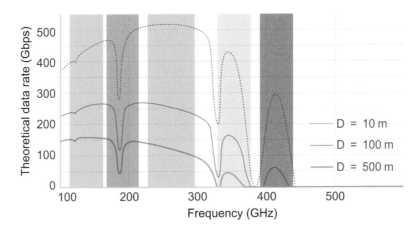

Figure 12.5 Evaluation of the theoretical throughput of a THz transmission link (D is the transmission distance).

transmission. We evaluated the theoretical throughput of a THz transmission link, taking into account the transmit power and component noise of current technologies and the impact of atmospheric attenuation [10]. The results are presented in Figure 12.5

From these results, we can conclude that the number of potential frequency-band windows with better propagation characteristics is relatively high (e.g., 140, 220, and 300 GHz). These frequency-band windows can be used for mid- (e.g., 200 meters) or short-distance (less than 10 meters) transmission by avoiding the frequency regions with high atmospheric absorption. Furthermore, wavelengths in the THz spectrum are much smaller than in the mmWave spectrum, and more antennas can be packed into the same area on a chip to overcome propagation attenuation, thereby improving the coverage of the THz spectrum.

Feasibility for Integrated Ultra-High Resolution Sensing

As mentioned earlier in this chapter, THz sensing can bring together much higher sensing accuracy and resolution owing to the ultra-wide bandwidth. Furthermore, THz

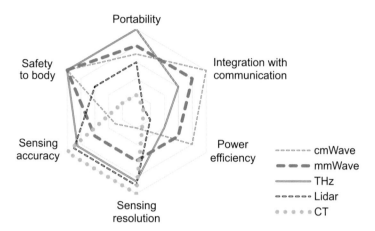

Figure 12.6 Comparison of sensing capabilities.

imaging shows enhanced spatial resolution compared with the lower-frequency bands thanks to shorter wavelengths and smaller equipment. This significantly widens the application scope of THz products in mobile communication devices. In the future, for example, smart terminals integrated with THz sensing technology will have electromagnetic imaging capabilities to obtain information such as the number of calories in food, or to detect hidden objects (Chapter 3 provides further details).

Unlike other bands of electromagnetic imaging, the THz band has a unique ability for less harmful detection, as it features non-ionizing, non-invasive, and spectral fingerprinting capabilities [11]. THz spectrometry in 6G has many potential applications in the healthcare, industrial, food quality, and environment sensing fields, as the vibration and rotation frequencies of most molecules are within the THz band. Thanks to its ability to provide continuous, real-time information via dynamic, non-invasive, passive, and contactless measurements, THz spectrometry will gain significant interest. Of particular note is that THz spectrometry achieves comparable results to professional CT or MRI machines, but it is much safer and more portable, as illustrated in Figure 12.6.

Although THz bands have greater bandwidth and better antenna aperture conditions, mmWave bands are more adaptable to imperfect environmental conditions such as dust, fog, and so on, making them the option of choice for outdoor sensing applications such as mapping and electromagnetic image reconstruction.

Challenges and Future Work

To support the rapid development of THz communication technologies, research and development leading to breakthroughs in THz devices are needed, such as: electronic, photonic, and hybrid transceiver design; large-scale antenna arrays; on-chip or on-die arrays; and new material array technologies. At the same time, further research is needed into the design of high-power high-frequency devices, new antenna and RF transistor materials, transceiver architectures, channel modeling, array signal processing, and issues of energy efficiency.

Although the ITU-R has allocated spectrum in excess of 230 GHz for mobile services in the frequency range 100–450 GHz, regulations and policies for the IMT industry are not clear and have yet to be globally unified. The meetings at both the ITU and WRC levels need to work together in order to promote further consensus.

References

[1] ITU-R, "IMT Vision – framework and overall objectives of the future development of IMT for 2020 and beyond," Recommendation ITU-R M.2083-0, Sept. 2015.

[2] GSA Report, "5G spectrum for terrestrial networks: Licensing developments worldwide," 2020, accessed Feb. 2020. [Online]. Available: https://gsacom.com/paper/5g-spectrum-report-for-terrestrial-networks-executive-summary-feb-2020/

[3] ITU. World Radiocommunication Conference, "Final Acts WRC-15," 2015. [Online]. Available: https://www.itu.int/pub/R-ACT-WRC.12-2015

[4] ITU. World Radiocommunication Conference, "Final Acts WRC-19," 2019. [Online]. Available: https://www.itu.int/pub/R-ACT-WRC.14-2019

[5] ITU-R, "IMT future technology trends towards 2030," ITU-R, M. [Future technology trends] (ongoing developing), 2020.

[6] White paper, "5G spectrum: Public policy position," 2020. [Online]. Available: https://www-file.huawei.com/-/media/corporate/pdf/public-policy/public_policy_position_5g_spectrum_2020_v2.pdf?la=en-gb

[7] *Radio Regulations (Edition of 2016)*. ITU-R publications, International Telecommunications Union, Geneva, Switzerland, 2016.

[8] *Results of the First Session of the Conference Preparatory Meeting for WRC-23 (CPM23-1)*. BR Administrative Circular CA/251, Geneva, Switzerland, 2019.

[9] J. Hasch, E. Topak, R. Schnabel, T. Zwick, R. Weigel, and C. Waldschmidt, "Millimeter-wave technology for automotive radar sensors in the 77 GHz frequency band," *IEEE Transactions on Microwave Theory and Techniques*, vol. 60, no. 3, pp. 845–860, 2012.

[10] ITU-R, "Attenuation by atmospheric gases and related effects," Recommendation ITU-R, p. 672–12, Aug. 2019.

[11] I. F. Akyildiz, J. M. Jornet, and C. Han, "Terahertz band: Next frontier for wireless communications," *Physical Communication*, vol. 12, pp. 16–32, 2014.

13 New Channels

Radio wave propagation is a fundamental part of wireless communications. Before constructing and operating real-world systems, we must understand the principles of radio propagation and develop the associated channel models. These models represent the key propagation processes and allow for reliable evaluation of and comparison between different systems. Each generation of the wireless system utilizes different frequency bands for commercialization purposes. Therefore, the channel model associated with these bands is integral to the development of technologies. Up to 5G, the frequency range spans from 0.4 GHz to 100 GHz [1, 2].

The 3GPP TR 38.901 standards include important channel modeling techniques for meeting requirements that come with the new scenarios and spectrum in 5G. For instance, the link budget is an important metric for evaluating path loss in bands above 6 GHz. This is the reason why we take frequency dependency into account in the path loss model beyond 6 GHz [3]. Another example is massive MIMO, which necessitates the beamforming effects as it becomes a key 5G technique. With the beamforming and tracking, spatial consistency is defined as an additional feature for dealing with the angular change in the channel model [4, 5]. The standards have also identified and modeled many other new propagation features, including blockage, outdoor-to-indoor penetration loss [6], bandwidth dependency [7], and oxygen/molecular absorption losses. The introduction of these new channel features ensures that we can comprehensively and accurately evaluate the impact of propagation space when new 5G technologies are designed.

Furthermore, we need to build upon the correct channel model when designing a 6G system and technology. This is because 6G is bound to bring significant challenges to channel modeling owing to its novel designs, such as new spectrum, scenarios, and antennas which will be discussed in other chapters. As mentioned in the paper [8], THz uses a different path loss coefficient compared to mmWave. Also, a spatial non-stationary channel geared towards the extremely large aperture array (ELAA) has been introduced [9]. With the development of new technologies, the angle-dependent phase shifter model included in the reconfigurable intelligent surface (RIS) is part of the propagation model [10]. In addition, in sensing scenarios such as reflective environmental sensing, instead of the propagation formula used in traditional communication, a radar-based reflection propagation formula must be included. These challenges impact channel modeling, and are not only relevant to the modification of formulas and parameters. In addition, traditional modeling methodologies may not be able to meet the requirements of these features.

13.1 New Requirements of 6G Channel Modeling

Looking at the channel model from a historical perspective, the physical channel models that use the multipath bidirectional propagation of electromagnetic waves to extract physical environment characteristics can be divided into three types: deterministic model, geometry-based stochastic model (GBSM), and non-geometry stochastic model [11]. In a deterministic model, the physical propagation parameters are completely fixed. Therefore, the real physical channels in specific scenarios can be reconstructed. Deterministic models include CEM, ray tracing, and measurement models. GBSM is a wireless channel modeled through the distribution of scattering clusters, which is randomly generated by a specified probability density function. Due to its stochastic nature, it depicts the propagation of a group of physical environments better than the deterministic models do, and is quite fitting for statistical evaluation, such as system simulation. The 3GPP model is a typical GBSM. The quasi-deterministic model is derived using a combination of deterministic models and stochastic models. The dominant paths are calculated by the deterministic model, and the scattering paths are generated by stochastic models.

Based on measurements and ray tracing across the frequency bands from 6 GHz to 100 GHz, studies in 5G have formed models in higher-frequency bands of up to 100 GHz, as demonstrated in Table 13.1. The main methodologies for channel modeling in these standards are stochastic modeling and quasi-deterministic modeling. These methodologies capture the essence of statistical fading models, along with some of the actual characterized propagation for certain antenna designs such as cross-polarization.

Table 13.1 5G channel standardization methodologies.

Channel model	Organization	Methodology
3GPP 38.901	3GPP	Stochastic model
METIS	METIS	Stochastic model, map-based model, or hybrid model
MiWEBA	MiWEBA	Quasi-deterministic channel model
ITU-R M	ITU	Stochastic model
COST2100	COST	Stochastic model
IEEE 802.11	IEEE	Quasi-deterministic or ray tracing
NYU WIRELESS	NYU	Stochastic
QuaDRiGa	Fraunhofer HHI	Stochastic or quasi-deterministic
5G mmWave Channel Model Alliance	NIST	Stochastic
mmMAGIC	mmMAGIC	Stochastic or rigorous validation
IMT-2020	IMT-2020	Stochastic

Statistical models can describe the channel more simply and efficiently, and are beneficial to large-scale simulation owing to their low computational complexity. We have also discovered that stochastic models cannot express the deterministic parameters related to a specific system or scenario, such as geometrical information related to multipath channel parameters or locations of communication devices and scatters. However, deterministic channel expression has not been a requirement for link and system designs over the past decades. The need for this modeling mainly arises from network planning, where the geometric layout of cellular sites influences base station configurations.

In the new 6G design, certain techniques (such as RIS, localization, and imaging) are highly related to specific environments, which cannot be depicted by stochastic models. We therefore anticipate the arrival of deterministic channel modeling methodologies that may lead to more precise evaluation.

For example, with the development of new antenna and integration technologies, we predict that large-scale arrays such as ELAA significantly impact channel modeling and performance evaluation. As shown in some papers [12], large-scale arrays bring new challenges in modeling, such as near-field spherical waves and non-stationary channels [13]. In the past, we simply modeled the far field, which could be approximated using plenary waves. However, the near field is now too large to ignore and we need to take spherical waves into account. These propagation features bring new opportunities for improving the communication capacity. The spatio-temporal characteristics of multi-antenna channels are the key factors that determine the performance of spatio-temporal processing. Therefore, research on large-array antenna channels must focus on the spatio-temporal characteristics of multi-antenna channels. It should be noted that these characteristics are highly dependent on the surrounding environment, and that is especially true for scattering distributions, which are difficult to describe through stochastic models. In addition, these features will greatly increase the modeling complexity. In light of this, future studies must focus on first identifying how these channel characteristics affect communication performance, after which the degree to which these characteristics are modeled in the channel can be determined.

Furthermore, as described in Chapter 3, new use cases (such as sensing) are introduced as new application scenarios of 6G. Its algorithm design and performance strongly depend on the location of the target and surrounding environment. Therefore, deterministic models related to a geographical location are preferred. Not only that; a typical application of sensing and imaging needs to take propagation effects (such as diffraction) into account when the size of an object is approximately equal to a wavelength, and it is difficult to model this through the conventional geometric optics method. As shown in Figure 13.1, the computational electromagnetic (CEM) methodology is expected to describe physical phenomena such as diffraction. Some technologies used in CEM are described in Chapter 8, such as the finite element method, the finite-difference time-domain method, and the method of moments.

From the preceding discussion, we can see that a single scheme alone may not meet the evaluation requirements of all applications. That is to say, models with higher

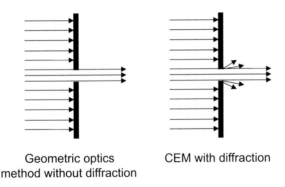

Geometric optics
method without diffraction

CEM with diffraction

Figure 13.1 Deterministic methodologies comparison of diffraction.

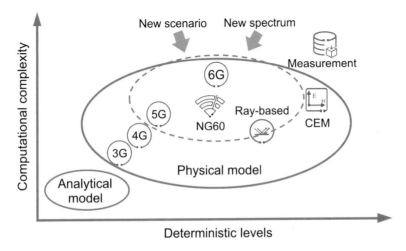

Figure 13.2 Evolution of channel modeling methodologies. The ellipse with broken lines refers to 6G.

accuracy and deterministic levels will introduce higher computational complexity, which in turn will far exceed the system evaluation capability. We therefore present the evolution of propagation models in the respective standards from 1G to 5G. As shown in Figure 13.2, research on the channel model historically tended to focus on improving the deterministic levels under the constraint of complexity. This might therefore be the potential direction of an evolving hybrid model, in which a plurality of mechanisms is included, but the weights of the mechanisms vary according to different application scenarios and evaluation criteria.

13.2 Channel Measurement in 6G

13.2.1 Channel Measurement in New Spectrum

Compared with the frequency band below 100 GHz, the THz frequency band has a higher free space path loss. Besides, THz signals excite gas molecules in the

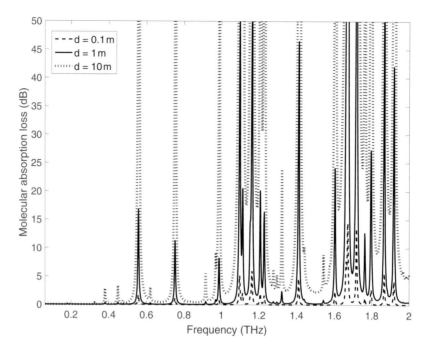

Figure 13.3 Molecular absorption loss in the THz band.

atmosphere. To add to that, some of the signal power will be converted into the kinetic energy of gas molecules, and this is referred to as molecular absorption. It is highly frequency selective as different gas molecules have various resonant frequencies that correspond to different molecular absorption peaks, as shown in Figure 13.3. The absorption effect of water vapor is more prevalent in the THz frequencies, whereas the absorption effect of oxygen is prevalent in the mmWave band. According to the Beer–Lambert law, the molecular absorption loss increases exponentially as the transmission distance increases. As a result, the molecular absorption effect introduces highly frequency-selective and distance-relative molecular absorption loss to THz links.

On the other hand, the THz channel also exhibits different propagation characteristics from the millimeter wave channel. Typically, the multipath component consists of LOS paths, specular reflection paths, and scattering paths. The high K-factor (averages of 13 dB in a meeting room at 140 GHz according to Huawei's measurements [8]) measures the power ratio between the LOS and NLOS paths. The THz channel measurements suggest that the LOS path is more prevalent in the THz channel. A surface that is smooth at lower frequencies will turn rough in the THz band as the THz waves have wavelengths that are considerably shorter, and comparable to the deviations of the surface height. This roughness leads to reflection loss and also disperses the reflection power to scattering paths, which ultimately weakens the strengths of multipath and makes the channel sparse. That said, some measurements indicate that in low-THz bands, e.g., 140 GHz and 220 GHz, multipath still exists indoors.

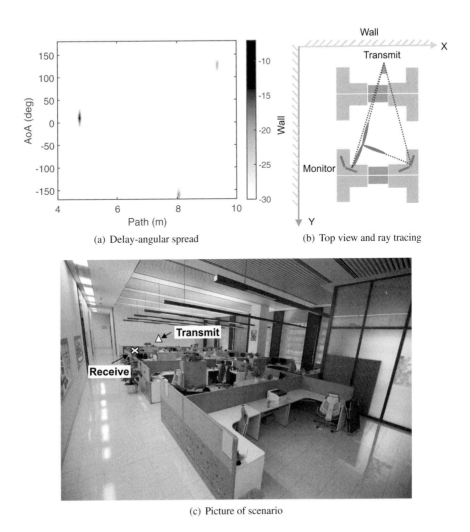

(a) Delay-angular spread

(b) Top view and ray tracing

(c) Picture of scenario

Figure 13.4 InH 140GHz measurement results and scenario.

For example, the received power of the reflected multipath is only 6–7 dB lower than that of the LOS path in an office room at 140 GHz owing to the rich reflection from the desk and liquid crystal monitors (as shown in Figure 13.4).

Moreover, different multipath distributions bring new features to the large- and small-scale parameters in the THz band. For example, to evaluate the path loss model, a measurement was obtained for a typical indoor hotspot (InH), i.e., the hallway of an indoor office, at 140 GHz, as shown in Figure 13.5 and Table 13.2. Compared to the 3GPP TR 38.901 InH path loss formula extended to 140 GHz, the measurement shows a path loss exponent (PLE) that is slightly higher than that in the 3GPP model and lower than the free space propagation PLE. This means that a weak wave-guide effect can also be observed in the sub-THz band. For the small-scale parameters obtained in typical InH scenarios, the measurements also indicate a noteworthy difference from

Table 13.2 Path loss at 140 GHz in InH scenarios.

	Path loss (dB)	Shadow fading std. (dB)
Measurement by Huawei	$PL_{InH_{140}} =$ $32.4 + 18.5\log_{10}(d_{3D}) + 20\log(f_{140})$	$\delta_{SF} = 0.8$
38.901 extended to 140 GHz	$PL_{InH_{140}} =$ $32.4 + 17.3\log_{10}(d_{3D}) + 20\log(f_{140})$	$\delta_{SF} = 3$

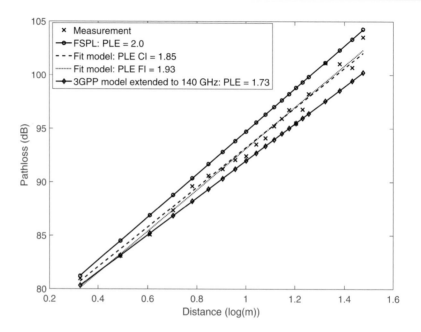

Figure 13.5 Path loss model comparison of measurement results, free space, and 3GPP TR 38.901.

the existing channel model standards [14], as shown in Table 13.3. Therefore, more accurate and representative parameterized channel models for new spectrum above 100 GHz need to be researched in depth and modeled.

13.2.2   Channel Measurement in New Scenarios

As described in Section 13.1, the new requirements of 6G mobile communications have led to new technologies being researched, such as ELAA, RIS, mmWave device-to-device communications, integrated mmWave non-terrestrial communication, and mmWave/THz sensing.

ELAA Channel Modeling

Naturally, the new application scenarios and antenna architecture also bring new challenges to channel modeling. For example, ELAA can effectively increase channel capacity and user peak rate. However, as the aperture of an antenna array increases,

Table 13.3 Small-scale parameters at 140 GHz in InH scenarios.

Scenario		Office LOS	
		3GPP model extended to 140 GHz	Measurement
Delay spread (DS)	$\mu_{\lg DS}$	−7.71	−8.70
$\lg DS = \log_{10}(DS/1\text{s})$	$\delta_{\lg DS}$	0.18	0.50
AOA spread (ASA)	$\mu_{\lg DS}$	1.37	1.29
$\lg ASA = \log_{10}(ASA/1°)$	$\delta_{\lg DS}$	0.37	0.30
ZOA spread (ZSA)	$\mu_{\lg ZSA}$	0.88	0.73
$\lg ASA = \log_{10}(ASA/1°)$	$\delta_{\lg ZSA}$	0.18	0.15

so does the near-field distance, and this is worth noting as some users will now reside within the near-field range. For this reason, the plane wave channel model used in 5G may no longer be applicable. As such, we need to comprehensively model and study the propagation characteristics of the new spherical wave propagation channel and the channel estimation technique.

Another characteristic of the ELAA propagation channel is the non-stationary feature in spatial domains. Phrased differently, as the aperture of antenna arrays increases, the different sub-arrays may have different propagation channel features, such as power, clusters, angle of arrival, and rank.

NTN Channel Modeling

Some new application scenarios, such as integrated non-terrestrial communication, have in turn brought great challenges to the channel sounding system. A case in point is the fact that conventional channel measurement requires high SNR and high synchronous accuracy, which are both difficult to achieve in non-terrestrial network (NTN) scenarios. The former is required because a low SNR makes it very difficult for the sounder to capture spatial multipath information. In addition, integrated non-terrestrial channel sounding systems also need to consider the impact of atmospheric conditions, such as wind, cloud, rain, and snow.

Millimeter wave/THz Sensing

In addition to communications applications, sensing applications are also an important direction of 6G research. Owing to the use of a higher spectrum, communication devices can achieve larger bandwidth and higher antenna integration. As a result, these devices can provide ultra-high-precision spatial and temporal resolution, further improving the channel detection capability. This is especially true in the THz frequency band. In other words, because of the short wavelength, it can deliver a millimeter-level imaging capability. That being said, we still cannot use the current channel modeling methodology to accurately evaluate imaging performance. For example, as shown in Figure 13.6, for a small hole that is comparable in size to the wavelength, the conventional geometric optics method cannot be used to simulate

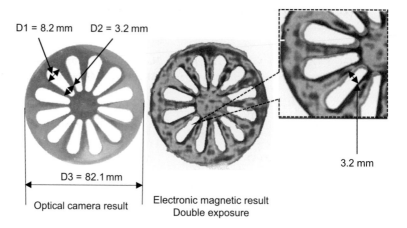

D1 = 8.2 mm   D2 = 3.2 mm

3.2 mm

D3 = 82.1 mm

Optical camera result

Electronic magnetic result
Double exposure

Figure 13.6 Example of millimeter-level imaging in the sub-THz band for a small hole with minimum size of 3.2 mm.

the diffraction phenomenon of an electromagnetic wave, and consequently cannot accurately evaluate the imaging performance. Within this context, a potential research direction that we can focus on is the CEM methodology, which can efficiently simulate the field distribution of electromagnetic waves and recover the channel propagation environment with high precision. This will ultimately help improve the accuracy with which the imaging resolution is evaluated. As shown in Figure 13.6, a 140 GHz reflective channel measurement equipment is used to detect metal objects with small holes (the minimum size is 3.2 mm). A CEM-based inverse scattering algorithm was used to solve the electromagnetic image. It is basically consistent with the optical image, as shown in the right figure. The minimum resolution is 3.2 mm. These imaging results justify the feasibility of millimeter-level imaging at sub-THz band.

References

[1] K. Haneda *et al.*, "5G channel model for bands up to 100 GHz," 2015, accessed Dec. 6, 2015. [Online]. Available: http://www.5gworkshops.com/2015/5G_Channel_Model_for_bands_up_to100_GHz(2015-12-6).pdf

[2] K. Haneda, J. Zhang, L. Tan, G. Liu, Y. Zheng, H. Asplund, J. Li, Y. Wang, D. Steer, C. Li *et al.*, "5G 3GPP-like channel models for outdoor urban microcellular and macrocellular environments," in *Proc. 2016 IEEE 83rd Vehicular Technology Conference (VTC Spring)*. IEEE, 2016, pp. 1–7.

[3] H. Yan, Z. Yu, Y. Du, J. He, X. Zou, D. Steer, and G. Wang, "Comparison of large scale parameters of mmwave wireless channel in 3 frequency bands," in *Proc. 2016 International Symposium on Antennas and Propagation (ISAP)*. IEEE, 2016, pp. 606–607.

[4] Y. Wang, Z. Shi, M. Du, and W. Tong, "A millimeter wave spatial channel model with variant angles and variant path loss," in *Proc. 2016 IEEE Wireless Communications and Networking Conference*. IEEE, 2016, pp. 1–6.

[5] K. Zeng, Z. Yu, J. He, G. Wang, Y. Xin, and W. Tong, "Mutual interference measurement for millimeter-wave D2D communications in indoor office environment," in *Proc. 2017 IEEE Globecom Workshops*. IEEE, 2017, pp. 1–6.

[6] Y. Du, C. Cao, X. Zou, J. He, H. Yan, G. Wang, and D. Steer, "Measurement and modeling of penetration loss in the range from 2 GHz to 74 GHz," in *Proc. 2016 IEEE Globecom Workshops*. IEEE, 2016, pp. 1–6.

[7] N. Iqbal, J. Luo, C. Schneider, D. Dupleich, S. Haefner, R. Müller, and R. S. Thomas, "Frequency and bandwidth dependence of millimeter wave ultra-wide-band channels," in *Proc. 2017 11th European Conference on Antennas and Propagation (EUCAP)*. IEEE, 2017, pp. 141–145.

[8] Z. Yu, Y. Chen, G. Wang, W. Gao, and C. Han, "Wideband channel measurements and temporal-spatial analysis for terahertz indoor communications," in *Proc. 2020 IEEE International Conference on Communications Workshops*. IEEE, 2020, pp. 1–6.

[9] S. Wu, C.-X. Wang, H. Haas, M. M. Alwakeel, B. Ai *et al.*, "A non-stationary wideband channel model for massive MIMO communication systems," *IEEE Transactions on Wireless Communications*, vol. 14, no. 3, pp. 1434–1446, 2014.

[10] W. Chen, L. Bai, W. Tang, S. Jin, W. X. Jiang, and T. J. Cui, "Angle-dependent phase shifter model for reconfigurable intelligent surfaces: Does the angle-reciprocity hold?" *IEEE Communications Letters*, 2020.

[11] P. Almers, E. Bonek, A. Burr, N. Czink, M. Debbah, V. Degli-Esposti, H. Hofstetter, P. Kyösti, D. Laurenson, G. Matz *et al.*, "Survey of channel and radio propagation models for wireless MIMO systems," *EURASIP Journal on Wireless Communications and Networking*, vol. 2007, no. 1, p. 019070, 2007.

[12] S. Wu, C.-X. Wang, H. Haas, M. M. Alwakeel, B. Ai *et al.*, "A non-stationary wideband channel model for massive MIMO communication systems," *IEEE Transactions on Wireless Communications*, vol. 14, no. 3, pp. 1434–1446, 2014.

[13] J. Chen, X. Yin, X. Cai, and S. Wang, "Measurement-based massive MIMO channel modeling for outdoor LoS and NLoS environments," *IEEE Access*, vol. 5, pp. 2126–2140, 2017.

[14] 3GPP, "Study on channel model for frequencies from 0.5 to 100 GHz," 3rd Generation Partnership Project (3GPP), Technical Specification (TS) 38.901, 12 2019, version 16.1.0. [Online]. Available: https://portal.3gpp.org/desktopmodules/Specifications/SpecificationDetails.aspx?specificationId=3173

14 New Materials

The tremendous evolution witnessed in digital communication over the years can be summed up as due to the remarkable progress in semiconductor technologies. With 6G no longer just a distant reality, new material technologies will continue to facilitate such transitions. In this chapter, we will review several material technologies that are key to 6G. First, we briefly expand on the silicon and III-V material evolution as enabling technologies for several applications related to THz frequencies. Then, we describe novel reconfigurable materials that offer flexible control over devices. As THz moves towards higher spectrum, advanced photonic materials come into play. More specifically, photonic crystals enable the low-cost silicon integration of optical components with low loss. Additionally, photovoltaics materials convert light into electrical signals and enable photodetectors to serve as a key component in the THz photonic system. Last but not least, plasmonic materials support surface plasmons, which can increase the interaction between light and materials. This is especially useful when we need to enhance the THz photonic/optoelectronic system performance.

14.1 Silicon Advancement

Based on the existing mature silicon platform, which inherently has low cost, high yield, small geometry, and low power, new process features are added to enable new capabilities. A fitting case in point is the SiGe-BiCMOS development process that started in the 1990s. The SiGe-BiCMOS platform can now successfully perform many applications, such as imaging, spectroscopy, and communication, which are not possible in the standard CMOS.

Efficient THz imaging requires faster imaging acquisition, and we can achieve this by utilizing focal plane arrays. Multi-pixel imaging arrays have been demonstrated for such purposes using SiGe technology [1]. Also, a compact THz spectroscopy system was demonstrated using SiGe BiCMOS technology up to 1 THz [2]. Further, to perform super-resolution imaging (imaging beyond the diffraction limit), a near-field imaging sensor was developed in [3] using a SiGe BiCMOS process, and a spatial resolution of 10 micrometers was achieved. Both CMOS and SiGe technologies have also been demonstrated in many THz communication circuits.

Silicon advancement has contributed to the development of the photonic systems [4, 5]. The progress made by silicon photonics over the past few decades is definitely worth mentioning. It was enabled by the optimization of the standard CMOS process, e.g., adding Ge growth steps to allow photodetector and modulator integration besides the common CMOS steps. Both planar and multi-layer architectures have been demonstrated for photonic and electronic integration. On top of that, various optical interconnections have been designed to enable low loss and the efficient use of space. Compared with the complexity of planar waveguides, multi-layer integration is preferred since it allows for more efficient use of space and reduces complexity. For these particular reasons, it is a promising candidate for many optoelectronic applications.

As previously mentioned, silicon technologies have been used to continuously drive next-generation applications in communication, imaging, computing, etc. Advanced processes allow for a more efficient and compact hybrid integration of both photonic and electronic components on the same silicon. It is expected that the progress achieved in terms of silicon will lead us to witness more advanced compact and low-cost photonic and electronic devices in the near future.

14.2 Heterogeneous III-V Material Platform

Electronic and photonic materials have shaped modern technologies across various disciplines. As fabrication processes leap forward, materials are continuously being developed or improved for more applications. In the past decades, Moore's law has enabled us to enhance the development in complexity and performance of silicon devices. However, silicon has fundamental limitations for photonic use owing to its indirect bandgap. Type III-V semiconductors with a direct bandgap were therefore proposed to meet this need. In that regard, III-V semiconductors, e.g., indium phosphide (InP) and gallium arsenide (GaAs), have indeed been used for the upper THz frequencies ($>1\,THz$) and photonic integrated circuits, obtaining satisfactory results. Be that as it may, they also entail high costs and have therefore not been implemented on a wide scale across the market. Nevertheless, the interest in photonics has gained traction and led to its use for many applications such as sensing, radar, and communication.

To overcome the fundamental silicon limitations while also leveraging photonic advantages, heterogeneous integration with III-V semiconductors entered the picture and combined both III-V and silicon advantages. The process of integrating III-V materials on the same silicon wafer in a standard lithography process has shown great potential in many photonic applications.

Many high-performing photonic devices [6], such as chip-scale tunable lasers, modulators, and optical amplifiers, have been developed featuring superior performance compared with their III-V counterparts. For example, compared with the silicon photonics-enabled counterparts, III-V heterogeneous integrated modulators can offer superior performance in terms of a large refractive-index-induced phase tuning, high electron mobility, and low carrier plasma absorption. In addition, III-V-enabled quantum wells offer additional advantages in performance elevation.

The photonic integrated circuit also benefits from this heterogeneous platform. More specifically, the bonding of different III-V layers on the same single die allows each component to achieve performance optimization. Put differently, optimized materials and designs for each component can be chosen and integrated on the same silicon die. In that regard, a heterogeneous network-on-chip system that comprises a complete set of photonic-communication components, including lasers, modulators, and amplifiers on different III-V layer stacks, has been demonstrated in [7].

We can also add other non-III-V materials using the heterogeneous integration approach. It has been verified that non-reciprocal magnetic materials such as Ce:YIG [8] can be integrated on a silicon chip, thereby providing more complex blocks (such as isolators and circulators) in a photonic integrated circuit.

14.3   Reconfigurable Material

Electrical tuning of material properties allows for devices with more functionalities, smaller dimensions, and reduced costs. As such, various tuning materials have been proposed and embedded in systems for flexible and dynamic control. For example, by adding reconfigurable materials to an intelligent surface and then interfacing the intelligent surface with a digital control circuit, an RIS can be realized through programmability.

Graphene, a 2D carbon material, can support the propagation of surface plasmon polaritons (SPPs). By biasing with different voltage levels, graphene can exhibit various levels of conductivity and therefore tunable electromagnetic behavior. Various graphene devices have been proposed using this variable conductivity, including switches, phase shifters, and antennas. The article [9] provided an exhaustive survey regarding graphene's potential applications in THz communication, biological sensing, and more. The articles [10, 11] demonstrated that a graphene leaky wave antenna can perform beam steering. Tunable reflection phases were achieved in reconfigurable graphene reflectarrays [12] by the application of different biasing voltages. A graphene switch was presented in [13] that operates by biasing segments of graphene sections with different voltages. Standard graphene deposition reactors are approximately worth US$1 million (low cost and not controlled).

Liquid crystals show both liquid and crystal properties. A nematic liquid crystal can change an incident wave's properties as the wave passes through the liquid, and this can be done by changing the intensity of the applied electric fields. The nematic liquid crystal can be very useful in a reconfigurable electronic system. Within this context, [14, 15] proposed a reconfigurable reflectarray and metasurface using a liquid crystal for beam-steering purposes. A liquid crystal (LC)-based phase shifter and phased array antenna were presented in [16], which also discussed the possibility of commercialization.

Phase change material (PCM) heats or cools materials to offer reconfigurability by switching between the amorphous (insulating) and crystalline (conductive) states. PCM switches are fast, compact, and have low insertion losses. Compared with micro-electro-mechanical systems (MEMS), PCM shows better reliability and performance at mmWave frequencies. In [17–19], PCM GeTe RF switches for mmWave applications and a true-time-delay phase shifter were first demonstrated. A monolithically integrated PCM-based phase shifter [20] was presented for miniaturized reconfigurable mmWave phased array beam-steering applications.

14.4   Photonic Crystal

A photonic crystal (PC) is an artificial periodic structure built by aligning different dielectric materials in a reoccurring sequence as a superlattice. By introducing a localized defect structure into the periodicity, a PC can manipulate the light as a light trap (PC cavity) or control its flow (PC waveguide). Compared with their metal counterparts, PC waveguides are made purely from dielectric materials, which means that the ohmic loss can be minimized. Further, the PC structure can be efficiently integrated on a silicon platform while remaining low cost.

Many PC devices have been fabricated and used for various photonic applications [21], such as modulators, high-resolution filters, sensors, and lasers. Some recent PC applications include biochemical sensors, LIDAR sensing, and THz communication. A PC nanolaser was presented in [22] for biomolecule and chemical detection. PC photonic phased arrays have also demonstrated their capability in LIDAR sensing [22]. Moreover, PCs have shown their potential in THz communication. The article [23] demonstrated a PC-based THz communication platform on silicon. This involved demonstrating a high data rate with an integrated resonant-tunneling diode detector and PC waveguide.

More recently, the topological photonic crystal has been explored as a new method of manipulating light [24]. By taking advantage of the topological phases of matter, light can be guided in a predetermined direction even when defects exist, as an effect of topologically protected boundary states. Additionally, various topological PC platforms have been explored for robust unidirectional light propagation, such as non-linear PC systems [25], non-Hermitian systems [26], and higher-order topological systems [27]. These systems will facilitate the protection of light propagation in nanoscale photonic devices.

14.5   Photovoltaics Material and Photodetector

When incident photon radiation reaches a photovoltaic (PV) material, electronic currents are generated in the material. Some of the conventional PV materials are silicon-based crystalline structures. With the advancement of nanotechnology, PV materials, such as nanoscale quantum dots, carbon nanotubes, and graphene, have stepped onto the scene, bringing many promising applications.

The most straightforward application of photovoltaic technology has probably been in solar cells, but we should not disregard the CMOS sensor used ubiquitously in digital cameras and many other applications. Traditionally, semiconductor materials like silicon were used to manufacture solar cells. Subsequently, metal was added to improve the photo conversion efficiency (PCE) and this led to the development of photodetectors, e.g., the Schottky detector. Further, the implementation of nano-scale light trapping layers such as metallic nanoparticless has been proven to increase the light concentration and scattering, which in turn increase the PCE. Moreover, metal gratings can also help excite surface plasmonic waves, thereby increasing the interaction between photons and electrons.

Photovoltaic-material-enabled photodetectors have been developed for many applications. For example, quantum dots have been used to increase the light absorption for health monitoring applications [28]. It was demonstrated that the carbon nanotube-based photodetector array [29] can perform fine-resolution imaging. A wristband based on the graphene photodetector array has been fabricated and demonstrated for health monitoring [30]. As photonic-system demands increase, the use of PV materials will continue to result in compact as well as high-performing photonic devices and advanced applications.

14.6   Plasmonic Material

Plasmonic materials take advantage of light energy to generate resonant oscillations of charge carriers – plasmons. As previously mentioned, plasmonic nanostructures have been used in photovoltaic devices to enhance the PCE. Two common structures are metal nanoparticles and patterned/grated metal electrodes. The former increases the light scattering by exciting localized surface plasmon polariton (SPP) resonances, while the latter increases the light path length coupled by SPP [31]. It is worth noting that Au and Ag are the common metals used in both structures. Up to 30% PCE enhancement was reported with Au nanoparticles in [32]. On the other hand, [33] reported a PCE enhancement of 19% using an Ag grating electrode.

Plasmonic materials have been studied as elementary excitations in theoretical solid state physics textbooks since the 1970s [34], but they are now being realized in laboratories, with the plasmonic metamaterial technology advancing rapidly over the last decade. Metamaterials can be utilized to achieve desired material properties by manipulating periodically distributed artificial structures. They were originally used at microwave frequencies. In order to work at optical frequencies, the size of the metamaterials needs to be nanometers or even smaller, and this is difficult to achieve in terms of fabrication. The article [35] offered a comprehensive survey on how to realize negative refraction at optical frequencies. In [36], a fishnet structure consisting of two layers of metal meshes with a dielectric spacer between them was presented to realize both negative permittivity and permeability at near-infrared frequencies. Also, a metal–dielectric–metal stacked waveguide was demonstrated with a negative refractive index at visible frequencies [37].

Plasmonic metasurfaces consist of nanoscale plasmonic structures with varied geometries and separations printed on a thin slab. By controlling the phase profile of the nanostructures, metasurfaces can use different configurations of incident light to generate desired responses. Apart from the more common functionalities, such as beam steering, orbital angular momentum (OAM) beam generation, and reflection beam control, plasmonic metasurfaces have recently proved to be capable of controlling optical surface waves in the near-field region [38]. When the patterned nanostructures interact with light, part of the re-radiated energy is converted into SPPs in the near field. By carefully designing the shape and alignment of the nanostructures, SPP excitation can be controlled to meet requirements, such as near-field sensing and biochemical detection.

Surface plasmon polaritons can also be supported in graphene. As previously mentioned, graphene can work as a reconfigurable material due to the fact that it possesses the tunable electrical property. Moreover, graphene can exhibit plasmonic behavior at THz and infrared frequencies. In graphene, the SPP excitation is confined in the transverse dimension. Therefore, graphene plasmons can provide significantly smaller wavelengths and stronger light–matter interactions compared with its metal counterparts. Furthermore, the hybridization of graphene and other conventional plasmonic nanostructures or metamaterials can be utilized to provide even further enhanced graphene-based devices. This is why it is a promising future direction from multiple perspectives.

Plasmonic materials with SPP wave properties have attracted interest across numerous fields. For example, graphene antennas have shown that their unique tunable plasmonic resonance behavior and potential can be used in many THz applications. By utilizing the SPP excitation from plasmonic metasurfaces, intelligent multi-functional surfaces can perform not only beam control but also biosensing applications. As we gain a deeper understanding of the SPP nature as well as how SPP-supported materials behave, more interesting applications will come along.

References

[1] U. R. Pfeiffer, R. Jain, J. Grzyb, S. Malz, P. Hillger, and P. Rodríguez-Vízquez, "Current status of terahertz integrated circuits-from components to systems," in *Proc. 2018 IEEE BiCMOS and Compound Semiconductor Integrated Circuits and Technology Symposium (BCICTS)*. IEEE, 2018, pp. 1–7.

[2] K. Statnikov, J. Grzyb, B. Heinemann, and U. R. Pfeiffer, "160-GHz to 1-THz multi-color active imaging with a lens-coupled SiGe HBT chip-set," *IEEE Transactions on Microwave Theory and Techniques*, vol. 63, no. 2, pp. 520–532, 2015.

[3] P. Hillger, R. Jain, J. Grzyb, W. Förster, B. Heinemann, G. MacGrogan, P. Mounaix, T. Zimmer, and U. R. Pfeiffer, "A 128-pixel system-on-a-chip for real-time super-resolution terahertz near-field imaging," *IEEE Journal of Solid-State Circuits*, vol. 53, no. 12, pp. 3599–3612, 2018.

[4] M. U. Khan, Y. Xing, Y. Ye, and W. Bogaerts, "Photonic integrated circuit design in a foundry+ fabless ecosystem," *IEEE Journal of Selected Topics in Quantum Electronics*, Article no. 8201014, 2019.

[5] R. Helkey, A. A. Saleh, J. Buckwalter, and J. E. Bowers, "High-performance photonic integrated circuits on silicon," *IEEE Journal of Selected Topics in Quantum Electronics*, Article no. 8300215, 2019.

[6] T. Komljenovic, D. Huang, P. Pintus, M. A. Tran, M. L. Davenport, and J. E. Bowers, "Photonic integrated circuits using heterogeneous integration on silicon," *Proceedings of the IEEE*, vol. 106, no. 12, pp. 2246–2257, 2018.

[7] C. Zhang, S. Zhang, J. D. Peters, and J. E. Bowers, "8×8×40 Gbps fully integrated silicon photonic network on chip," *Optica*, vol. 3, no. 7, pp. 785–786, 2016.

[8] T. Shintaku and T. Uno, "Optical waveguide isolator based on nonreciprocal radiation," *Journal of Applied Physics*, vol. 76, no. 12, pp. 8155–8159, 1994.

[9] D. Correas-Serrano and J. S. Gomez-Diaz, "Graphene-based antennas for terahertz systems: A review," *arXiv preprint arXiv:1704.00371*, 2017.

[10] J. Gómez-Díaz, M. Esquius-Morote, and J. Perruisseau-Carrier, "Plane wave excitation–detection of non-resonant plasmons along finite-width graphene strips," *Optics Express*, vol. 21, no. 21, pp. 24 856–24 872, 2013.

[11] M. Esquius-Morote, J. S. Gómez-Dı, J. Perruisseau-Carrier *et al.*, "Sinusoidally modulated graphene leaky-wave antenna for electronic beamscanning at THz," *IEEE Transactions on Terahertz Science and Technology*, vol. 4, no. 1, pp. 116–122, 2014.

[12] E. Carrasco, M. Tamagnone, and J. Perruisseau-Carrier, "Tunable graphene reflective cells for THz reflectarrays and generalized law of reflection," *Applied Physics Letters*, vol. 102, no. 10, p. 104103, 2013.

[13] J.-S. Gómez-Díaz and J. Perruisseau-Carrier, "Graphene-based plasmonic switches at near infrared frequencies," *Optics Express*, vol. 21, no. 13, pp. 15 490–15 504, 2013.

[14] S. Bildik, S. Dieter, C. Fritzsch, W. Menzel, and R. Jakoby, "Reconfigurable folded reflectarray antenna based upon liquid crystal technology," *IEEE Transactions on Antennas and Propagation*, vol. 63, no. 1, pp. 122–132, 2014.

[15] S. Foo, "Liquid-crystal reconfigurable metasurface reflectors," in *Proc. 2017 IEEE International Symposium on Antennas and Propagation & USNC/URSI National Radio Science Meeting*. IEEE, 2017, pp. 2069–2070.

[16] T. Ting, "Technology of liquid crystal based antenna," *Optics Express*, pp. 17 138–17 153, 2019.

[17] T. Singh and R. R. Mansour, "Characterization, optimization, and fabrication of phase change material germanium telluride based miniaturized DC–67 GHz RF switches," *IEEE Transactions on Microwave Theory and Techniques*, vol. 67, no. 8, pp. 3237–3250, 2019.

[18] N. El-Hinnawy, P. Borodulin, B. P. Wagner, M. R. King, E. B. Jones, R. S. Howell, M. J. Lee, and R. M. Young, "Low-loss latching microwave switch using thermally pulsed nonvolatile chalcogenide phase change materials," *Applied Physics Letters*, vol. 105, no. 1, p. 013501, 2014.

[19] T. Singh and R. R. Mansour, "Miniaturized reconfigurable 28 GHz PCM based 4-bit latching variable attenuator for 5G mmWave applications," in *Proc. IEEE MTT-S International Microwave Symposium (IMS)*. IEEE, 2020, pp. 53–56.

[20] T. Singh and R. R. Mansour, "Loss compensated PCM GeTe-based latching wideband 3-bit switched true-time-delay phase shifters for mmWave phased arrays," *IEEE Transactions on Microwave Theory and Techniques*, vol. 68, no. 9, pp. 3745–3755, 2020.

[21] T. Asano and S. Noda, "Photonic crystal devices in silicon photonics," *Proceedings of the IEEE*, vol. 106, no. 12, pp. 2183–2195, 2018.

[22] T. Baba, "Photonic crystal devices for sensing," in *Proc. 2019 Conference on Lasers and Electro-Optics (CLEO)*, 2019, pp. 1–2.

[23] W. Withayachumnankul, M. Fujita, and T. Nagatsuma, "Integrated silicon photonic crystals toward terahertz communications," *Advanced Optical Materials*, vol. 6, no. 16, p. 1800401, 2018.

[24] H. Wang, S. K. Gupta, B. Xie, and M. Lu, "Topological photonic crystals: a review," *Frontiers of Optoelectronics*, pp. 1–23, 2020.

[25] X. Zhou, Y. Wang, D. Leykam, and Y. D. Chong, "Optical isolation with nonlinear topological photonics," *New Journal of Physics*, vol. 19, no. 9, p. 095002, 2017.

[26] S. Yao and Z. Wang, "Edge states and topological invariants of non-hermitian systems," *Physical Review Letters*, vol. 121, no. 8, p. 086803, 2018.

[27] H. Hu, B. Huang, E. Zhao, and W. V. Liu, "Dynamical singularities of floquet higher-order topological insulators," *Physical Review Letters*, vol. 124, no. 5, p. 057001, 2020.

[28] E. O. Polat, G. Mercier, I. Nikitskiy, E. Puma, T. Galan, S. Gupta, M. Montagut, J. J. Piqueras, M. Bouwens, T. Durduran *et al.*, "Flexible graphene photodetectors for wearable fitness monitoring," *Science Advances*, vol. 5, no. 9, p. eaaw7846, 2019.

[29] D. Suzuki, S. Oda, and Y. Kawano, "A flexible and wearable terahertz scanner," *Nature Photonics*, vol. 10, no. 12, pp. 809–813, 2016.

[30] M. Zhang and J. T. Yeow, "A flexible, scalable, and self-powered mid-infrared detector based on transparent PEDOT: PSS/graphene composite," *Carbon*, vol. 156, pp. 339–345, 2020.

[31] S. Ahn, D. Rourke, and W. Park, "Plasmonic nanostructures for organic photovoltaic devices," *Journal of Optics*, vol. 18, no. 3, p. 033001, 2016.

[32] C. C. Wang, W. C. Choy, C. Duan, D. D. Fung, E. Wei, F.-X. Xie, F. Huang, and Y. Cao, "Optical and electrical effects of gold nanoparticles in the active layer of polymer solar cells," *Journal of Materials Chemistry*, vol. 22, no. 3, pp. 1206–1211, 2012.

[33] X. Li, W. E. Sha, W. C. Choy, D. D. Fung, and F. Xie, "Efficient inverted polymer solar cells with directly patterned active layer and silver back grating," *Journal of Physical Chemistry C*, vol. 116, no. 12, pp. 7200–7206, 2012.

[34] O. Madelung, *Introduction to solid-state theory*, Vol. 2. Springer Science & Business Media, 2012.

[35] K. Yao and Y. Liu, "Plasmonic metamaterials," *Nanotechnology Reviews*, vol. 3, no. 2, pp. 177–210, 2014.

[36] S. Zhang, W. Fan, N. Panoiu, K. Malloy, R. Osgood, and S. Brueck, "Experimental demonstration of near-infrared negative-index metamaterials," *Physical Review Letters*, vol. 95, no. 13, p. 137404, 2005.

[37] S. P. Burgos, R. De Waele, A. Polman, and H. A. Atwater, "A single-layer wide-angle negative-index metamaterial at visible frequencies," *Nature Materials*, vol. 9, no. 5, pp. 407–412, 2010.

[38] S. Sun, Q. He, S. Xiao, Q. Xu, X. Li, and L. Zhou, "Gradient-index meta-surfaces as a bridge linking propagating waves and surface waves," *Nature Materials*, vol. 11, no. 5, pp. 426–431, 2012.

15 New Antennas

Conventional antennas are designed to connect to RF systems through coaxial cables or microstrip lines, whereas THz antennas are quite different owing to the fact that transmission line loss is considerable at THz frequencies. At the lower THz frequency band (100–500 GHz), THz silicon-based integrated circuit platforms have been demonstrated with integrated antenna-on-chip and antenna-in-package [1, 2]. Beyond 500 GHz, it is possible to utilize III-V/silicon technology to drive the conventional antenna up to 1 THz.

On the other hand, since the THz system can be realized in the photonic platform, a photoconductive antenna (PCA) or electro-optic crystal can be used to generate THz waves by driving photocurrent through them. Under illumination from laser beams at specified frequencies, photoconduction current can be generated in certain semiconductor substrates, e.g., InP and GaAs. The photocurrent can then radiate as a THz wave in space by employing biased antenna electrodes.

Nano-photodetectors have a similar working principle, the difference being that they measure the generated photocurrent using a read-out circuit intended for imaging/sensing. Due to their small size and low-power consumption, nano-photodetectors show enormous potential in mobile and wearable applications.

With carefully designed scatterer shapes and spacing, intelligent surfaces, such as reflectarrays and transmitarrays, serve as planar "reflectors" or "lenses." Reconfigurable components are usually embedded within such surfaces to allow for flexible control under different incident illuminations. Metasurfaces leverage the intelligent function by providing user-designed responses under a certain level of illumination. You can use a digital controller to further program these responses, and these intelligent surfaces are guaranteed to deliver compact, highly efficient, as well as low-cost, THz communication systems.

Considerable progress has been made in the 5G massive MIMO communication system since it was introduced in 2010 [3]. As 6G is bound to step onto the scene, new research perspectives should be raised to further enhance the massive MIMO system performance. In that respect, orbital angular momentum (OAM) is being proposed to provide extra DoF, with the ultimate goal being performance improvement. It has also been reported that several joint OAM and massive MIMO wireless communication frameworks can obtain multiplicative spectral gain [4, 5].

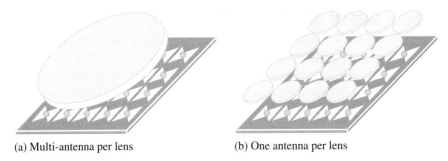

(a) Multi-antenna per lens   (b) One antenna per lens

Figure 15.1 Photoconductive lens antennas.

15.1 Photoconductive Lens Antenna

As mentioned previously, a PCA can convert photonic energy to THz radiation. However, the photo conversion efficiency is usually low, which in turn results in low radiated power (μW). That said, we can utilize a dielectric lens to help collimate radiated beams and ultimately increase antenna gain.

In a typical photoconductive lens antenna, the lens is fed by a thin high-permittivity substrate with printed planar antenna electrodes. Common antenna shapes include dipole, bowtie, spiral, and log-periodic. An external bias voltage is imposed across the antenna to help accelerate the transient photocurrents, and a THz wave is eventually radiated into space. To reduce mismatch and maximize antenna performance, we need to design an optimal lens size and shape.

Advanced techniques have been explored to enhance the photo conversion efficiency. For example, nanostructures [6] have been deposited in the photoconductive gap to increase the wave scattering or excite plasmonic waves [7], thereby increasing the photo coupling efficiency. Quantum dots have also been used as coatings to increase photo absorption [8].

Focal plane arrays have been proposed to accelerate the imaging speed. One scenario involves connected antenna arrays that are printed on the feeding substrate and used to feed a large aperture lens. The arrays can be configured to achieve beam steering or shaping functionality, and this configuration is intended for general imaging as well as telecommunication systems. The other scenario applies a closely spaced lens on top of antenna elements to form a tightly packed array. This structure is more suited for space or astronomy applications, where multi-pixel imaging functionality is desired. These two scenarios [9] are shown in Figure 15.1.

15.2 Reflectarray and Transmitarray

Both reflectarrays and transmitarrays are applicable to many applications, including communication range extension, wireless power transfer, spatial modulation, and high-gain antennas. Both arrays have a low profile, high efficiency, and are

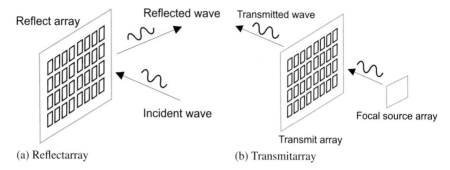

Figure 15.2 Reflectarray and transmitarray.

reconfigurable. This means that they can be easily integrated into a wide variety of systems, and are therefore promising candidates for utilization in next-generation communication.

Reflectarrays use a variety of electromagnetic scatterers printed on a surface. When illuminated by a feeder antenna, reflectarrays behave like traditional reflector antennas. More specifically, as shown in Figure 15.2, scatterers can take the form of microstrip patches (or other types of resonant structures). Each scatterer is carefully designed to produce a phase shift, simulating curved reflector surface and therefore reflecting a predetermined beam.

The scatterers can be co-polarized without changing the incident field polarization. They can also be dual-polarized or cross-polarized to control incident polarizations separately or change them. We can also add reconfigurability in order to reflect beams in different directions at different times. Advanced tunable materials, such as liquid crystal and graphene, can be applied and independently controlled to produce a reconfigurable phase profile.

Similarly, a transmitarray consists of a planar substrate with a number of printed resonators of predetermined phase distribution. When the focal source array illuminates a transmitarray, the incident wave passes through the transmitarray plane and is then converted into the desired beam pattern. In most cases, the transmitarray behaves like a planar dielectric lens. It is worth noting that the beam-steering functionality can be implemented at both the source and array planes. For example, focal sources can be configured to achieve the desired source function. Then, at the transmitarray plane, the resonators are manipulated to give the desired radiation properties.

15.3   Metasurfaces

Metasurfaces deliver attractive radiating solutions for many emerging applications. Due to their control flexibility, metasurfaces are "programmable" through a digital platform [10]. To add to that, they feature a compact size that enables low-cost integration in various platforms. When printed on a flexible substrate, metasurfaces can be designed as wearable devices potentially used for communication, imaging, and more advanced applications. For example, [11] demonstrated a programmable smart metasurface glass that allows for the full penetration, partial reflection, and full reflection of incident radio waves. This glass can be helpful to improve the channel performance in wireless communication systems.

In the microscopic view, each unit cell element can have a frequency response that varies in terms of amplitude and phase values. To achieve independent control, tunable elements can be applied on each unit cell. Macroscopically, unit cells can be designed to form an electromagnetically interconnected network that works constructively towards a specific functionality, e.g., wave absorption, surface wave cancellation, antenna decoupling, beam shaping, etc.

For metasurface antennas, unit cells are designed carefully to match the amplitudes and phase with those of the desired beam pattern in free space. Each unit cell serves as a small radiator of a specific pattern, which in turn allows the combined beam from all the elements to be shaped as needed. Both transmitter and receiver antennas can be integrated on the same substrate, and we can implement separate waveguide structures to connect the radiators to the input or output signal port.

Metasurface holography is probably one of the most intriguing applications. In holography, amplitude and phase information scattered from the desired object is recorded and encoded on a photographic film. We then apply an imaging reconstruction algorithm to generate a fictitious 3D computed hologram, which looks like the real object. Metasurfaces can be used to record this type of hologram information because the unit cells are designed to record holograms by acting like "pixels." Put differently, each element can record a specific amplitude and phase response. When subsequently illuminated by waves with specific configurations, these responses can be superimposed to generate a 3D hologram of the original object. The working principle is shown in Figure 15.3.

Another interesting application is cloaking. It involves the placing of a thin metasurface layer at a certain distance from the object to be cloaked, which in turn makes it possible to create "anti-phase" destructive interference that cancels out the scattering. The end result is an "invisible" object at the specific frequency. By adding active metasurface elements, we can broaden the bandwidth of operation.

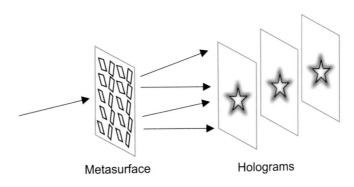

Metasurface   Holograms

Figure 15.3 Holograms created by a metasurface.

15.4   Nano-Photodetectors

As mentioned before, imaging platforms above 1 THz remain challenging owing to limitations in the silicon performance. Current "beyond 1 THz" imaging systems are mostly based on large optical systems and are too bulky for integration. A portable and convenient measurement system is therefore needed. With the advancement of nanotechnology, this can be realized using nano-photodetectors.

The recent developments in nanotechnology have opened up new possibilities for THz imaging. To put it into perspective, nano-photodetectors are small, scalable, cost effective, and power efficient. As such, they can be easily integrated into a variety of imaging systems and deliver satisfactory performance levels.

The carbon nanotube (CNT) sensing array is a potential candidate in that regard. More precisely, by using the photothermoelectric effect, CNTs can convert the illuminating THz irradiation to a measurable photocurrent at the read-out circuit. Articles [12, 13] reported that a CNT scanner array for THz imaging and the spatial resolution can reach hundreds of micrometers at 1.4 THz. Further, the proposed CNT array can be fabricated on a flexible substrate for potential wearable use.

The graphene photodetector works in a similar fashion to the CNT sensor. It can transform the illuminating THz irradiation to a measurable photocurrent. Within this context, [14] reported that a flexible wearable graphene photodetector-array wristband can be used to monitor health. To add to this, graphene's transparency means that it could possibly be used for screen antennas, intelligent glasses, wearable devices, etc.

15.5   Antenna-on-Chip and Antenna-in-Package

Advances in semiconductor technologies, such as CMOS and SiGe BiCMOS, have made it possible for us to achieve a THz integrated circuit. Articles [15–17] showed that it is entirely possible to fabricate ICs up to 700 GHz through SiGe HBT technology. Estimates indicate that the performance limit of SiGe HBT may reach or even exceed 1 THz soon. Silicon THz IC holds several advantages, such as low cost, compact size, high yield, and ease of integration.

A convenient way to implement the THz antenna involves directly integrating it with the front-end circuit on a silicon substrate. However, on-chip antenna design is challenging due to the surface waves generated in substrates. The surface waves will interfere with the antenna radiation and result in poor performance. A backside-radiated lens antenna was proposed to improve the performance and demonstrated in THz imaging and sensing applications in [18, 19].

Terahertz near-field imaging (NFI) is an emerging application. For NFI, the imaging spatial resolution is not restrained by the diffraction limit and can be used for microscopic-scale or nano-scale imaging. NFI requires the antenna to have a strong near-field coupling performance. For example, resonant structures, such as split-ring-resonators (SRRs), were proven capable of biometric human fingerprint imaging in [20].

The antenna-in-package facility provides us with yet another way of achieving integration. However, at THz frequencies, the interconnect loss between the antenna and the monolithic microwave integrated circuit (MMIC) is high. Fortunately, effective packaging techniques are currently being developed to address this issue and minimize losses. A 300 GHz substrate integrated waveguide horn antenna was proposed in a multi-layer LTCC package [21] and demonstrated in the KIOSK downloading system.

15.6   Orbital Angular Momentum

Traditionally, the capacity of a radio link has been increased through the multiplexing of space, frequency, time, code, and polarization. From an antenna perspective, a MIMO antenna with different polarizations has been developed to enhance data performance. Recently, a new dimension of multiplexing – orbital angular momentum (OAM) – has attracted much attention. In OAM, the antenna can generate orthogonal modes and each is associated with a different orbital momentum. For example, the signal can have a phase factor $e^{-j\phi}, e^{-j2\phi}$, and so on. Each mode can carry different information; therefore, multiple OAM modes can co-exist and transmit data simultaneously through a single communication link. By extracting the desired OAM mode properly at the receiver side, the spectral efficiency can be optimized.

A lot of research work has been devoted to OAM communication systems. For instance, in [22] it was proposed to generate OAM waves by steering phase modes using a circular array with several phase shifters and combiners. The article [23] demonstrated an OAM–MIMO multiplexing system that utilizes multiple uniform circular antennas to generate five OAM modes. The articles [24, 25] have demonstrated that it is also possible to achieve a high data rate through OAM communication systems.

OAM technology has great potential in LOS wireless communications, such as cellular backhaul and interconnections within a data center. Further, combined OAM and massive MIMO communication will greatly boost the data rate and achieve considerably higher spectrum efficiency. For example, [5] reported an OAM–MIMO multiplexing communication system that achieves a data rate of 100 Gbps at a distance of 10 m. The proposed multiplexing method uses 11 OAM modes at 28 GHz.

OAM is a promising transmission technology and is being researched for 6G applications. Due to the transmission antenna characteristics of OAM array, the practical frequency bands that allow for 6G to utilize OAM systems will be above 20 GHz, and, as such, mmWave bands are good candidates. Several efforts have been made towards compact antenna architecture and the design of a mmWave OAM array [26, 27]. The main issue regarding OAM application involves enabling mobility usage scenarios. Within this context, technology breakthroughs are needed to enable us to use simple user-equipment side antennas to demodulate OAM signals, especially when it comes to supporting mobility applications [28].

References

[1] P. Hillger, J. Grzyb, R. Jain, and U. R. Pfeiffer, "Terahertz imaging and sensing applications with silicon-based technologies," *IEEE Transactions on Terahertz Science and Technology*, vol. 9, no. 1, pp. 1–19, 2018.

[2] Y. Zhang and J. Mao, "An overview of the development of antenna-in-package technology for highly integrated wireless devices," *Proceedings of the IEEE*, vol. 107, no. 11, pp. 2265–2280, 2019.

[3] T. L. Marzetta, "Noncooperative cellular wireless with unlimited numbers of base station antennas," *IEEE Transactions on Wireless Communications*, vol. 9, no. 11, pp. 3590–3600, 2010.

[4] W. Cheng, H. Zhang, L. Liang, H. Jing, and Z. Li, "Orbital-angular-momentum embedded massive MIMO: Achieving multiplicative spectrum-efficiency for mmWave communications," *IEEE Access*, vol. 6, pp. 2732–2745, 2017.

[5] D. Lee, H. Sasaki, H. Fukumoto, Y. Yagi, and T. Shimizu, "An evaluation of orbital angular momentum multiplexing technology," *Applied Sciences*, vol. 9, no. 9, p. 1729, 2019.

[6] S.-G. Park, Y. Choi, Y.-J. Oh, and K.-H. Jeong, "Terahertz photoconductive antenna with metal nanoislands," *Optics Express*, vol. 20, no. 23, pp. 25 530–25 535, 2012.

[7] H. Tanoto, J. Teng, Q. Wu, M. Sun, Z. Chen, S. Maier, B. Wang, C. Chum, G. Si, A. Danner *et al.*, "Nano-antenna in a photoconductive photomixer for highly efficient continuous wave terahertz emission," *Scientific Reports*, vol. 3, p. 2824, 2013.

[8] E. O. Polat, G. Mercier, I. Nikitskiy, E. Puma, T. Galan, S. Gupta, M. Montagut, J. J. Piqueras, M. Bouwens, T. Durduran *et al.*, "Flexible graphene photodetectors for wearable fitness monitoring," *Science Advances*, vol. 5, no. 9, p. eaaw7846, 2019.

[9] O. Yurduseven, "Wideband integrated lens antennas for terahertz deep space investigation," Ph.D. dissertation, Delft University of Technology, 2016.

[10] T. J. Cui, M. Q. Qi, X. Wan, J. Zhao, and Q. Cheng, "Coding metamaterials, digital metamaterials and programmable metamaterials," *Light: Science & Applications*, vol. 3, no. 10, p. e218, 2014.

[11] NTT DoCoMo, "DOCOMO conducts worlds first successful trial of transparent dynamic metasurface," Tokyo, 2020.

[12] D. Suzuki, S. Oda, and Y. Kawano, "A flexible and wearable terahertz scanner," *Nature Photonics*, vol. 10, no. 12, pp. 809–813, 2016.

[13] D. Suzuki, Y. Ochiai, and Y. Kawano, "Thermal device design for a carbon nanotube terahertz camera," *ACS Omega*, vol. 3, no. 3, pp. 3540–3547, 2018.

[14] M. Zhang and J. T. Yeow, "A flexible, scalable, and self-powered mid-infrared detector based on transparent PEDOT: PSS/graphene composite," *Carbon*, vol. 156, pp. 339–345, 2020.

[15] J. Grzyb, B. Heinemann, and U. R. Pfeiffer, "Solid-state terahertz superresolution imaging device in 130-nm SiGe BiCMOS technology," *IEEE Transactions on Microwave Theory and Techniques*, vol. 65, no. 11, pp. 4357–4372, 2017.

[16] J. Grzyb, B. Heinemann, and U. R. Pfeiffer, "A 0.55 THz near-field sensor with a μm-range lateral resolution fully integrated in 130 nm SiGe BiCMOS," *IEEE Journal of Solid-State Circuits*, vol. 51, no. 12, pp. 3063–3077, 2016.

[17] P. Hillger, R. Jain, J. Grzyb, W. Förster, B. Heinemann, G. MacGrogan, P. Mounaix, T. Zimmer, and U. R. Pfeiffer, "A 128-pixel system-on-a-chip for real-time

super-resolution terahertz near-field imaging," *IEEE Journal of Solid-State Circuits*, vol. 53, no. 12, pp. 3599–3612, 2018.

[18] J. Grzyb and U. Pfeiffer, "Thz direct detector and heterodyne receiver arrays in silicon nanoscale technologies," *Journal of Infrared, Millimeter, and Terahertz Waves*, vol. 36, no. 10, pp. 998–1032, 2015.

[19] D. F. Filipovic, S. S. Gearhart, and G. M. Rebeiz, "Double-slot antennas on extended hemispherical and elliptical silicon dielectric lenses," *IEEE Transactions on Microwave Theory and Techniques*, vol. 41, no. 10, pp. 1738–1749, 1993.

[20] P. Hillger, R. Jain, J. Grzyb, W. Förster, B. Heinemann, G. MacGrogan, P. Mounaix, T. Zimmer, and U. R. Pfeiffer, "A 128-pixel system-on-a-chip for real-time super-resolution terahertz near-field imaging," *IEEE Journal of Solid-State Circuits*, vol. 53, no. 12, pp. 3599–3612, 2018.

[21] T. Tajima, T. Kosugi, H.-J. Song, H. Hamada, A. El Moutaouakil, H. Sugiyama, H. Matsuzaki, M. Yaita, and O. Kagami, "Terahertz MMICs and antenna-in-package technology at 300 GHz for KIOSK download system," *Journal of Infrared, Millimeter, and Terahertz Waves*, vol. 37, no. 12, pp. 1213–1224, 2016.

[22] M. Klemes, H. Boutayeb, and F. Hyjazie, "Orbital angular momentum (OAM) modes for 2-D beam-steering of circular arrays," in *Proc. 2016 IEEE Canadian Conference on Electrical and Computer Engineering (CCECE)*. IEEE, 2016, pp. 1–5.

[23] H. Sasaki, D. Lee, H. Fukumoto, Y. Yagi, T. Kaho, H. Shiba, and T. Shimizu, "Experiment on over-100-Gbps wireless transmission with OAM-MIMO multiplexing system in 28-GHz band," in *Proc. 2018 IEEE Global Communications Conference (GLOBECOM)*. IEEE, 2018, pp. 1–6.

[24] Y. Ren, L. Li, G. Xie, Y. Yan, Y. Cao, H. Huang, N. Ahmed, Z. Zhao, P. Liao, C. Zhang *et al.*, "Line-of-sight millimeter-wave communications using orbital angular momentum multiplexing combined with conventional spatial multiplexing," *IEEE Transactions on Wireless Communications*, vol. 16, no. 5, pp. 3151–3161, 2017.

[25] A. M. Yao and M. J. Padgett, "Orbital angular momentum: Origins, behavior and applications," *Advances in Optics and Photonics*, vol. 3, no. 2, pp. 161–204, 2011.

[26] M. Klemes, H. Boutayeb, and F. Hyjazie, "Minimal-hardware 2-D steering of arbitrarily large circular arrays (combining axial patterns of phase-modes)," in *Proc. 2016 IEEE International Symposium on Phased Array Systems and Technology (PAST)*. IEEE, 2016, pp. 1–8.

[27] Z. Zhao, G. Xie, L. Li, H. Song, C. Liu, K. Pang, R. Zhang, C. Bao, Z. Wang, S. Sajuyigbe *et al.*, "Performance of using antenna arrays to generate and receive mm-wave orbital-angular-momentum beams," in *Proc. IEEE Global Communications Conference (GLOBECOM)*. IEEE, 2017, pp. 1–6.

[28] M. Klemes, "Reception of OAM radio waves using pseudo-doppler interpolation techniques: A frequency-domain approach," *Applied Sciences*, vol. 9, no. 6, p. 1082, 2019.

16 THz Technology

Recent development in semiconductor technology has bridged the "THz band gap" (caused by the lack of THz hardware enablers) and stimulated the progress of various THz applications. In terms of spectrum, THz lies between the mmWave and infrared frequencies. THz signals can penetrate dielectric materials at different depths, enabling new ways to perform imaging. Owing to the effects of the diffraction limit theory, the spatial imaging resolution in the THz spectrum will be much higher compared with its mmWave counterpart. THz radiation is also non-ionizing since its frequency is lower than the UV range, meaning that it is likely to be applied extensively in biomedical applications. Furthermore, various materials exhibit unique THz responses in the 0.5–3 THz band, enabling such responses to be used as spectral "fingerprints" for material detection and characterization.

The increasing demand for higher data rates and lower latency has meant that higher frequencies and wider bandwidths have become more and more important in communication system development. Various architectures for the THz communication system have been explored, and testbeds have been demonstrated based on two different approaches: electronic, where the frequencies are multiplied up to THz; and photonic, where the optical frequencies are divided to THz. Of note is the fact that most of these systems are being developed mainly for short-range indoor communication, due in part to the high atmospheric attenuation in the THz spectrum. However, this can be avoided to some extent by choosing a "THz window" where the atmospheric loss is low; for example, 140, 220, and 300 GHz, as discussed in Chapter 12.

In this section, we summarize the state-of-the-art THz technology. We first discuss THz components and compare their performance in different technologies. We then analyze THz communication and imaging system performance, and conclude by discussing the challenges involved in THz and the current progress of related research.

16.1   THz Components

As mentioned earlier, the THz system can be implemented using either an electronic or a photonic approach. While the photonic approach mainly targets the upper frequencies in the THz band, the electronic approach can provide solutions in lower bands. The maximum frequency at which a THz electronic circuit can operate depends on the maximum frequency f_{max} of a transistor using a given solid-state process technology.

In the conventional CMOS and BiCMOS technologies, the f_{max} of a transistor is between 200 GHz and 350 GHz [1]. With the SiGe BiCMOS technology, the transistor can reach an f_{max} of 0.5 THz or even 0.75 THz [2]. Type III-V semiconductors, such as InP-based high electron mobility transistors (HEMTs) or heterojunction bipolar transistors (HBTs), can increase f_{max} to more than 1 THz [3]. Technology roadmaps have been developed for both InP- and SiGe-based technologies [1]. THz electronics are expected to surpass 1 THz and reach 2 THz using both SiGe CMOS and III-V technologies in the near future.

16.1.1 Electronic Approach

Owing to the fundamental limit of transistor frequency f_{max}, the electronic approach mainly targets systems at lower THz frequencies. Moving beyond these lower frequencies requires the system to be carefully designed in order to mitigate the device non-linearity and harmonic effects that come into play.

In [4], a comparison of state-of-the-art THz sources in both CMOS and SiGe technologies was made. In this section a non-exhaustive search is performed on InP HEMT/HBT THz sources. Table 16.1 summarizes the performance of InP THz sources beyond 300 GHz, and Figure 16.1 and Figure 16.2 plot comparisons of THz sources in different technologies. As shown in the figures, most THz sources operate at frequencies from 0.2 THz to 0.5 THz, with an output power level from −10 dBm to +10 dBm. With a similar output power level, InP THz sources show better direct-current(DC)-to-RF efficiency (ratio of output power to consumed DC power). For the frequency range beyond 0.5 THz, all three technology-based THz sources were reported. CMOS and SiGe THz sources show a lower output power level because the operating frequency exceeds f_{max}, and the performance is undermined by the

Table 16.1 Comparison of state-of-the-art InP-based THz sources.

Technology	Frequency (GHz)	Process	Pout	Pdc	Ref.
InP HBT	300	250 nm	+5.3 dBm	87.4 mW	[6]
InP DHBT	300	250 nm	+4.8 dBm	88 mW	[7]
InP DHBT	300	250 nm	−5 dBm	46.2 mW	[8]
InP DHBT	300	250 nm	+1.5 dBm	148 mW	[9]
InP DHBT	300	130 nm	+4.7 dBm	75.6 mW	[10]
InP DHBT	303	800 nm	−6.2 dBm	37.6 mW	[11]
InP DHBT	306	800 nm	−1.6 dBm	36 mW	[12]
InP DHBT	325	800 nm	−7 dBm	40 mW	[13]
InP DHBT	330	250 nm	−6.5 dBm	13.5 mW	[14]
InP HBT	413	250 nm	−5.6 dBm	<115 mW	[15]
InP DHBT	480	300 nm	−11 dBm	15 mW	[5]
InP HBT	487	250 nm	−8.9 dBm	<115 mW	[15]
InP HBT	573	250 nm	−19.2 dBm	<115 mW	[15]
InP HBT	591	250 nm	−17.4 dBm	49.3 mW	[6]
InP HBT	645	250 nm	−17.4 dBm	49.3 mW	[16]
InP HEMT	670	25 nm	+2.55 dBm	1.7 W	[17]

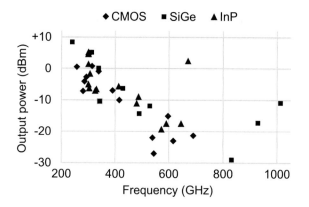

Figure 16.1 Comparison of the output power of state-of-the-art THz sources in CMOS, SiGe, and InP technologies. Data for CMOS and SiGe come from [4], while the data for InP come from Table 16.1.

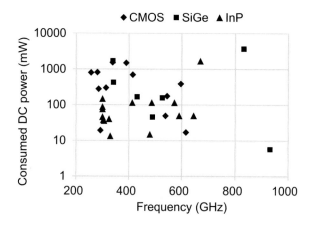

Figure 16.2 Comparison of the DC power consumed by state-of-the-art THz sources in CMOS, SiGe, and InP technologies. Data for CMOS and SiGe come from [4], while the data for InP come from Table 16.1.

harmonic effect. High output power is achievable with InP THz sources [5, 6], which is essential for THz beam steering and shaping applications.

Prior works have compared a variety of THz power amplifiers using CMOS, SiGe, GaAs, and InP technologies [18–20]. A comprehensive comparison of various PA technologies is summarized in [20], though the frequency is mostly under 100 GHz. The article [18] compared state-of-the-art PA performances at 110–180 GHz using both CMOS and SiGe technologies, showing that up to 20 dBm output power can be generated with a power-added efficiency (PAE) generally under 10%. InP-based PAs around 200 GHz were compared in [19], which showed that 20–30 dBm output power can be generated with a PAE under 10%. In this section, a non-exhaustive literature review was performed on PAs above 300 GHz. Table 16.2 and Figure 16.3 compare PA performances in different technologies. Both GaAs and InP-based PAs were demonstrated around 300 GHz, though InP PAs showed a higher output power level. Further works [21, 22] have extended the operating frequencies of InP PAs up to 850 GHz.

Table 16.2 Comparison of state-of-the-art THz PAs in different technologies.

Technology	Process (nm)	Freq. (GHz)	Gain (dB)	Psat (dBm)	Pdc (mW)	PAE	Ref.
CMOS SOI	32	210	15	4.6	40	6.00%	[23]
SiGe BiCMOS	130	230	12.5	12	740	1.00%	[24]
SiGe BiCMOS	130	215	25	9.6	—	0.50%	[25]
GaN	50	190	12	14.1	—	1.20%	[26]
InGaAs mHEMT	35	320	13.5	8.6	—	—	[27]
InGaAs mHEMT	35	320	12	7	—	—	[27]
InGaAs mHEMT	35	310	7	8.5	521	—	[28]
InGaAs mHEMT	35	294	15	4.8	—	—	[29]
InP HEMT	50	340	15	10	—	—	[30]
InP HBT	250	300	12	9.2	848	1.10%	[31]
InP HBT	250	300	13.4	13.5	—	—	[32]
InP HBT	130	325	10	9.4	243	2.20%	[33]
InP HBT	130	325	9.4	11.4	243	1.09%	[33]
InP DHBT	250	325	11	1.13	—	0.60%	[34]
InP HEMT	80	300	20	12	—	—	[35]
InP HEMT	80	300	14	9.5	—	—	[36]
InP DHBT	130	670	24	−4	—	—	[37]
InP DHBT	130	655	20	−0.7	—	—	[37]
InP HBT	130	585	20	2.8	455	—	[22]
InP HEMT	25	850	17	−0.3	60	—	[21]

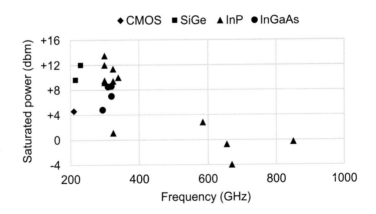

Figure 16.3 Comparison of state-of-the-art THz PAs in different technologies. Data for CMOS and SiGe come from [4], while the data for InP come from Table 16.2.

THz receivers can be classified as homodyne/heterodyne and direct detector receivers. The article [4] compared a variety of THz receivers based on both CMOS and SiGe technologies, ranging from 200 GHz up to nearly 1 THz. Here, we have performed a non-exhaustive search of InP and GaAs THz receivers. Table 16.3, Figure 16.4, and Figure 16.5 compare the performance of THz receivers in different technologies. As shown in Figure 16.4, both InP and SiGe receivers were reported beyond 500 GHz. Gains of up to 25 dB were reported at 670 GHz using the InP

Table 16.3 Comparison of state-of-the-art THz receivers in different technologies.

Technology	Process (nm)	Freq. (GHz)	Gain (dB)	Noise figure (dB)	Power diss. (mW)	Ref.
InP HBT	130	577	20	16	—	[43]
InP HEMT	25	850	—	12	1160	[40]
InP HEMT	25	670	25	10.3	1800	[38]
InP HEMT	80	300	—	15	—	[39]
InP DHBT	250	300	26	16.3	482	[41]

Technology	Process (nm)	Freq. (GHz)	Responsivity (kV/W)	NEP (pW/Hz$^{0.5}$)	Ref.
InP DHBT	250	280	350	0.13	[44]
InP HBT	250	300	40	35	[45]
GaAs HEMT	—	271	42	135	[46]
GaAs HEMT	—	632	1.6	1250	[46]
GaAs HEMT	—	650	70	300	[47]

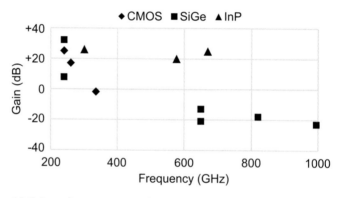

(a) Gain performance comparison

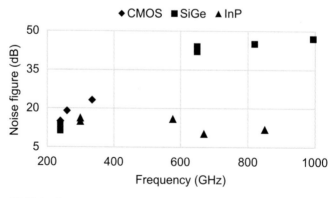

(b) Noise figure comparison

Figure 16.4 Comparison of state-of-the-art THz homodyne/heterodyne receivers in different technologies. Data for CMOS and SiGe come from [4], while the data for InP come from Table 16.3.

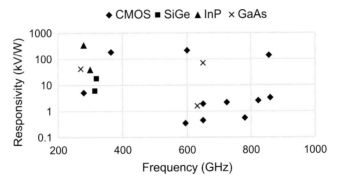

(a) Responsivity performance comparison

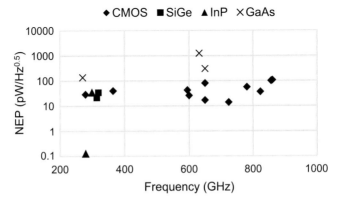

(b) NEP performance comparison

Figure 16.5 Comparison of state-of-the-art THz direct receivers in different technologies. Data for CMOS and SiGe come from [4], while the data for InP come from Table 16.3.

HEMT technology [38]. Several InP HEMT transceivers have been reported for communication use [39–41]. Silicon–germanium receivers are mostly found in THz imaging and sensing applications [42].

Most direct detectors are integrated with antennas and therefore known as antenna-coupled direct detectors. Backside-radiating on-chip antennas and external hyper-hemispherical silicon lens antennas are two of the most common forms of integration. Responsivity and noise-equivalent power (NEP) are used to evaluate the gain and sensitivity performance of these detectors. Compared with homodyne/heterodyne receivers, direct detectors show much higher noise in the operating bands. As a result, they are mostly used in imaging and sensing applications, where the noise requirement can be somewhat relaxed.

16.1.2   Hybrid and Photonic Approaches

As mentioned in the previous chapter, the indirect bandgap in silicon prevents on-chip integration of active optical sources, such as lasers. In order to address this limitation, a

III-V/silicon heterogeneous platform is required, which will enable the full integration of various optical components, such as lasers, optical amplifiers, and modulators, on a single silicon wafer. This is a promising development for the next generation of photonic integrated circuits.

Two approaches are applied when integrating III-V semiconductor components on silicon: the direct growth of III-V materials on the silicon substrate and heterogeneous integration by wafer bonding. Both approaches have demonstrated their maturity in delivering commercial products.

The first demonstration of heterogeneous III-V/silicon distributed feedback lasers was achieved in 2008 [48]. Since then, significant progress has been made [49, 50]. By controlling electron confinement in quantum wells between heterostructure layers (e.g., InGaAs and GaAs), III-V/silicon lasers can provide enhanced performance, such as faster modulation and higher efficiency, compared with conventional lasers. A quantum dot (QD) laser uses QDs in an active region and can be directly grown on a silicon substrate. Using QDs in the active region allows for three-dimensional confinement of charge carriers, leading to atom-like energy states. This property can help to offer better performance at high temperatures, lower threshold current, and longer lifetime compared with quantum wells and conventional lasers.

Optical amplifiers, as the name suggests, amplify optical signals in a semiconductor gain medium. Type III-V materials allow for faster switching since the carrier lifetimes are shorter due to their direct bandgaps. The first demonstration of a III-V/silicon optical amplifier was in 2007 [51], following which significant progress has been made. A variety of recent advances in heterogeneous III-V/silicon optical amplifiers was reviewed in [52]. For example, [53] reported a III-V/silicon optical amplifier with a gain of up to 28 dB and [52] demonstrated a III-V/silicon optical amplifier with a saturated output power of up to 16.8 dBm. The heterogeneous optical amplifier has enabled new applications of photonic ICs, for example, beam-steering devices like LIDAR.

Optical modulators transform electrical signals into the light domain. Conventional silicon optical modulators take advantage of the plasma dispersion effect in silicon devices, where the material refractive index changes depending on the variation of free charge density. This change leads to phase variations, which are subsequently converted to amplitude modulation using an interferometer or resonator. The use of III-V materials in the heterogeneous platform can help enhance the modulator performance, offering a large refractive-index change, high electron mobility, and more [49]. Further utilization of quantum wells allows for the quantum-induced Stark effect (the material absorption coefficient is changed by the electric field variation) and can offer rapid switching solutions for optical modulators [54].

A photodetector in a photonic platform is functionally equivalent to an antenna. It needs to be sufficiently sensitive to pick up power that meets the bit error rate (BER) requirement at a given data rate. A popular material candidate is Ge because of its high bandwidth, good responsivity, and ability to be easily integrated in CMOS-based platforms. A III-V-based photodetector can generally offer comparable or better performance. A III-V photodetector with responsivity of up to 0.7 A/W was demonstrated with a bandwidth of over 67 GHz in [55].

16.2   THz Systems

The development of various THz components in a highly integrated platform has enabled many THz applications. For example, SiGe-based platforms have been used in THz wireless communication, imaging, and sensing applications, mainly at frequencies below 300 GHz [4]. Beyond that, III-V technologies such as those involving InP have shown that they are capable of enabling high-performing THz systems, as discussed in previous sections. The heterogeneous III-V/silicon integrated platform is a promising option for delivering both commercial and industrial THz portable systems.

16.2.1   THz Communication Systems

Terahertz wireless communication has exhibited great potential in many short-range communication applications at some of the THz "windows" (e.g., 140, 220, and 300 GHz). The IEEE 802.15.3d task force has investigated the spectrum 252–325 GHz and defined use cases such as kiosk downloading, intra-chip/intra-board radio communication, wireless communication in data centers, and mobile fronthaul and backhaul links [56]. In this section, we summarize state-of-the-art THz system performance in the 275–450 GHz band.

As mentioned in previous sections, THz spectrum can experience high path loss due to strong atmospheric attenuation. Table 16.4 estimates the link budget at 275–450 GHz, with antenna transmit (Tx) and receive (Rx) gains of 30 dB as well as a Tx power of 0 dBm, link distance of 10 meters, and noise figure of 15 dB. Because the path loss is high (~100 dB), a highly directional pencil beam is needed as compensation.

Table 16.5 summarizes a non-exhaustive search of recent state-of-the-art THz wireless communication systems and their performance from 275 GHz to 450 GHz. Both electronic and optoelectronic approaches have demonstrated data rates of up to 100 Gbps, though the communication distance is mostly less than two meters. The paper [35] demonstrated a system using the InP–HEMT technology that achieved a data rate of up to 120 Gbit/s at a distance of 9.8 meters. It also demonstrated an optoelectronic system consisting of a photonic uni-traveling-carrier diode transmitter and an active electronic receiver based on the InGaAs HEMT technology, which was reportedly able to achieve a data rate of up to 100 Gbit/s at a distance of 15 meters.

Table 16.4 Link budget estimation at 275–450 GHz.

Tx power (dBm)	Freq. (GHz)	Distance (m)	PL (dB)	Atmos. absorp. (dB/km)	Noise figure (dB)	Ant. Tx/Rx gain (dB)	Rx power (dBm)
0	275–296	10	−101.5	10	15	30	−56.6
0	306–313	10	−102.3	16	15	30	−57.4
0	318–333	10	−102.7	20	15	30	−57.9
0	356–450	10	−104.5	10	15	30	−59.6

Table 16.5 A state-of-the-art summary of THz link performance.

Technology	Freq. (GHz)	Data rate (Gbps)	Distance (m)	Tx power (dBm)	Modulation	Ref.
GaAs mHEMT	300	64	1	−4	QPSK	[57]
INP HEMT	300	20	0.8	+3	ASK	[58]
CMOS	300	56	0.05	−5.5	16QAM	[59]
CMOS	300	105	—	−5.5	32QAM	[60]
CMOS	300	20	0.1	—	16QAM	[61]
InP-HEMT	300	100	2.22	—	16QAM	[62]
InP-HEMT	300	120	9.8	—	16QAM	[35]
InGaAs	300	60	0.5	−7	16QAM	[63]
Optoelectronic	280	100	0.5	−10	16QAM	[64]
Optoelectronic	300	10	0.3	−20	OOK	[65]
Optoelectronic	300	100	0.5	−16.1	16QAM	[66]
Optoelectronic	300	100	15	−8	32QAM	[67]
Optoelectronic	330	50	1	−10.5	ASK	[68]
Optoelectronic	350	100	2	−12	16QAM	[69]
Optoelectronic	350	100	2	+14	16QAM	[69]
Optoelectronic	385	32	0.5	−12	QPSK	[70]
Optoelectronic	400	60	0.5	−17	QPSK	[71]
Optoelectronic	400	60	0.5	−21	QPSK	[72]
Optoelectronic	450	132	1.8	+16	64QAM	[73]
Optoelectronic	350–475	120	0.5	−15	QPSK	[74]
Optoelectronic	400	106	0.5	—	QPSK	[75]
Optoelectronic	400	160	0.5	−17.5	QPSK	[76]

16.2.2   THz Imaging and Sensing Systems

The THz spectrum can be used in many imaging and sensing applications, such as material characterization, biomedical imaging, and biochemical sensing. THz spectroscopy takes advantage of the sample's characteristic response to illumination by THz radiation, meaning that it can be used for such things as material characterization and security imaging. THz radar imaging uses reflected signals in order to realize gesture recognition with enhanced precision thanks to the shorter wavelength. THz near-field imaging can overcome the fundamental imaging diffraction limit and offer ultra-high-resolution sampling images, showing great promise in medical applications such as biomolecular imaging.

THz spectroscopic imaging can be categorized into time-domain spectroscopy (TDS) and continuous wave spectroscopy (CWS). In TDS, a THz pulse is generated and directed to the material sample. Transmitted or reflected signals are collected and then converted into the frequency domain. This recorded spectrum contains the sample's unique "fingerprint," which can be used to determine the sample material properties. To generate a broadband THz signal, a photoconductive lens antenna (which we describe later) can be used. A TDS system is generally bulky, as it contains auxiliary optical components.

THz CWS follows a similar principle, but, instead of generating a broadband pulse, CWS generates a narrowband spectrum to achieve high-frequency resolution. A silicon-based electronic platform can offer a compact solution to implement spectroscopy at a lower THz spectrum. The article [77] demonstrated a multi-band (six harmonics of 160 GHz) imaging chipset operating from 160 GHz to 1000 GHz using a 250-nm SiGe HBT process. Additional performance enhancements are expected with advances in III-V/silicon technology.

THz radar imaging takes advantage of radar's range-gating capabilities and can provide high-resolution/precision image quality. The demands for radar sensors in automotive systems, healthcare, mobile devices, and other applications have recently increased – a promising candidate to meet such demands is the silicon platform. It was reported in [78] that a single-chip frequency modulated continuous wave (FMCW) radar front-end module, operating at 210–270 GHz using 130-nm SiGe HBT technology, was able to achieve a spatial resolution of 2.57 mm.

THz NFI can perform imaging with ultra-high resolution, achieving results in the μm or even nm range. Conventionally, THz near-field imaging has been performed using near-field scanning optical microscopy. Although this resolution can achieve the nm range, the system is typically bulky and difficult to integrate. An on-chip SRR sensor was therefore considered for near-field imaging. For example, a near-field imaging system using 130-nm SiGe technology was reported in [79], where an SRR sensor was used to couple the near-field power (at a distance of several μm) and then fed to the read-out circuit for measurement. A spatial resolution of 10–12 μm was reported at 550 GHz.

16.3   Challenges

Both mmWave and THz wireless communication systems experience high path losses, although this is more severe in the THz spectrum. For example, the path loss at 28 GHz over a distance of 10 meters is 81 dB, but increases to 101 dB at 280 GHz. Highly directive antenna arrays are typically used to compensate for the high path loss. To generate a highly directional beam, the equiphase surface of the transmitted beams from all antenna elements should be perpendicular to the wave propagation direction. In most cases, analog phase shifters are needed to compensate for the phase delays caused by different antenna elements being spaced at certain distances (usually at half wavelengths). The generated phase shift is mostly determined by the carrier frequency. This becomes a problem in a broadband system, where multiple carriers that span a wide frequency range are used. The resulting beam may be dispersed with frequency variation, leading to array gain loss – an effect known as beam squint in mmWave systems. In THz systems, this effect becomes more severe because the bandwidth is even wider and the beam width is extremely narrow (such beams are known as pencil beams). The generated beams may split in different directions as the frequency varies, resulting in increasingly severe array gain loss. To differentiate it from mmWave systems, this effect is called beam split in THz systems. Recently, [80, 81] proposed the use of metasurfaces (which we discussed in Chapter 15) to mitigate this effect. The promising results provided a new path to design low-complexity communication systems with improved beam performance.

Although III-V/silicon semiconductor technology has demonstrated its ability to enable a scalable and low-cost platform for THz imaging and communication systems, a few key challenges remain. The first challenge originates from the material's broadband spectral characteristics and the limited tuning range of semiconductor sources, whereby the operating frequency of a THz source may not be sufficient to cover the entire spectral range of the material under detection. THz sources that can be tuned over a broad range are therefore needed for spectroscopic imaging. Recent works have shown promising progress; for example, [82] reported a THz source with a wide tunable range (0.04–0.99 THz). Furthermore, fast THz imaging acquisition requires tunable sources with suitable beamforming capabilities in order to avoid the clumsiness inherent in mechanical scanning. Compact integration of reconfigurable sources with large THz antenna arrays is needed, and these sources are expected to facilitate the implementation of advanced imaging algorithms such as compressive sensing.

The advances made in III-V/silicon semiconductors have enabled the heterogeneous integration of various high-performing components, such as electronic, photonic, magnetic, and graphene components. An integrated platform that is both compact and highly heterogeneous in addition to encompassing all such components is therefore desired to enable optimal performance. Furthermore, recent progress on plasmonic technology has shown promise for achieving on-chip silicon integration of plasmonic components [83, 84]. Work is expected to continue in the future, despite the challenging nature of this field, with the aim of achieving maturity in plasmonic-to-silicon transfer technology. Consequently, an optimized integration of these heterogeneous components on a single silicon wafer could lead to an electronic–photonic/plasmonic system capable of providing high performance along with an optimal balance between cost, efficiency, and programmability – a system for the next generation of commercial and industrial THz devices.

References

[1] P. Garcia, A. Chantre, S. Pruvost, P. Chevalier, S. T. Nicolson, D. Roy, S. P. Voinigescu, and C. Garnier, "Will BiCMOS stay competitive for mmW applications?," in *Proc. 2008 IEEE Custom Integrated Circuits Conference*. IEEE, 2008, pp. 387–394.

[2] B. Heinemann, H. Rücker, R. Barth, F. Bärwolf, J. Drews, G. Fischer, A. Fox, O. Fursenko, T. Grabolla, F. Herzel *et al.*, "SiGe HBT with fx/fmax of 505 GHz/720 GHz," in *Proc. 2016 IEEE International Electron Devices Meeting (IEDM)*. IEEE, 2016, pp. 3.1.1–3.1.4.

[3] M. Urteaga, Z. Griffith, M. Seo, J. Hacker, and M. J. Rodwell, "InP HBT technologies for THz integrated circuits," *Proceedings of the IEEE*, vol. 105, no. 6, pp. 1051–1067, 2017.

[4] U. R. Pfeiffer, R. Jain, J. Grzyb, S. Malz, P. Hillger, and P. Rodríguez-Vízquez, "Current status of terahertz integrated circuits – from components to systems," in *Proc. 2018 IEEE*

BiCMOS and Compound Semiconductor Integrated Circuits and Technology Symposium (BCICTS). IEEE, 2018, pp. 1–7.

[5] M. Hossain, N. Weimann, M. Brahem, O. Ostinelli, C. R. Bolognesi, W. Heinrich, and V. Krozer, "A 0.5 THz signal source with 11 dBm peak output power based on InP DHBT," in *Proc. 2019 49th European Microwave Conference (EuMC)*. IEEE, 2019, pp. 856–859.

[6] J. S. Rieh, J. Yun, D. Yoon, J. Kim, and H. Son, "Terahertz InP HBT oscillators," in *Proc. 2018 IEEE International Symposium on Radio-Frequency Integration Technology (RFIT)*. IEEE, 2018, pp. 1–3.

[7] J. Yun, D. Yoon, H. Kim, and J.-S. Rieh, "300-GHz InP HBT oscillators based on common-base cross-coupled topology," *IEEE Transactions on Microwave Theory and Techniques*, vol. 62, no. 12, pp. 3053–3064, 2014.

[8] J.-Y. Kim, H.-J. Song, K. Ajito, M. Yaita, and N. Kukutsu, "InP HBT voltage controlled oscillator for 300-GHz-band wireless communications," in *Proc. 2012 International SoC Design Conference (ISOCC)*. IEEE, 2012, pp. 262–265.

[9] D. Kim and S. Jeon, "A WR-3 band fundamental voltage-controlled oscillator with a wide frequency tuning range and high output power," *IEEE Transactions on Microwave Theory and Techniques*, vol. 67, no. 7, pp. 2759–2768, 2019.

[10] D. Kim and S. Jeon, "A 300-GHz high-power high-efficiency voltage-controlled oscillator with low power variation," *IEEE Microwave and Wireless Components Letters*, vol. 30, no. 5, pp. 496–499, 2020.

[11] T. K. Johansen, M. Hossain, S. Boppel, R. Doerner, V. Krozer, and W. Heinrich, "A 300 GHz active frequency tripler in transferred-substrate InP DHBT technology," in *Proc. 2019 14th European Microwave Integrated Circuits Conference (EuMIC)*. IEEE, 2019, pp. 180–183.

[12] M. Hossain, S. Boppel, W. Heinrich, and V. Krozer, "Efficient active multiplier-based signal source for 300 GHz system applications," *Electronics Letters*, vol. 55, no. 23, pp. 1220–1221, 2019.

[13] M. Hossain, K. Nosaeva, N. Weimann, V. Krozer, and W. Heinrich, "A 330 GHz active frequency quadrupler in InP DHBT transferred-substrate technology," in *Proc. 2016 IEEE MTT-S International Microwave Symposium (IMS)*. IEEE, 2016, pp. 1–4.

[14] D. Yoon, J. Yun, and J.-S. Rieh, "A 310–340-GHz coupled-line voltage-controlled oscillator based on 0.25-μm InP HBT technology," *IEEE Transactions on Terahertz Science and Technology*, vol. 5, no. 4, pp. 652–654, 2015.

[15] M. Seo, M. Urteaga, J. Hacker, A. Young, Z. Griffith, V. Jain, R. Pierson, P. Rowell, A. Skalare, A. Peralta *et al.*, "InP HBT IC technology for terahertz frequencies: Fundamental oscillators up to 0.57 THz," *IEEE Journal of Solid-State Circuits*, vol. 46, no. 10, pp. 2203–2214, 2011.

[16] J. Yun, J. Kim, D. Yoon, and J.-S. Rieh, "645-GHz InP heterojunction bipolar transistor harmonic oscillator," *Electronics Letters*, vol. 53, no. 22, pp. 1475–1477, 2017.

[17] A. Zamora, K. M. Leong, G. Mei, M. Lange, W. Yoshida, K. T. Nguyen, B. S. Gorospe, and W. R. Deal, "A high efficiency 670 GHz x36 InP HEMT multiplier chain," in *Proc. 2017 IEEE MTT-S International Microwave Symposium (IMS)*. IEEE, 2017, pp. 977–979.

[18] S. Daneshgar and J. F. Buckwalter, "Compact series power combining using subquarter-wavelength baluns in silicon germanium at 120 GHz," *IEEE Transactions on Microwave Theory and Techniques*, vol. 66, no. 11, pp. 4844–4859, 2018.

[19] M. Urteaga, Z. Griffith, M. Seo, J. Hacker, and M. J. Rodwell, "InP HBT technologies for THz integrated circuits," *Proceedings of the IEEE*, vol. 105, no. 6, pp. 1051–1067, 2017.

[20] H. Wang, F. Wang, H. Nguyen, S. Li, T. Huang, A. Ahmed, M. Smith, N. Mannem, and J. Lee, "Power amplifiers performance survey 2000–present," vol. 10, 2018.

[21] K. M. Leong, X. Mei, W. Yoshida, P.-H. Liu, Z. Zhou, M. Lange, L.-S. Lee, J. G. Padilla, A. Zamora, B. S. Gorospe *et al.*, "A 0.85 THz low noise amplifier using InP HEMT transistors," *IEEE Microwave and Wireless Components Letters*, vol. 25, no. 6, pp. 397–399, 2015.

[22] M. Seo, M. Urteaga, J. Hacker, A. Young, A. Skalare, R. Lin, and M. Rodwell, "A 600 GHz InP HBT amplifier using cross-coupled feedback stabilization and dual-differential power combining," in *2013 IEEE MTT-S International Microwave Symposium Digest (MTT)*. IEEE, 2013, pp. 1–3.

[23] Z. Wang, P.-Y. Chiang, P. Nazari, C.-C. Wang, Z. Chen, and P. Heydari, "A CMOS 210-GHz fundamental transceiver with OOK modulation," *IEEE Journal of Solid-State Circuits*, vol. 49, no. 3, pp. 564–580, 2014.

[24] M. H. Eissa and D. Kissinger, "4.5 A 13.5 dBm fully integrated 200-to-255 GHz power amplifier with a 4-way power combiner in SiGe: C BiCMOS," in *Proc. 2019 IEEE International Solid-State Circuits Conference (ISSCC)*. IEEE, 2019, pp. 82–84.

[25] N. Sarmah, K. Aufinger, R. Lachner, and U. R. Pfeiffer, "A 200–225 GHz SiGe power amplifier with peak Psat of 9.6 dBm using wideband power combination," in *Proc. ESSCIRC Conference 2016: 42nd European Solid-State Circuits Conference*. IEEE, 2016, pp. 193–196.

[26] M. Ćwikliński, P. Brückner, S. Leone, C. Friesicke, R. Lozar, H. Maßler, R. Quay, and O. Ambacher, "190-GHz G-band GaN amplifier MMICs with 40GHz of bandwidth," in *Proc. 2019 IEEE MTT-S International Microwave Symposium (IMS)*. IEEE, 2019, pp. 1257–1260.

[27] L. John, A. Tessmann, A. Leuther, P. Neininger, and T. Zwick, "Investigation of compact power amplifier cells at THz frequencies using InGaAs mHEMT technology," in *Proc. 2019 IEEE MTT-S International Microwave Symposium (IMS)*. IEEE, 2019, pp. 1261–1264.

[28] L. John, P. Neininger, C. Friesicke, A. Tessmann, A. Leuther, M. Schlechtweg, and T. Zwick, "A 280–310 GHz InAlAs/InGaAs mHEMT power amplifier MMIC with 6.7–8.3 dBm output power," *IEEE Microwave and Wireless Components Letters*, vol. 29, no. 2, pp. 143–145, 2018.

[29] A. Tessmann, A. Leuther, V. Hurm, H. Massler, S. Wagner, M. Kuri, M. Zink, M. Riessle, H.-P. Stulz, M. Schlechtweg *et al.*, "A broadband 220–320 GHz medium power amplifier module," in *Proc. 2014 IEEE Compound Semiconductor Integrated Circuit Symposium (CSICS)*. IEEE, 2014, pp. 1–4.

[30] V. Radisic, W. R. Deal, K. M. Leong, X. Mei, W. Yoshida, P.-H. Liu, J. Uyeda, A. Fung, L. Samoska, T. Gaier *et al.*, "A 10-mW submillimeter-wave solid-state power-amplifier module," *IEEE Transactions on Microwave Theory and Techniques*, vol. 58, no. 7, pp. 1903–1909, 2010.

[31] Z. Griffith, M. Urteaga, P. Rowell, and R. Pierson, "A 6–10 mW power amplifier at 290–307.5 GHz in 250 nm InP HBT," *IEEE Microwave and Wireless Components Letters*, vol. 25, no. 9, pp. 597–599, 2015.

[32] J. Kim, S. Jeon, M. Kim, M. Urteaga, and J. Jeong, "H-band power amplifier integrated circuits using 250-nm InP HBT technology," *IEEE Transactions on Terahertz Science and Technology*, vol. 5, no. 2, pp. 215–222, 2015.

[33] A. S. Ahmed, A. Simsek, M. Urteaga, and M. J. Rodwell, "8.6–13.6 mW series-connected power amplifiers designed at 325 GHz using 130 nm InP HBT technology," in *Proc. 2018 IEEE BiCMOS and Compound Semiconductor Integrated Circuits and Technology Symposium (BCICTS)*. IEEE, 2018, pp. 164–167.

[34] J. Hacker, M. Urteaga, D. Mensa, R. Pierson, M. Jones, Z. Griffith, and M. Rodwell, "250 nm InP DHBT monolithic amplifiers with 4.8 dB gain at 324 GHz," in *2008 IEEE MTT-S International Microwave Symposium Digest*. IEEE, 2008, pp. 403–406.

[35] H. Hamada, T. Tsutsumi, G. Itami, H. Sugiyama, H. Matsuzaki, K. Okada, and H. Nosaka, "300-GHz 120-Gb/s wireless transceiver with high-output-power and high-gain power amplifier based on 80-nm InP-HEMT technology," in *Proc. 2019 IEEE BiCMOS and Compound semiconductor Integrated Circuits and Technology Symposium (BCICTS)*. IEEE, 2019, pp. 1–4.

[36] H. Hamada, T. Kosugi, H.-J. Song, M. Yaita, A. El Moutaouakil, H. Matsuzaki, and A. Hirata, "300-GHz band 20-Gbps ASK transmitter module based on InP-HEMT MMICs," in *Proc. 2015 IEEE Compound Semiconductor Integrated Circuit Symposium (CSICS)*. IEEE, 2015, pp. 1–4.

[37] J. Hacker, M. Urteaga, M. Seo, A. Skalare, and R. Lin, "InP HBT amplifier MMICs operating to 0.67 THz," in *2013 IEEE MTT-S International Microwave Symposium Digest (MTT)*. IEEE, 2013, pp. 1–3.

[38] W. Deal, K. Leong, A. Zamora, W. Yoshida, M. Lange, B. Gorospe, K. Nguyen, and G. X. Mei, "A low-power 670-GHz InP HEMT receiver," *IEEE Transactions on Terahertz Science and Technology*, vol. 6, no. 6, pp. 862–864, 2016.

[39] H. Hamada, T. Tsutsumi, H. Matsuzaki, T. Fujimura, I. Abdo, A. Shirane, K. Okada, G. Itami, H.-J. Song, H. Sugiyama *et al.*, "300-GHz-band 120-Gb/s wireless front-end based on InP-HEMT PAs and mixers," *IEEE Journal of Solid-State Circuits*, vol. 55, no. 9, pp. 2316–2335, 2020.

[40] K. M. Leong, X. Mei, W. H. Yoshida, A. Zamora, J. G. Padilla, B. S. Gorospe, K. Nguyen, and W. R. Deal, "850 GHz receiver and transmitter front-ends using InP HEMT," *IEEE Transactions on Terahertz Science and Technology*, vol. 7, no. 4, pp. 466–475, 2017.

[41] S. Kim, J. Yun, D. Yoon, M. Kim, J.-S. Rieh, M. Urteaga, and S. Jeon, "300 GHz integrated heterodyne receiver and transmitter with on-chip fundamental local oscillator and mixers," *IEEE Transactions on Terahertz Science and Technology*, vol. 5, no. 1, pp. 92–101, 2014.

[42] K. Statnikov, J. Grzyb, B. Heinemann, and U. R. Pfeiffer, "160-GHz to 1-THz multi-color active imaging with a lens-coupled SiGe HBT chip-set," *IEEE Transactions on Microwave Theory and Techniques*, vol. 63, no. 2, pp. 520–532, 2015.

[43] M. Urteaga, Z. Griffith, M. Seo, J. Hacker, and M. J. Rodwell, "InP HBT technologies for THz integrated circuits," *Proceedings of the IEEE*, vol. 105, no. 6, pp. 1051–1067, 2017.

[44] C. Yi, M. Urteaga, S. H. Choi, and M. Kim, "A 280-GHz InP DHBT receiver detector containing a differential preamplifier," *IEEE Transactions on Terahertz Science and Technology*, vol. 7, no. 2, pp. 209–217, 2017.

[45] J. Yun, S. J. Oh, K. Song, D. Yoon, H. Y. Son, Y. Choi, Y.-M. Huh, and J.-S. Rieh, "Terahertz reflection-mode biological imaging based on InP HBT source and detector," *IEEE Transactions on Terahertz Science and Technology*, vol. 7, no. 3, pp. 274–283, 2017.

[46] E. Javadi, A. Lisauskas, M. Shahabadi, N. Masoumi, J. Zhang, J. Matukas, and H. G. Roskos, "Terahertz detection with a low-cost packaged GaAs high-electron-mobility transistor," *IEEE Transactions on Terahertz Science and Technology*, vol. 9, no. 1, pp. 27–37, 2018.

[47] D. M. Yermolaev, I. Khmyrova, E. Polushkin, A. Kovalchuk, V. Gavrilenko, K. Mare-myanin, N. Maleev, V. Ustinov, V. Zemlyakov, V. A. Bespalov *et al.*, "Detector for terahertz applications based on a serpentine array of integrated GaAs/InGaAs/AlGaAs-field-effect transistors," in *Proc. 2017 International Conference on Applied Electronics (AE)*. IEEE, 2017, pp. 1–4.

[48] A. W. Fang, E. Lively, Y.-H. Kuo, D. Liang, and J. E. Bowers, "A distributed feedback silicon evanescent laser," *Optics Express*, vol. 16, no. 7, pp. 4413–4419, 2008.

[49] T. Komljenovic, D. Huang, P. Pintus, M. A. Tran, M. L. Davenport, and J. E. Bowers, "Photonic integrated circuits using heterogeneous integration on silicon," *Proceedings of the IEEE*, vol. 106, no. 12, pp. 2246–2257, 2018.

[50] M. Tang, J.-S. Park, Z. Wang, S. Chen, P. Jurczak, A. Seeds, and H. Liu, "Integration of III-V lasers on Si for Si photonics," *Progress in Quantum Electronics*, vol. 66, pp. 1–18, 2019.

[51] H. Park, A. W. Fang, O. Cohen, R. Jones, M. J. Paniccia, and J. E. Bowers, "A hybrid AlGaInAs–silicon evanescent amplifier," *IEEE Photonics Technology Letters*, vol. 19, no. 4, pp. 230–232, 2007.

[52] M. L. Davenport, S. Skendžić, N. Volet, J. C. Hulme, M. J. Heck, and J. E. Bowers, "Het-erogeneous silicon/III–V semiconductor optical amplifiers," *IEEE Journal of Selected Topics in Quantum Electronics*, vol. 22, no. 6, pp. 78–88, 2016.

[53] P. Kaspar, G. de Valicourt, R. Brenot, M. A. Mestre, P. Jennevé, A. Accard, D. Make, F. Lelarge, G.-H. Duan, N. Pavarelli *et al.*, "Hybrid III-V/silicon SOA in optical network based on advanced modulation formats," *IEEE Photonics Technology Letters*, vol. 27, no. 22, pp. 2383–2386, 2015.

[54] D. A. Miller, "Device requirements for optical interconnects to silicon chips," *Proceedings of the IEEE*, vol. 97, no. 7, pp. 1166–1185, 2009.

[55] L. Shen, Y. Jiao, W. Yao, Z. Cao, J. van Engelen, G. Roelkens, M. Smit, and J. van der Tol, "High-bandwidth uni-traveling carrier waveguide photodetector on an InP-membrane-on-silicon platform," *Optics Express*, vol. 24, no. 8, pp. 8290–8301, 2016.

[56] K. Sengupta, T. Nagatsuma, and D. M. Mittleman, "Terahertz integrated electronic and hybrid electronic–photonic systems," *Nature Electronics*, vol. 1, no. 12, pp. 622–635, 2018.

[57] I. Kallfass, P. Harati, I. Dan, J. Antes, F. Boes, S. Rey, T. Merkle, S. Wagner, H. Massler, A. Tessmann *et al.*, "MMIC chipset for 300 GHz indoor wireless communication," in *Proc. 2015 IEEE International Conference on Microwaves, Communications, Antennas and Electronic Systems (COMCAS)*. IEEE, 2015, pp. 1–4.

[58] H.-J. Song, T. Kosugi, H. Hamada, T. Tajima, A. El Moutaouakil, H. Matsuzaki, Y. Kawano, T. Takahashi, Y. Nakasha, N. Hara *et al.*, "Demonstration of 20-Gbps wireless data transmission at 300 GHz for KIOSK instant data downloading applications with InP MMICs," in *Proc. 2016 IEEE MTT-S International Microwave Symposium (IMS)*. IEEE, 2016, pp. 1–4.

[59] K. Takano, K. Katayama, S. Amakawa, T. Yoshida, and M. Fujishima, "56-Gbit/s 16-QAM wireless link with 300-GHz-band CMOS transmitter," in *Proc. 2017 IEEE MTT-S International Microwave Symposium (IMS)*. IEEE, 2017, pp. 793–796.

[60] K. Takano, S. Amakawa, K. Katayama, S. Hara, R. Dong, A. Kasamatsu, I. Hosako, K. Mizuno, K. Takahashi, T. Yoshida *et al.*, "17.9 A 105 Gb/s 300 GHz CMOS transmitter," in *Proc. 2017 IEEE International Solid-State Circuits Conference (ISSCC)*. IEEE, 2017, pp. 308–309.

[61] S. Hara, K. Takano, K. Katayama, R. Dong, S. Lee, I. Watanabe, N. Sekine, A. Kasamatsu, T. Yoshida, S. Amakawa *et al.*, "300-GHz CMOS transceiver for terahertz wireless communication," in *Proc. 2018 Asia-Pacific Microwave Conference (APMC)*. IEEE, 2018, pp. 429–431.

[62] H. Hamada, T. Fujimura, I. Abdo, K. Okada, H.-J. Song, H. Sugiyama, H. Matsuzaki, and H. Nosaka, "300-GHz. 100-Gb/s InP-HEMT wireless transceiver using a 300-GHz fundamental mixer," in *Proc. 2018 IEEE/MTT-S International Microwave Symposium-IMS*. IEEE, 2018, pp. 1480–1483.

[63] I. Dan, G. Ducournau, S. Hisatake, P. Szriftgiser, R.-P. Braun, and I. Kallfass, "A terahertz wireless communication link using a superheterodyne approach," *IEEE Transactions on Terahertz Science and Technology*, vol. 10, no. 1, pp. 32–43, 2019.

[64] V. Chinni, P. Latzel, M. Zégaoui, C. Coinon, X. Wallart, E. Peytavit, J. Lampin, K. Engenhardt, P. Szriftgiser, M. Zaknoune *et al.*, "Single-channel 100 Gbit/s transmission using III–V UTC-PDs for future IEEE 802.15. 3D wireless links in the 300 GHz band," *Electronics Letters*, vol. 54, no. 10, pp. 638–640, 2018.

[65] E. Lacombe, C. Belem-Goncalves, C. Luxey, F. Gianesello, C. Durand, D. Gloria, and G. Ducournau, "10-Gb/s indoor THz communications using industrial si photonics technology," *IEEE Microwave and Wireless Components Letters*, vol. 28, no. 4, pp. 362–364, 2018.

[66] C. Castro *et al.*, "32 GBd 16QAM wireless transmission in the 300 GHz band using a PIN diode for THz upconversion," in *Proc. Optical Fiber Communications Conference and Exhibition (OFC)*. Optical Society of America, 2019.

[67] I. Dan, P. Szriftgiser, E. Peytavit, J.-F. Lampin, M. Zegaoui, M. Zaknoune, G. Ducournau, and I. Kallfass, "A 300-GHz wireless link employing a photonic transmitter and an active electronic receiver with a transmission bandwidth of 54 GHz," *IEEE Transactions on Terahertz Science and Technology*, vol. 10, no. 3, pp. 271–281, 2020.

[68] T. Nagatsuma and G. Carpintero, "Recent progress and future prospect of photonics-enabled terahertz communications research," *IEICE Transactions on Electronics*, vol. 98, no. 12, pp. 1060–1070, 2015.

[69] K. Liu, S. Jia, S. Wang, X. Pang, W. Li, S. Zheng, H. Chi, X. Jin, X. Zhang, and X. Yu, "100 Gbit/s THz photonic wireless transmission in the 350-GHz band with extended reach," *IEEE Photonics Technology Letters*, vol. 30, no. 11, pp. 1064–1067, 2018.

[70] G. Ducournau, K. Engenhardt, P. Szriftgiser, D. Bacquet, M. Zaknoune, R. Kassi, E. Lecomte, and J.-F. Lampin, "32 Gbit/s QPSK transmission at 385 GHz using coherent fibre-optic technologies and THz double heterodyne detection," *Electronics Letters*, vol. 51, no. 12, pp. 915–917, 2015.

[71] X. Yu, R. Asif, M. Piels, D. Zibar, M. Galili, T. Morioka, P. U. Jepsen, and L. K. Oxenløwe, "400-GHz wireless transmission of 60-Gb/s Nyquist-QPSK signals using UTC-PD and heterodyne mixer," *IEEE Transactions on Terahertz Science and Technology*, vol. 6, no. 6, pp. 765–770, 2016.

[72] X. Yu, R. Asif, M. Piels, D. Zibar, M. Galili, T. Morioka, P. U. Jepsen, and L. K. Oxenløwe, "60 Gbit/s 400 GHz wireless transmission," in *Proc. 2015 International Conference on Photonics in Switching (PS)*. IEEE, 2015, pp. 4–6.

[73] X. Li, J. Yu, L. Zhao, W. Zhou, K. Wang, M. Kong, G.-K. Chang, Y. Zhang, X. Pan, and X. Xin, "132-Gb/s photonics-aided single-carrier wireless terahertz-wave signal transmission at 450 GHz enabled by 64QAM modulation and probabilistic shaping," in

Proc. Optical Fiber Communication Conference. Optical Society of America, 2019, pp. M4F–4.

[74] S. Jia, X. Yu, H. Hu, J. Yu, T. Morioka, P. U. Jepsen, and L. K. Oxenløwe, "120 Gb/s multi-channel THz wireless transmission and THz receiver performance analysis," *IEEE Photonics Technology Letters*, vol. 29, no. 3, pp. 310–313, 2017.

[75] S. Jia, X. Pang, O. Ozolins, X. Yu, H. Hu, J. Yu, P. Guan, F. Da Ros, S. Popov, G. Jacobsen *et al.*, "0.4 THz photonic-wireless link with 106 Gb/s single channel bitrate," *Journal of Lightwave Technology*, vol. 36, no. 2, pp. 610–616, 2018.

[76] X. Yu, S. Jia, H. Hu, M. Galili, T. Morioka, P. U. Jepsen, and L. K. Oxenløwe, "Ultra-broadband THz photonic wireless transmission," in *Proc. 2018 23rd Opto-Electronics and Communications Conference (OECC)*. IEEE, 2018, pp. 1–2.

[77] K. Statnikov, J. Grzyb, B. Heinemann, and U. R. Pfeiffer, "160-GHz to 1-THz multi-color active imaging with a lens-coupled SiGe HBT chip-set," *IEEE Transactions on Microwave Theory and Techniques*, vol. 63, no. 2, pp. 520–532, 2015.

[78] J. Grzyb, K. Statnikov, N. Sarmah, B. Heinemann, and U. R. Pfeiffer, "A 210–270-GHz circularly polarized FMCW radar with a single-lens-coupled SiGe HBT chip," *IEEE Transactions on Terahertz Science and Technology*, vol. 6, no. 6, pp. 771–783, 2016.

[79] P. Hillger, R. Jain, J. Grzyb, W. Förster, B. Heinemann, G. MacGrogan, P. Mounaix, T. Zimmer, and U. R. Pfeiffer, "A 128-pixel system-on-a-chip for real-time super-resolution terahertz near-field imaging," *IEEE Journal of Solid-State Circuits*, vol. 53, no. 12, pp. 3599–3612, 2018.

[80] A. Mehdipour, J. W. Wong, and G. V. Eleftheriades, "Beam-squinting reduction of leaky-wave antennas using Huygens metasurfaces," *IEEE Transactions on Antennas and Propagation*, vol. 63, no. 3, pp. 978–992, 2015.

[81] M. Faenzi, G. Minatti, D. Gonzalez-Ovejero, F. Caminita, E. Martini, C. Della Giovam-paola, and S. Maci, "Metasurface antennas: New models, applications and realizations," *Scientific Reports*, vol. 9, no. 1, pp. 1–14, 2019.

[82] X. Wu and K. Sengupta, "Single-chip source-free terahertz spectroscope across 0.04–0.99 THz: Combining sub-wavelength near-field sensing and regression analysis," *Optics Express*, vol. 26, no. 6, pp. 7163–7175, 2018.

[83] T. Harter, S. Muehlbrandt, S. Ummethala, A. Schmid, S. Nellen, L. Hahn, W. Freude, and C. Koos, "Silicon–plasmonic integrated circuits for terahertz signal generation and coherent detection," *Nature Photonics*, vol. 12, no. 10, pp. 625–633, 2018.

[84] S. Moazeni, "CMOS and plasmonics get close," *Nature Electronics*, vol. 3, no. 6, pp. 302–303, 2020.

17 Post Moore's Law Computing

Moore's law has been used over the past few decades to successfully predict the trend of development in the technology industry. But now, with the advances in silicon lithography and miniaturization of electronics, Moore's law is expected to flatten by 2025 [1]. In the near term, silicon-based platforms will continue to drive IC performance improvement and functional expansion, with heterogeneous integration of III-V semiconductors, photonic/plasmonic and other advanced materials as mentioned in Chapter 14. New computing architectures, such as specialized computational hardware, and optimized algorithms will help boost the computing performance on the software side. In the long term, two computing technologies are promising to achieve a fundamental leap forward in terms of computing performance. One is brain-inspired technology, such as neuromorphic computing and deep learning, which enable us to handle information more efficiently like a biological brain. The other is quantum computing, which takes advantages of the polynomial superposition nature of qubits; this can significantly lower the computational burden. This chapter discusses these technologies in detail.

17.1 Post Moore's Law Era

In 1965, Gordon Moore posited that the number of transistors in a given area of use would double every two years [2]. In 1975, the observation was confirmed and the trend held true until the mid-2000s. Since then, this trend has slowed down, leading to many believing that the industry is in the post Moore's law era.

Over the past 50 years, significant innovations have been made in shrinking the size of components. For example, the gate length of metal-oxide–semiconductor field-effect transistors (MOSFETs) has decreased from 10 μm in the early 1970s to 5 nm today [2]. MOSFETs, once a gate length of 50 nm was achieved, were split into two different families: the planar fully depleted silicon-on-insulator structure, and the three-dimensional (3D) fin field-effect transistor (FinFET) structure. Nowadays, advances in extreme ultraviolet lithography and many improvements in various processes have given rise to atomic-level transistors. Relying exclusively on Moore's law is no longer possible; new concepts of devices, architectures, and computational paradigms will be required.

Over the next few years, silicon-based devices will continue to evolve. For example, multi-layer electronic IC schemes have been developed to facilitate the vertical growth of electronic circuits, and heterogeneous integration of silicon with III-V semiconductors and other functional materials will continue to boost the performance of ICs. Furthermore, silicon photonics – integrating photonic components – will expand functionality into the photonic domain. Such efforts will continue to evolve as nanotechnology advancement, efficient energy management, and advanced circuit design.

Mass production of 7-nm FinFETs was achieved in 2018, and 5 nm was achieved only two years later in 2020. Although the size reduction does not seem significant – from 7 nm to 5 nm, it is in fact a 30% reduction, which is considerable. Following this trend, we can expect to achieve atomic-level transistors by 2022 to 2024. At the atomic scale, quantum effects will start to become dominant, enabling us to implement novel techniques that are not possible to use today and rendering conventional circuit logic obsolete. By 2030, 2-nm technology is expected to be ready for implementing the first batch of 6G devices. A long-term roadmap for the semiconductor chip technology is critical to successful 6G commercialization.

In the long term, fundamental advances are needed in how we control and manipulate information in an energy-efficient manner. To achieve a paradigm shift in the world of computing, technologies that can break the CMOS limit are essential. Historical trends in transistor development show that it takes about 10 years for basic device physics to reach mainstream use. Over the next two decades, ongoing research is expected to bear fruit, culminating in chip process advances, a new computing paradigm, and associated real-world applications.

Economic considerations are also noteworthy. The new silicon fabrication for 5 nm is expected to cost US$12 billion. The number of suppliers that can build such facilities and have the expertise to run them is extremely small. The non-recurring engineering cost for application-specific integrated circuits (ASICs) in these processes is tens of millions of dollars, meaning that new fabrication processes make economic sense only for those with extremely high volume.

17.2   Neuromorphic Computing

With the emergence of AI, computational technologies that mimic the functions of a brain (e.g., neural networks and machine learning) have revolutionized the computing industry. AI manipulates data similarly to a biological brain and can perform self-learning to recognize hidden patterns in large volumes of data. Initially, AI was designed to perform some fundamental functions such as speech recognition and imaging processing, but it has since evolved to perform more advanced functions, such as playing chess, composing music, and even writing articles.

Although the recent progress in AI has enabled us to automate data processing, we still desire a computing platform that can process information like a human brain in order to realize true machine intelligence. A human brain consists of billions of neurons connected by trillions of synapses in three-dimensional space. By exciting synaptic potentials, information is transferred among neurons with a considerably low energy consumption. For a neural network, its interconnected nature enables it to process multiple information streams in parallel. This allows computations to be performed much faster while consuming very little power. A computing model that imitates the brain is therefore desirable to enable both smart and energy-efficient machine intelligent chips.

From the perspective of algorithms, a deep learning algorithm mimics a human brain in how it processes information. Leveraging the huge amount of data available, such an algorithm trains itself by extracting higher-level features from the original data using a multi-layer network that mimics neural activity transmission. Each layer learns to extract more abstract and representative features from its previous layer and sends these features to the next layer for more refined extraction. Eventually, the original data is recognized at the output layer.

In hardware realizations, early implementations of neuromorphic computing used CMOS transistor circuits to mimic the activities of neurons and synapses [3]. This led to the advent of "neuromorphic" silicon neurons (SiNs) based on the digital very-large-scale integration (VLSI) technology. Many types of SiNs have subsequently been proposed to imitate real neurons with multiple levels of complexity [4]. The article [4] presents a wide range of circuits commonly used to design SiNs. For example, synaptic transmission can be modeled mathematically by using a simple first-order differential equation. To realize such a function, a temporal integration circuit can be used. However, implementing a system that achieves extensive connections based on this circuit is difficult because the circuit is bulky.

Digital implementation with field programmable gate arrays (FPGAs) or ASICs was proposed to better adapt to the nature of neuromorphic algorithms and achieve higher computing efficiency. Optimized memory interfaces and processing, such as reconfigurable memory interfaces and high-bandwidth memory, have been applied in neuromorphic computing systems to enhance the data processing.

Recently, non-volatile memory has attracted a great deal of research, as it provides alternative ways to store data [5]. One example of this is the memristor, which records the historic profile of excitations in devices – this is a unique characteristic. Memristor, the electronic equivalent of biological memory in simple terms, can bring a new dimension in data processing when applied to neuromorphic computing. For example, it enables historic data searches to be recorded and applied in subsequent searches more efficiently, helping deep learning neural networks to perform tasks more intelligently.

Neuromorphic computing imitates information handling at a biological level and processes information on a compact and energy-efficient platform. This technology, inspired by the brain, is a promising candidate to achieve true machine intelligence, and it will continue to advance as the AI market grows.

17.3   Quantum Computing

Quantum computing is a new computing paradigm. Compared with the current binary operations in a conventional computer, a quantum computer processes information based on quantum mechanics. Quantum bits, called qubits, are used to represent information in quantum mechanics and are the fundamental computational unit in a quantum computer. Compared with the discrete nature of binary bits, qubits can take coherent states from 0 to 1 on a continuous sphere, i.e., the Bloch sphere. By taking advantage of the concept of superposition, qubits can represent information of 2^n (where n is the number of qubits) states at the same time. Because a qubit is essentially a polynomial superposition of multiple states simultaneously, a quantum computer can process all 2^n states in parallel, the result of which is a significant increase in computation speed.

Over the past decade, many quantum algorithms have been developed to demonstrate different levels of complexity in both quantum and conventional computing platforms [6]. For example, a quantum Fourier transform takes $O(n \log n)$ steps, while classical algorithms take $O(n2^n)$. We can conclude from this that the computational burden is lowered in a polynomial way.

A great deal of research has been conducted into the hardware implementation of qubits. Initially, liquid-state nuclear-magnetic resonance (NMR) and ion traps were studied as possible ways to implement qubits [7, 8]. Recently, however, focus has shifted towards semiconductor-based qubits. One example is the silicon qubit, which uses one or more QDs coupled capacitively between two electrically tunable electrodes on a double-gate silicon transistor. Different quantum states can be modeled based on different capacitive coupling levels between these QDs. Figure 17.1 shows an example of a double QD system. Of particular note is the fact that silicon qubits have longer spin coherence times while also inheriting all the advantages of silicon. Since the first silicon qubit emerged [9, 10], many research labs and foundries are collaborating to drive the fabrication of scalable quantum devices. For example, in Canada, the Certified Management Consultant (CMC) recently announced that it is working to launch a quantum computing program with IBM Q-HUB [11].

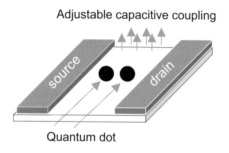

Figure 17.1   A double QD system.

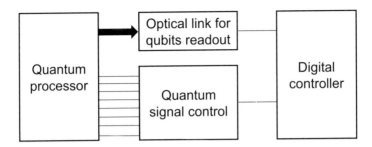

Figure 17.2 A block diagram of a peripheral system for qubit control/read-out [12].

A peripheral system is needed for qubit control and read-out. The functions of a controller can include quantum signal amplification, down-conversion, digitization, and more. Figure 17.2 shows a block diagram of such a system [12]. Optical links are used for qubit read-out, and multiplexing and demultiplexing are used to facilitate multi-stream quantum signal transmission between qubits and the controller.

The controller circuit usually operates at room temperature, while the quantum computing system operates at a cryo-cooled temperature. The large temperature gradient between them can induce thermal noise, which will cause quantum errors. Redundant qubits are therefore used for error correction, but this increases latency and reduces the quantum computation efficiency. To address this issue, many researchers have proposed CMOS-based controller systems working at deep cryogenic temperatures (less than 10 kelvin), i.e., cryo-CMOS circuits. Thermal noise can be significantly reduced when both the quantum processor and controller circuit are at a cryogenic temperature. This can also reduce the interconnect complexity between them. The articles [12, 13] presented several cryo-CMOS components, including qubit read-out, passive circulators, and low-noise amplifiers (LNAs). Then a chip consisting of those components was fabricated in a 160-nm CMOS technology as an example of the controller system above.

Cryo-CMOS allows both the quantum controller and processor to work under the same temperature and eventually the potential full integration of both on the same chip. Although the current understanding of how CMOS circuits behave at cryogenic temperatures is still limited, significant research has been conducted into cryo-physics, new models, and system-level simulation. As we gain a better understanding of cryo-CMOS circuit models and quantum system-level performance, we expect that silicon quantum computers featuring high performance, low power consumption, and scalable yield will become a reality in the near future.

17.4   New Computing Architectures

While the end of Moore's law can stimulate the growth of new computing paradigms, new improvements are still necessary in the current digital computing platform. Although conventional parallel computing can increase the computation speed roughly proportionally to the number of threads or cores, it cannot optimize task allocation, meaning that individual processors might not be allocated tasks for which they were designed. The advent of graphical processing units (GPUs) has enabled specialized task handling by its targeted hardware, and other specialized processors/accelerators have been developed to perform certain scientific tasks. For example, a specialized supercomputer for molecular dynamics simulations, achieving 180 times the performance of a conventional high-performing computer (HPC), was reported in [1]. Specialization delivers significant improvements in computing performance by optimizing the hardware use more efficiently. In the short term, specialized hardware can continue to provide the desired computing performance increase.

Multiple specialized processor/accelerator types can be co-located in one heterogeneous computing platform. For example, a central processing unit (CPU) and a GPU can be integrated on the same IC to combine the advantages of both. The CPU focuses on operating-system-related tasks, whereas the GPU handles 3D graphics-related tasks and intensive scientific computing tasks. Numerous specialized processors/accelerators have been reported recently, including cryptography coprocessors, tensor processing units, and other deep learning accelerators. The integration of CPU and specialized hardware allows much more efficient task handling.

From an economic perspective, increased costs and long lead time are associated with cutting-edge node process technology. While big companies have the financial resources to purchase customized high-end hardware, small and medium-size companies do not, and so cannot fully benefit from this advance. To fully leverage the concept of heterogeneous computing, the flexible use of hardware was proposed, specifically, the chiplet [1]. In contrast to a single silicon wafer integrating all specialized processors/accelerators and CPU heterogeneously in order to meet specific needs (the cost is usually high due to customization), the chiplet concept breaks the system into functional blocks that each perform the minimal task possible. As a result, this allows manufacturers to piece together all the blocks needed to achieve customized use at a much lower cost and faster turn-around.

Heterogeneous computing platforms require new software implementation. The computer architectures in use today separate data storage and processing units, resulting in a considerable slow-down in computational speed as data moves between them. This issue is known as the "von Neumann bottleneck" or "memory wall." To address this issue, new software must minimize data movement operations in addition to performing higher-order arithmetic operations.

Redesigning current algorithms and programming environments to reflect the advantages offered by specialized hardware in targeted applications is a key issue of new software implementation. Apart from algorithm optimization, the cost associated with data movement is another concern. The publication [1] stated that the overall computation complexity should not be calculated only on the basis of the number of floating-point operations; rather, it should also consider the complexity of compulsory data movement required by the algorithm. Consequently, it is desirable for new software to utilize an optimized memory/data access topology, for example, non-uniform memory access.

In terms of hardware, a complete redesign of hardware is needed to address the memory bottleneck issue. For example, the reduction or replacement of metal interconnects is desired, as they limit the energy efficiency. Silicon photonics is a promising candidate to address this, where low-loss photonic interconnects and waveguides can be applied to reduce the energy needed for data movement. One such example of the use of integrated photonic technologies is the ARPAe ENLITENED project [14], which developed novel photonic-based network topologies for information transfer in a data center environment.

References

[1] J. Shalf, "The future of computing beyond Moore's law," *Philosophical Transactions of the Royal Society A*, vol. 378, no. 2166, p. 20190061, 2020.

[2] Wikipedia, "Moore's law," 2020, accessed Sept. 29, 2020. [Online]. Available: https://en .wikipedia.org/wiki/Moore's_law

[3] C. Mead, *Analog VLSI and neural systems*. Addison-Wesley Longman Publishing Co., 1989.

[4] G. Indiveri, B. Linares-Barranco, T. J. Hamilton, A. Van Schaik, R. Etienne-Cummings, T. Delbruck, S.-C. Liu, P. Dudek, P. Häfliger, S. Renaud *et al.*, "Neuromorphic silicon neuron circuits," *Frontiers in Neuroscience*, vol. 5, p. 73, 2011.

[5] Y. Chen, H. H. Li, C. Wu, C. Song, S. Li, C. Min, H.-P. Cheng, W. Wen, and X. Liu, "Neuromorphic computing's yesterday, today, and tomorrow – an evolutional view," *Integration*, vol. 61, pp. 49–61, 2018.

[6] A. Steane, "Quantum computing," *Reports on Progress in Physics*, vol. 61, no. 2, p. 117, 1998.

[7] L. M. Vandersypen and I. L. Chuang, "NMR techniques for quantum control and computation," *Reviews of Modern Physics*, vol. 76, no. 4, p. 1037, 2005.

[8] P. Schindler, D. Nigg, T. Monz, J. T. Barreiro, E. Martinez, S. X. Wang, S. Quint, M. F. Brandl, V. Nebendahl, C. F. Roos *et al.*, "A quantum information processor with trapped ions," *New Journal of Physics*, vol. 15, no. 12, p. 123012, 2013.

[9] R. Maurand, X. Jehl, D. Kotekar-Patil, A. Corna, H. Bohuslavskyi, R. Lavit'eville, L. Hutin, S. Barraud, M. Vinet, M. Sanquer *et al.*, "A CMOS silicon spin qubit," *Nature Communications*, vol. 7, p. 13575, 2016.

[10] L. Hutin, R. Maurand, D. Kotekar-Patil, A. Corna, H. Bohuslavskyi, X. Jehl, S. Barraud, S. De Franceschi, M. Sanquer, and M. Vinet, "Si CMOS platform for quantum information processing," in *Proc. 2016 IEEE Symposium on VLSI Technology*. IEEE, 2016, pp. 1–2.

[11] CMC, "CMC becomes member of IBM Q-HUB at Université de Sherbrooke," 2020, accessed Sept. 29, 2020. [Online]. Available: https://www.cmc.ca/cmc-becomes-member-of-ibm-q-hub-at-universite-de-sherbrooke/

[12] E. Charbon, "Cryo-CMOS electronics for quantum computing applications," in *Proc. ESSDERC 2019–49th European Solid-State Device Research Conference (ESSDERC)*. IEEE, 2019, pp. 1–6.

[13] S. Schaal, A. Rossi, V. N. Ciriano-Tejel, T.-Y. Yang, S. Barraud, J. J. Morton, and M. F. Gonzalez-Zalba, "A CMOS dynamic random access architecture for radio-frequency readout of quantum devices," *Nature Electronics*, vol. 2, no. 6, pp. 236–242, 2019.

[14] ARPA-E, "ENLITENED," 2020, accessed Sept. 29, 2020. [Online]. Available: https:// arpa-e.energy.gov/?q=arpa-e-programs/enlitened

18 New Devices

Wireless technology has had a profound impact on society. Today, most people consider mobile devices to be a fundamental part of everyday life. Smartphones, which have completely replaced their predecessor, the humble feature phone, are no longer simply a nice-to-have item; rather, they are a necessity. The smartphone revolution started with a vision of using wireless computers to access the Internet via mobile broadband services. Since then, as mobile apps and the OTT service model have become almost ubiquitous, we have witnessed an explosion in the use of mobile apps. There is an extensive array of apps now available to us for almost everything: calendar, camera, calculator, travel agent, map, wallet, news, and public transport are just a few. Other apps can even intelligently recommend restaurants, shopping malls, parking lots, and online classes. In this chapter we will discuss the trends for mobile devices and the interfaces between human beings and devices.

18.1 Future Mobile Devices

It is envisioned that the next mobile device revolution – just as smartphones replaced feature phones – will happen during the lifetime of 6G. By 2030, future devices are expected to adopt new capabilities enabled by the 6G communication system, enriching their functionality with sensing and imaging, haptic/tactile communication, holographic display, AI, and other capabilities.

These innovative capabilities are redefining devices and the role they play in our life. Mobile devices are evolving to have the following capabilities:

- Human-level perception (e.g., unlimited bandwidth for human-level visual/audio perception and high-fidelity human-to-human communications [1])
- Ambient sensing (e.g., the capability for proximity multispectral imaging [2] and high-precision positioning)
- Human–cyber interaction (e.g., holographic near-eye displays for human interaction with the cyber world [3])
- Energy harvesting (e.g., wireless charging, and simultaneous wireless information and power transfer)

Such capabilities will transform the current "intelligent assistant" world into a "cyber-physical fusion" world, as shown in Figure 18.1. Today's smartphones act as

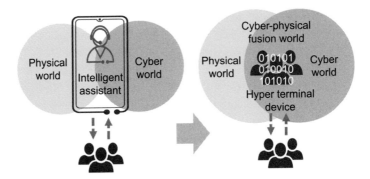

Figure 18.1 Evolution from intelligent assistant to "digital self," which is a virtual representation of an end user or device.

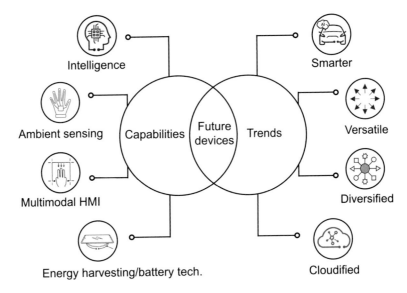

Figure 18.2 Four capabilities enable four development trends of future devices.

gateways connecting the physical world to the cyber world, mainly providing Internet access. In the future, we expect that devices will evolve to wearable hyper terminal devices in the cyber-physical fusion world – the so-called "digital self." These devices will have some capabilities that surpass the human equivalents, such as intelligent recognition and environmental perception.

Enabled by the preceding capabilities, the development of mobile devices is being shaped by four major trends, as shown in Figure 18.2.

Smarter – not only making smartphones smarter, but also augmenting realities to automate everything.

Today's smartphones are millions of times more powerful than the computer National Aeronautics and Space Administration (NASA) used during the Apollo 11 mission [4] – they are becoming smarter as computing capabilities increase.

Computing power has steadily increased in line with Moore's law, benefiting the development of AI and ML in recent years. Furthermore, the advances made in semiconductor manufacturing pave the way for higher computing performance, better energy efficiency, smaller chips, and higher transistor density.

As smartphones become more and more powerful, they are able to implement AI capabilities. Today, an increasing number of smartphones are equipped with specialized built-in neural processing units for AI computing. The AI/ML algorithms can be used to perform many computationally intensive tasks, such as AR, facial recognition, and voice and speech recognition. To facilitate the development of innovative applications, there are numerous AI APIs available, such as Huawei's HiAI, Facebook's Caffe2Go, Google's TensorFlow Lite, and Apple's CoreML.

At the same time, computationally intensive tasks can be offloaded from mobile devices to edge clouds; this is known as the edge computing paradigm. It leverages the ultra-high data rate, ultra-low latency, and ultra-high reliability provided by next-generation networks. In the future, devices will become more intelligent while also ensuring privacy protection by leveraging distributed computing and learning, thanks to edge computing, cloud computing, and local CPU, GPU, and dedicated AI acceleration hardware.

The advances made in short-range communication technologies and AI algorithms have enabled groups of drones, vehicles, and robots to interact locally with each other and with the surrounding environment. By using decentralized and self-organized control (e.g., swarm intelligence and collaborative robotics), these groups of devices are able to achieve their objectives. As 6G short-range communication technologies and AI/ML algorithms continue to develop, devices in the future will have greater intelligence, improving service experience and productivity by automating more aspects of our life.

Versatile – not only providing connectivity, but also offering novel sensing capabilities to open new possibilities for future mobile applications.

In the future, wireless connectivity will be a fundamental capability for devices as it becomes essential for every person, home, organization, and industry. As introduced in Chapter 2 regarding extremely immersive human-centric experience use cases, the 6G network can achieve throughput in the Tbit/s range while ensuring a transmission latency of less than 1 ms. This supports the smooth exchange of multi-sensory VR information, including immersive video, audio, and even haptics. With multi-sensory capabilities, devices in the future could be integrated in humans, for example, to overcome disabilities. Integrating such capabilities so that they work together with the human body has the potential to advance the human race, forming cybernetic organisms.

In addition to communication, novel sensing capabilities will be supported by integrated communication and sensing technologies (Figure 18.3). This creates the potential for mobile devices to support many new functions in the future. For example, THz radio communication can be used for imaging and spectroscopy, providing a personal X-ray at the molecular level, as mentioned in Chapter 3. In addition, healthcare

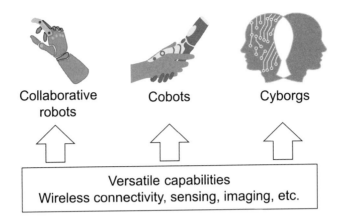

Figure 18.3 Versatile capabilities of future devices supported by 6G.

monitoring (for heartbeat, falls, etc.), trespassing, and gesture recognition through touchless interaction with devices can also be realized by using wireless sensing and machine learning.

Moreover, interactions between humans and machines, and those between the physical and cyber worlds, will be further enriched. For example, haptic sensor-actuator systems will incorporate VR to provide more cues for users to mentally construct a virtual world beyond spatial and temporal limitations, and equipment controlled remotely by brainwaves will significantly extend the scope of personal activities. Human bond communication (HBC) aims to enhance humans' communication sensations by incorporating five sensory features. It will enable the exchange of more expressive and holistic sensory information by leveraging highly reliable, responsive, and intelligent connectivity.

To facilitate these versatile applications, a significantly extended battery life is essential. By using new battery materials and architectures (e.g., graphene, lithium iron phosphate, and polymer batteries), humans with integrated devices will not need to recharge the batteries daily using wired chargers. Instead, various types of energy harvesting technologies will be available, including wireless power transfer, which can charge mobile devices efficiently and safely over long distances (e.g., across a room).

Diversified – not only smartphones, but also various types of devices that act as sensors and actuators.

Today, smartphones are the dominant mobile devices used by individual consumers. Globally, the penetration rate of smartphones is 65% in 2020, and is predicted to exceed 80% by 2025 [5]. At the same time, wireless terminal technologies continue to evolve as wireless networks aim to offer higher speed, lower latency, and higher reliability. Furthermore, there is a paradigm shift from serving individual customers to serving both individual and diverse vertical/industry customers.

Smart devices, such as smartphones, tablets, laptops, and even smart cars, are becoming an indispensable part of our life thanks to the advances made in semiconductor technologies and software platforms. In the future, a wide range of

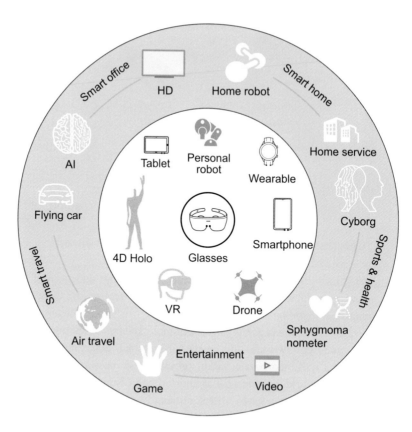

Figure 18.4 Example of human-centric applications and diversified devices.

human-centric and industrial devices will emerge integrating advanced sensors, new display technologies, and AI (Figure 18.4).

- **In terms of human-centric devices,** we predict the enhancing of wearable devices (VR headsets, AR glasses, smart watches, and exoskeleton robots), implantable medical devices, and more.
- **In terms of industrial devices,** automobiles, robots/cobots (collaborative robots), and smart factory equipment will be connected to 6G networks; a large number of low-cost, low-power (or even battery-free, referred to as "passive") IoT devices will be scattered everywhere, enabling smart city and home, smart healthcare, pollution monitoring, asset tracking, and other applications.

These devices will serve as the sensors and actuators of the cyber-physical system, integrating computation, networking and physical processes.

Owing to the explosive growth of diversified devices, there will be significantly higher requirements on interconnectivity. For example, smartphones will be connected to TVs for enhanced video watching experience, and wearable devices will be connected to smartphones for information collection. What kind of device will function as the anchor device in the future? Such an anchor device should possess strong wireless

Central cloud   Virtual device in   Physical
                      the edge cloud   device

Figure 18.5 The virtual device concept.

connectivity, battery life, and computing capabilities; good candidates include smartphones, VR headsets, exoskeleton robots, and smart cars. Through 6G networking, anchor devices and their connected devices will provide a seamless and consistent user experience.

Cloudified – not only physical devices, but also virtual devices that enable privacy protection and new business models.

As cloud computing and the IoE become mature, driven by 6G connectivity, each physical 6G device will have a virtual counterpart in the cloud acting as its proxy (Figure 18.5). By using decentralized technologies, users' personal data can be stored in virtual devices under the control of users – rather than in terminal devices or by centralized third parties – and therefore freeing users from concerns about potential privacy issues. Based on the concept of virtualized devices, shared 6G devices will emerge, typically residing in public places and used on demand. By integrating biometrics, AI, personalized automatic configuration, and privacy protection technologies, 6G devices will be able to provide easy-to-use features such as biometric-based authentication and hyper-personalized configurations. This will allow end users to access their desired services at any time and anywhere via the shared devices. The types of 6G shared devices will be extensive, covering rental vehicles, meeting rooms, cloud devices, and any other devices with input/output and computing capabilities that can be shared or rented in public places.

Furthermore, some 6G devices will provide cloud computing. Customers will be able to purchase 6G devices enabled with private cloud computing – such devices will help customers meet greater computing storage requirements for smarter voice assistance, private picture and video storage, and more. These devices can interact with the public cloud services through 6G connections to provide enhanced functionality. In addition, with virtual devices in the cloud, physical devices can offload computationally intensive tasks to the cloud, thereby extending battery life and avoiding overheating issues.

18.2 Future Brain and Device Interface

The brain–computer interface (BCI) concept dates back to the 1970s when Jacques Vidal attempted to establish a systematic approach for person–computer communication [6, 7]. Initially, researchers aimed for the BCI system to be used in the diagnosis

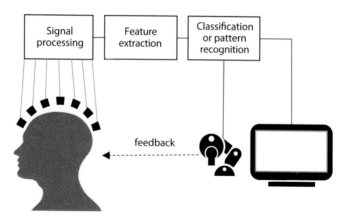

Figure 18.6 BCI system diagram.

of neurological disorders and brain diseases (e.g., the control of epileptic seizures). The system recorded and processed scalp electroencephalography (EEG) signals, and extracted features of interest from the data to diagnose epilepsy. In 2004, the article [8] demonstrated a BCI system that can control multi-dimensional movements using a non-invasive approach. This breakthrough ignited tremendous interest in BCI research across the science and industrial communities. As shown in Figure 18.6, at the least the typical BCI system consists of the following components: signal processing, feature extraction, classification or pattern recognition for translating data to machine commands, and application interfaces (e.g., computers or robotic arms). The application interfaces will finally feed sensory information back to the user, forming a bidirectional link.

In the past decade, BCI technology has evolved to realize feats that were once considered science fiction. For example, it was shown in [9] that a neural prosthesis can help locomotion recovery after spinal injury. With electrodes placed in the brain, it is possible for a paralytic patient to control a robotic prosthesis with their mind. Moreover, telekinesis was once considered a superpower but is now possible to a certain extent. In that regard, the article [10] showed that we can control the flight of a drone through brain waves. More recently, it was demonstrated in [11] that with neurochips implanted into an animal's brain, it is possible to accurately predict the animal's joint movements. This opens up a whole realm of possibilities in terms of curing blindness, paralysis, mental diseases, etc.

Both invasive and non-invasive methods have been used to record brain activity. The latter uses electrodes placed over the scalp, while the former uses intracortical and stereoelectroencephalography (SEEG) electrodes placed over the cortex. Compared with the non-invasive method, the invasive method can provide signals with a higher temporal–spatial resolution and lower noise. Within this context, the article [11] demonstrated that a coin-sized neurochip containing thousands of electrodes can be implanted into an animal's brain to collect intracortical EEG signals. As promising as this sounds, there are still some difficulties being faced in human application.

Further, advancements in deep learning have also facilitated the progress of BCI technology. Conventionally, EEG source localization (the inverse problem) was solved by iteratively computing the EEG forward problem until the simulated EEG signals matched measurements within desired tolerance levels. Deep learning can help solve the inverse problem by taking advantage of the large amount of data available [12, 13].

Besides the passive reading of brain signals, the bidirectional BCI link can provide sensory feedback from the controlled object (e.g., a prosthetic limb). Along those lines, [14, 15] showed that somatosensory feedback can be provided by electrically stimulating the peripheral and/or cortical nerves of patients. Cortical stimulation involves the stimulation of specific somatosensory cortex regions, which are linked to certain sensory functions and body parts. These regions are usually localized and demand micro-sized electrodes for precise stimulation. In addition, real-time feedback is desired to simulate natural biomechanics mediated by the central nervous system. This renders the connection speed critical (i.e., the bandwidth for brain and machine communication). To take it a step further, the authors of [16] showed that an ultra-dense packaged microelectrode array chip can record cortical signals in the brains of rodents. The size of the packaged neurochip is less than $23 \times 18.5 \times 2$ mm^3, and it is reported to be capable of creating stimulation on all channels for the purpose of neural modulation, though this is yet to be demonstrated. Implantable neurochips with new neural sensors are expected to further boost progress in this research direction [17].

Peripheral stimulation can be linked to the concept of a cyborg, which is mainly known from science-fiction movies where human enhancement is achieved by integrating robotic parts with biological bodies. Recent advancements in bionics have shown that the age of cyborg might be upon us. To put it into perspective, the publications [18, 19] demonstrated an agonist–antagonist myoneural interface (AMI) that allows brain commands to be sent to an end muscle for dynamic prosthetic control, and enables proprioceptive sensory feedback to the brain. In this AMI, surgeons first create AMI muscle sets in a patient's residual limb during amputation. When the patient undergoes an intention to move, electromyography (EMG) on the targeted AMI muscle is recorded and subsequently translated into machine commands to control the motor in the prosthetic limb. In the reverse direction, sensory signals are naturally generated in the AMI muscle when the prosthetic joint moves, and then sent to the brain through nerves, enabling the patient to feel the movement. The AMI was demonstrated in a TED talk [20], where the prosthetic limb was shown to run, climb, and dance. The AMI provides a promising framework, within which we can integrate robotics and biological humans. It seems that the era of cyborg–human may not be far away.

Over the next few decades, we expect brain-controlled neural prosthesis to become a reality, improving the quality of life for paralyzed and disabled patients. By controlling the robotic arms or prosthesis through the human mind, patients can walk, move objects, and engage in limb-related activities.

18.3   New Wearable

Wearable devices have flourished over the last few years owing to their functionalities. When attached onto human skin, wearable electronics can monitor various physiological functions, e.g., heart rate, electrocardiograms, skin temperature, body motion, and blood pressure. These devices have numerous applications, including in sports, military activities, health, and daily life. With 6G bound to step on the scene, new wearable devices are projected to bring new use cases and applications.

The article [21] showed that a scalable tactile glove with a knitted sensor array is capable of identifying individual object shapes, estimating their weights, and mapping the tactile patterns when grasping the objects. In the study, interactions between an entire hand and different objects were recorded through 548 sensors assembled on the tactile glove. A deep convolutional neural network was then trained to extract the tactile signatures of the human grasp. These signatures can be quite useful in the development of robots and prosthetics.

In [22] a wideband and flexible THz imaging wearable device based on CNT films under room temperature was demonstrated. The THz CNT scanner was able to perform passive imaging on both flat and bent samples. This technology will facilitate bringing convenient and portable THz devices for many applications, including security inspections and health monitoring.

Graphene-based wearables were demonstrated for applications that monitor health and wellness. Specifically, the article [23] showed that a graphene photodetector sensitized with semiconducting quantum dots (GQD-PD) can be used for flexible and transparent wearables. A number of prototype devices could be designed with the proposed GQD-PD. For example, a bracelet with GQD-PD and integrated light source was shown to monitor the heart rate in both reflectance and transmission modes. The authors of [24] presented a flexible and transparent wristband integrated with a graphene and poly(3,4-ethylenedioxythiophene) : poly(4-styrenesulfonate) (PEDOT:PSS)/graphene mid-infrared photodetector. The wristband can be used to monitor health by tracking variations in body emissions.

References

[1] W. Wang, X. Deng, L. Ding, and L. Zhang, *Brain-inspired intelligence and visual perception*. Springer, 2019.

[2] V. C. Coffey, "Multispectral imaging moves into the mainstream," *Optics and Photonics News*, vol. 23, no. 4, pp. 18–24, 2012.

[3] A. Maimone, A. Georgiou, and J. S. Kollin, "Holographic near-eye displays for virtual and augmented reality," *ACM Transactions on Graphics*, vol. 36, no. 4, pp. 1–16, 2017.

[4] "Your smartphone is millions of times more powerful than the Apollo 11 guidance computers." [Online]. Available: https://www.zmescience.com/science/news-science/smartphone-power-compared-to-apollo-432/

[5] GSM. Association *et al.*, "Global mobile trends 2020 new decade, new industry?" 2019.

[6] J. J. Vidal, "Toward direct brain-computer communication," *Annual Review of Biophysics and Bioengineering*, vol. 2, no. 1, pp. 157–180, 1973.

[7] Wikipedia, "Brain computer interface." [Online]. Available: https://en.wikipedia.org/wiki/Brain%E2%80%93computer_interface

[8] J. R. Wolpaw and D. J. McFarland, "Control of a two-dimensional movement signal by a noninvasive brain–computer interface in humans," *Proceedings of the National Academy of Sciences*, vol. 101, no. 51, pp. 17 849–17 854, 2004.

[9] D. Borton, M. Bonizzato, J. Beauparlant, J. DiGiovanna, E. M. Moraud, N. Wenger, P. Musienko, I. R. Minev, S. P. Lacour, J. D. R. Millán *et al.*, "Corticospinal neuroprostheses to restore locomotion after spinal cord injury," *Neuroscience Research*, vol. 78, pp. 21–29, 2014.

[10] CTV News, "'Pure thought': Edmonton graduates using brain waves to fly drones." [Online]. Available: https://www.ctvnews.ca/sci-tech/pure-thought-edmonton-graduates -using-brain-waves-to-fly-drones-1.4343715

[11] A. Regalado, "Elon Musk's Neuralink is neuroscience theater." [Online]. Available: https://www.technologyreview.com/2020/08/30/1007786/elon-musks-neuralink-demo -update-neuroscience-theater

[12] S. Cui, L. Duan, B. Gong, Y. Qiao, F. Xu, J. Chen, and C. Wang, "EEG source localization using spatio-temporal neural network," *China Communications*, vol. 16, no. 7, pp. 131–143, 2019.

[13] K. H. Jin, M. T. McCann, E. Froustey, and M. Unser, "Deep convolutional neural network for inverse problems in imaging," *IEEE Transactions on Image Processing*, vol. 26, no. 9, pp. 4509–4522, 2017.

[14] D. W. Tan, M. A. Schiefer, M. W. Keith, J. R. Anderson, and D. J. Tyler, "Stability and selectivity of a chronic, multi-contact cuff electrode for sensory stimulation in human amputees," *Journal of Neural Engineering*, vol. 12, no. 2, p. 026002, 2015.

[15] G. A. Tabot, J. F. Dammann, J. A. Berg, F. V. Tenore, J. L. Boback, R. J. Vogelstein, and S. J. Bensmaia, "Restoring the sense of touch with a prosthetic hand through a brain interface," *Proceedings of the National Academy of Sciences*, vol. 110, no. 45, pp. 18 279–18 284, 2013.

[16] E. Musk *et al.*, "An integrated brain–machine interface platform with thousands of channels," *Journal of Medical Internet Research*, vol. 21, no. 10, p. e16194, 2019.

[17] C. Li and W. Zhao, "Progress in the brain–computer interface: An interview with Bin He," *National Science Review*, vol. 7, no. 2, pp. 480–483, 2020.

[18] T. R. Clites, M. J. Carty, J. B. Ullauri, M. E. Carney, L. M. Mooney, J.-F. Duval, S. S. Srinivasan, and H. M. Herr, "Proprioception from a neurally controlled lower-extremity prosthesis," *Science Translational Medicine*, vol. 10, no. 443, p. eaap8373, 2018.

[19] S. Srinivasan, M. Carty, P. Calvaresi, T. Clites, B. Maimon, C. Taylor, A. Zorzos, and H. Herr, "On prosthetic control: A regenerative agonist–antagonist myoneural interface," *Science Robotics*, vol. 2, no. 6, 2017.

[20] H. Herr, "New bionics let us run, climb and dance." [Online]. Available: https://ed.ted .com/lessons/g8KC49mB

[21] S. Sundaram, P. Kellnhofer, Y. Li, J.-Y. Zhu, A. Torralba, and W. Matusik, "Learning the signatures of the human grasp using a scalable tactile glove," *Nature*, vol. 569, no. 7758, pp. 698–702, 2019.

[22] D. Suzuki, S. Oda, and Y. Kawano, "A flexible and wearable terahertz scanner," *Nature Photonics*, vol. 10, no. 12, pp. 809–813, 2016.

[23] E. O. Polat, G. Mercier, I. Nikitskiy, E. Puma, T. Galan, S. Gupta, M. Montagut, J. J. Piqueras, M. Bouwens, T. Durduran *et al.*, "Flexible graphene photodetectors for wearable fitness monitoring," *Science Advances*, vol. 5, no. 9, p. eaaw7846, 2019.

[24] M. Zhang and J. T. Yeow, "A flexible, scalable, and self-powered mid-infrared detector based on transparent PEDOT : PSS/graphene composite," *Carbon*, vol. 156, pp. 339–345, 2020.

Summary of Part IV

The forthcoming 6G era has sparked a great deal of research interest into the hardware technologies needed to brace the infrastructure for the upcoming paradigm shift. As mmWave spectrum continues to mature, new 6G spectrum – including several THz candidate frequency bands – will flourish. The combination of both mmWave and THz frequency bands is expected to deliver data rates in the Tbit/s range and high-resolution imaging. With the new spectrum, use cases, and hardware technologies (such as intelligent surface), an updated channel model that can accurately reflect radio wave propagation is needed. From a hardware perspective, the heterogeneous III-V platform can further boost silicon system performance by optimizing each component on a single die. Novel materials such as photonic crystals, photovoltaics, and plasmonics can be added onto silicon to achieve higher performance. Antenna on-chip and in-package technologies, along with compact lens technology such as RIS, can enable greater control of antenna performance while also reducing the system size. THz technology will enable new communication and imaging methods, but the implementation of THz systems based on electronics, optoelectronics, and photonics will depend on the applications and operating frequencies. The performance of these THz systems was discussed in Part IV. A framework covering joint orbital angular momentum and massive MIMO wireless communication (a promising transmission technology for 6G) has been shown to improve the system performance and achieve multiplicative spectral gain. Finally, as Moore's law flattens and silicon technology approaches the limit, two new computing paradigms – neuromorphic computing and quantum computing – are promising to bring a fundamental leap forward in terms of computing performance.

Part V

Enabling Technologies for 6G Air Interface Design

Introduction to Part V

The design philosophy of the 5G radio access network was inspired by the vision for the IoE, which is manifest in diversified usage scenarios such as eMBB, URLLC, and mMTC. The three representative requirements of these usage scenarios are 20 Gbit/s at 100 MHz bandwidth, 99.999% reliability at 1 ms air interface latency, and 1 million devices per square kilometers with wider coverage, respectively.

To provide an optimized method of supporting versatile application scenarios and a wide spectrum range, a unified new air interface featuring both flexibility and adaptability has been employed in 5G. This air interface can natively enable radio access network (RAN) slicing to facilitate efficient service multiplexing, effectively improve spectral efficiency, increase connectivity, and reduce latency. To avoid the limitations of the geographic "cells" that introduce devices detrimental "cell-edge experience," 5G radio access decouples user-equipment (UE)-specific physical signals from "cell" identifiers (IDs), providing networks with more freedom to select an optimal "beam" or "beams" to serve users, realizing "user-centric no-cell (UCNC)" architecture to improve cell-edge user experience with natively supported cooperation transmission schemes.

For 6G, as elaborated in Chapter 1, we envision a new era featuring connected people, connected things, and connected intelligence with new services such as sensing and networked AI in addition to enhanced 5G usage scenarios. Within this context, 6G air interface should be able to support new KPIs and much higher or stricter KPIs than that of 5G, as shown in Chapter 1. 6G needs to support an even higher spectrum range and wider bandwidth in order to deliver extremely high-speed data services and high-resolution sensing. To meet these new challenging goals, 6G air interface designs require revolutionary breakthroughs. The design philosophy needs rethinking at the onset of the 6G air interface design. We envision several paradigm shifts as follows.

From soft air interface to intelligent air interface.

5G air interface abandons the "one-size-fits-all" approach. Its flexibility and configurability (i.e., "softness") enable us to optimize air interface for different usage scenarios, such as eMBB, URLLC, and mMTC within a unified framework.

The 6G air interface design is powered by a combination of model- and data-driven AI and expected to enable the tailored optimization of air interface from provisional configuration to self-learning. The personalized air interface can customize the transmission scheme and parameters at the UE/service level to maximize experience without sacrificing system capacity. It can be easily scaled to support the near-zero-latency URLLC. In addition, the new simple and agile signaling mechanism will minimize the signaling overhead and latency.

From retrofit AI to native AI.

AI and machine learning technologies have the potential to be used to continuously enhance 5G capabilities. For 6G, AI will be a built-in feature of air interface, enabling intelligent physical layer (PHY) and media access control (MAC). It will not be limited to network management optimization (such as load balancing and power saving), replacing some non-linear or non-convex algorithms in the transceiver modules, or compensating for deficiencies in the non-linear models.

Intelligence will make 6G PHY more powerful and efficient. It will facilitate the optimization of PHY building block and procedural designs, including the possible re-architecturing of the transceiver process. On top of that, it will help provide new sensing and positioning capabilities, which in turn will significantly change the air interface component designs. AI-assisted sensing and positioning will also make low-cost and highly accurate beamforming and tracking possible. Intelligent MAC will provide a smart controller based on single- or multi-agent reinforced learning, including cooperative machine learning for network and UE nodes. For example, with multi-parameter joint optimization and individual or joint procedure training, enormous performance gains can be obtained in terms of system capacity, UE experience, and power consumption.

From add-on power saving to power saving by design.

Minimizing power consumption for both network nodes and terminal devices should be a paramount design target of 6G air interface. Unlike the 5G power saving mechanism where power saving is an add-on feature or optional mode, power saving in 6G will be a built-in feature and default operation mode. With intelligent power utilization management, an on-demand power consumption strategy, and the help of other new enabling technologies (such as the sensing/positioning-assisted channel sounding scheme), we anticipate that the 6G network and terminals will feature significantly improved power utilization efficiency.

From connectivity only to integrated connectivity and sensing.

Sensing not only provides new functionalities and therefore new business opportunities, but also assists communications. For example, a communications network can serve as radar network with high resolution and wide coverage. A communications network can also be viewed as a sensing network that could provide high resolution and wide coverage and generate useful information (such as locations, Doppler, beam directions, and images) for assisting communications. In addition, the sensing-based imaging capability of terminal devices offers new device functions.

A new design requirement for 6G involves building a single network with both sensing and communication functions, which are to be integrated under the same air interface design framework. We hope that a carefully designed communication and sensing network will offer full sensing capabilities, while also meeting all communication KPIs more effectively.

From passive on-demand beam management to proactive UE-centric beam operations.

Beam-based transmission is important for high frequencies, such as mmWave. With highly directional antennas, major efforts are needed in order to generate and maintain the precise alignment of the transmitter and receiver beams. 5G beam management aims to adapt beam direction to the change of the transmission environment and UE mobility. However, it faces issues relating to extra training overhead, access latency, and high power consumption, especially for higher spectrum.

6G expects more challenging beam management due to the exploration of higher-frequency ranges. Fortunately, with the help of new technologies such as sensing, advanced positioning, and AI, we can replace the conventional beam sweeping, beam failure detection, and beam recovery mechanisms with proactive UE-centric beam generation, tracking, and adjustment schemes. In addition, "handover-free" mobility can be realized at the physical layer, at the very least. These new intelligent UE-centric beamforming and beam management technologies will maximize UE experience and overall system performance.

Moreover, the emerging RISs and new types of mobile antennas, such as those equipped with unmanned aerial vehicles (UAVs), make it possible for us to shift from passively dealing with channel conditions to actively controlling them. With channel-aware antenna array deployment assisted by RISs and moving distributed antennas, the radio transmission environment can be changed to create the desired transmission channel condition, thereby achieving optimal performance.

From tracking channel change to predicting channel change.

Accurate channel information is imperative to achieving highly reliable wireless communications. Currently, channel acquisition is based on reference-signal (RS)-assisted channel sounding. As such, it is difficult to obtain real-time channel information owing to the measurement and report delay as well as the concern about channel measurement overhead. It is also worth noting that channel aging deteriorates performance, especially for high-speed mobile UEs.

Sensing and positioning-assisted channel sounding powered by AI can transform RS-based channel acquisition to environment-aware channel acquisition. With the information obtained from sensing/localization, we can simplify the beam search process dramatically. Proactive channel tracking and predication can provide real-time channel information and solve channel aging. In addition, the new channel acquisition technology minimizes both channel acquisition overhead and power consumption for network and terminal devices.

From terrestrial plus satellite systems to integrated terrestrial and non-terrestrial systems.

Satellite systems have been introduced into recent 5G releases as extensions of terrestrial communication systems. It is expected that the integrated terrestrial and non-terrestrial systems will achieve full-earth coverage and on-demand capacity in 6G. With 6G, tightly integrated terrestrial and non-terrestrial systems, satellite constellations, UAVs, HAPSs, drones, etc. will be viewed as new types of moving network nodes, which require new design consideration. Combining the designs of the terrestrial and non-terrestrial systems will allow more efficient multi-connection joint operations, more flexible functionality sharing, and faster cross-connection switching. These new mechanisms will go a long way in helping 6G achieve global coverage and seamless global mobility with low power consumption.

From multi-carrier operations to super-flexible spectrum utilization.

5G supports sub-6G and mmWave carrier aggregation, and also allows cross-operation of TDD and FDD carriers. Intelligent spectrum utilization and channel resource management are important 6G design aspects. More higher-frequency spectra with wider bandwidth (for example, the high end of mmWave frequency bands up to THz) will be explored to support the unprecedented data rates required by 6G. However, higher frequencies suffer from severe path loss and atmospheric absorption. In light of this, when designing 6G air interface, we must consider how to effectively utilize these new spectra jointly with other lower-frequency bands. Moreover, more mature full duplex in the 6G era is being eagerly anticipated, even though it has already been promoted in 5G. We should consider developing a simplified mechanism to allow fast cross-carrier switching and flexible bidirectional spectrum resource assignment in 6G. Also, a unified frame structure definition and signaling for FDD, TDD, and full duplex is expected to simplify system operations and support the co-existence of UEs with different duplex capabilities.

From analog- and RF-transparent to analog- and RF-aware systems.

Baseband signal processing and algorithms are usually designed without carefully considering the characteristics of the analog and RF components, owing to the difficulty in modeling the impairments and non-linearity of the analogy and RF components. This is acceptable with lower frequencies, especially with linearization effects such as the digital pre-distortion of power amplifier. In 6G, the baseband physical layer design is expected to account for RF impairments or restrictions, especially with higher-frequency spectra such as THz. With the native AI capability, joint RF and baseband design and optimization may also be possible.

19 Intelligent Air Interface Framework

19.1 Background and Motivations

New radio (NR) has made a significant progress towards developing new radio access technologies to support different usage scenarios. NR air interfaces have shown significant improvement in terms of performance, flexibility, scalability, and efficiency over the 4G long-term evolution (LTE) network.

New radio soft air interface design provides a unified framework to support below 6 GHz and beyond 6 GHz frequency bands, such as mmWave, for both licensed and unlicensed access. The configurable unified air interface also supports both Uu link, between the RAN and users, and sidelinks, between devices. The flexibility provided by the scalable numerology allows transmission parameter optimization for different spectrum bands and for different services.

The unified air interface in NR is also self-contained in the frequency domain. The frequency domain's self-contained design can support more flexible RAN slicing through channel resource sharing between difference services in both frequency and time. This domain's self-contained signal design principle also provides forward compatibility, enabling seamless compatibility with future NR releases. Moreover, time-domain self-contained design allows fast feedback for low latency services.

From the discussions of technology trends in Chapter 1 and of use cases and KPIs in Part II of the book, it is to be expected that the 6G radio network will be much more complex than the 5G network in terms of functionalities. The new air interface design could be extremely challenging as the new applications, requirements, and KPIs are tackled at the same time as attempting to minimize cost. Because of this, a revolutionary 6G air interface is needed. The 6G air interface framework should be more intelligent and greener compared with NR. It must be able to meet all 6G requirements in the most efficient way in terms of deployment efficiency, cost, power consumption, and complexity. To achieve these goals, the 6G air interface framework may require a fundamental and effective design from inception. In order to create such a design, all the relevant enabling air interface technology components must be considered, including AI, new spectrum, non-terrestrial and sensing communications.

The objective of this chapter is to provide an overview of intelligent air interface frameworks. Section 19.2 contains a comprehensive survey of the current state of AI and ML technologies in wireless communication. Section 19.2.1 will then discuss the new design expectation and potential research direction of air interfaces.

19.2   Overview of Existing Technologies

The need to optimize OPEX and CAPEX both for network operators and for device vendors has become urgent. From the perspective of network operators, the most vital factors for optimizing OPEX and CAPEX are to utilize fragmented spectrum in the most efficient way and to optimize energy efficiency for a sustainable network. The latter has been a critical success factor for network operation. Moreover, device battery life is an important aspect of user experience, which will influence the adoption of diverse services. It is thus critical to minimize UE power consumption to ensure better user experience in 6G. To this end, we provide a brief introduction of spectrum operation and power saving mechanisms in NR air interface design in Section 19.2.1.

In addition, artificial intelligence technologies, especially machine learning, have been promising technologies for introduction into telecommunications to improve system performance and efficiency. In Sections 19.2.2 and 19.2.3, we will provide a survey of three key aspects of the state of AI/ML methods in wireless communication: AI/ML methods in the physical layer and in the media access layer, and AI algorithms and learning architectures.

In relation to PHY, research is ongoing on how to utilize AI/ML to optimize component design and improve the algorithm performance, such an AI/ML for channel coding and MIMO. For MAC, research has shown that engagement is needed with the problem of how to utilize AI/ML capability with learning and prediction to solve complex optimization problems with improved strategies and an optimal solution. Several examples are discussed in Section 19.2.3, such as the use of DRL to optimize the functionality in MAC in aspects like scheduling and power control.

19.2.1   Spectrum Utilization and Energy Efficiency in NR

Generally, there are flexible mechanisms for joint spectrum operation in NR, which can enable multiple available spectrum resources in an efficient way for an operator. To utilize these spectrum resources effectively, carrier aggregation (CA) and dual connectivity (DC) are used in NR to increase the bandwidth for UEs.

In CA, multiple component carriers (CCs) are assigned to the same UE. In DC, a UE can simultaneously transmit and receive data on multiple CCs from two cell groups via Master and Secondary base stations.

Flexible spectrum aggregation is supported in NR CA, including aggregating sub-6 GHz and above 6 GHz, as well as aggregating FDD and TDD. In NR DC, multiple network architectures are supported to enable tighter interworking between Master base station and Secondary base station, including LTE-NR DC and NR-NR DC. In LTE-NR DC, a UE is firstly connected to the LTE radio and core networks, then connected to the NR radio by radio resource control (RRC) reconfiguration.

Energy efficiency is a KPI on both the UE side and the network side in wireless networks. Flexible and scalable system design in 5G NR enables different standardized power saving techniques to adapt to various traffic loads and traffic types in the networks. Specifically, the NR standard supports flexible reference signal design (avoiding always-on reference signal) and flexible muting of resources, and ensures forward compatibility for energy-efficient network implementation, the discontinuous reception (DRX) mechanism, and the inactive RRC state.

Regarding the RRC inactive state which was introduced in NR Rel-15, the motivation of introducing the RRC inactive state in addition to RRC connected and RRC idle states is to allow faster and more efficient resumption to the RRC connected state. This will allow data transmission to be done with less signaling overhead, lower latency, and lower power consumption. With this RRC inactive state, the state transition to RRC connected state in NR becomes more energy efficient.

19.2.2   AI/ML for the PHY Layer

Research into how to apply AI/ML algorithms to physical layer components has been recently conducted [1–3]. Below, some examples are briefly introduced, including examples related to AI/ML for channel modeling and estimation, channel coding, modulation, MIMO, and waveform design.

The AI/ML method can be used to extract features of wireless channel in the time, frequency, and spatial domain. For example, in [4], a neural network model is built up based on RNNs, such as LSTM and gated recurrent unit, to learn the temporal correlation property of a wireless channel. The trained model can predict channel changing, which provides even more accurate channel information than pilot-based channel estimation in deep fading scenarios. To extract more features from the wireless channel, a multiple-domain embedding method was introduced in [5], where the embedded data are fed into a transformer network for channel model pre-training. The pre-training channel model can be used in many downstream tasks, such as channel prediction, channel charting, and positioning.

Although channel coding is usually designed under sound theoretical guidelines, these are not always available. A typical example is that the performance of polar codes with successive cancellation list (SCL) decoding is hard to analyze. In [6], a RL-based framework was used to search polar codes with a given decoding list size in an SCL decoder.

Neural networks have also been employed directly to serve as decoders. The benefits of a neural network channel decoder include complexity reduction as in [7] and better compensation of non-linearity as in [8]. Joint source and channel coding (JSCC) is another application field of AI/ML methods. An auto-encoder-based JSCC framework was introduced in [9], where a better information recovery quality was obtained compared with conventional methods.

The first challenge for AI/ML methods for modulation is to recognize modulation types from unknown radio signals. In [10], CNNs are used to do the recognition among multiple digital and analog modulations. Neural-network-based demodulators have also been introduced. For example, in [11], fully connected networks are employed to approximate the logarithmic maximum a posteriori (log-MAP) used in soft demodulation. To exploit the shaping gains, the auto encoder framework is also used to design modulation schemes where both geometric and probabilistic shaping are obtained [12].

AI/ML methods have an extensive use in MIMO systems. First, channel acquisition is facilitated through AI/ML methods in several ways. For example, in [13] channels at one set of antennas and one frequency band are mapped to the channels at another set of antennas at a different frequency band through fully connected networks, which shows the possibility of directly acquiring downlink channel from uplink channel sounding even in FDD systems. In [14] and [15], auto encoder-based NNs were trained to compress channel state information (CSI), which is then reported to the transmitter with less overhead. The auto-encoder framework is also used to perform precoding design especially when imperfect CSI can be guaranteed [16], where a more robust performance can be expected. The MIMO receiver can also be designed on the basis of AI/ML methods. For example, in [17], a specially designed DNN based on the deep image prior network was used for MIMO channel estimation. Moreover, the deep unfolding method is normally used for MIMO detection works such as [18], where better performance and lower complexity can be achieved.

For waveforms, [19] proposed to use a deep complex convolutional network to convert orthogonal frequency-division multiplexing (OFDM) waveform with quadrature amplitude modulation (QAM) directly into bits under noisy and Rayleigh fading channels. Meanwhile, the aim of AI/ML-based waveform generation is mainly to focus on reducing the peak-to-average power ratio (PAPR) of OFDM signals. In [20], an auto-encoder-based framework was proposed that would map the modulated symbols onto OFDM subcarriers at the transmitter and demap them at the receiver. As a result, the PAPR of OFDM symbols can be reduced more than it can in the traditional way. In [21], to generate the OFDM signal, a set of subcarriers was chosen from the whole bandwidth by a proposed DNN order to map the designed modulation symbols, which could dramatically reduce the PAPR of the signal.

19.2.3   AI/ML for the MAC Layer

For the MAC layer, some works such as [22] rely on either supervised learning or unsupervised learning. For supervised learning, optimization problems are first formulated to provide the data label. For unsupervised learning, optimization problems are formulated to provide the optimization objective. However, decision-making problems in the MAC layer are quite sensitive to channel environment changes, such as channel and localization. Deep reinforcement learning (DRL) makes decisions and adjusts decision-making policy dynamically, which makes it useful for the MAC layer. A compressed survey of DRL in communications and networking is provided in [23]. Here we summarize below some typical cases.

The traditional adaptive modulation and coding (AMC) schemes are mostly reactive ones, which adjust the modulation and coding scheme (MCS) on the basis of feedback from the receivers. In [24], a DRL agent was used to decide MCS settings by learning from the experience of and interactions with other agents, which resulted in a better decision made proactively by the agent.

Resource allocation is another important functionality in the MAC layer. The resource to be allocated can be access opportunity, transmission opportunity, power, or spectrum, amongst others. In [25], an access-opportunity DRL agent helped devices to choose their own access patterns, that is to say, which time slot to turn on or off, and also to determine the probability of each chosen pattern. The simulations showed that a DRL-based method outperforms fixed Tx/Rx mode switch protocols.

A DRL agent was used as a MAC scheduler in a base station to decide the allocation of transmission opportunities in [26] and [27], where the optimal performance was achieved for the full-buffer traffic scenario and there was a gain of up to 30% over a proportional fairness scheme for the non-full-buffer traffic scenario.

In [28], a fingerprint-based deep Q-network method was proposed to solve the spectrum sharing problem among vehicle-to-vehicle (V2V) and vehicle-to-infrastructure (V2I) links. The DRL agent was trained centrally and distributed to each V2V link to make decisions locally. The system throughput is improved without complicated centralized control. Similar spectrum sharing problems were considered in millimeter wave ultra-dense networks [29] and LTE-based heterogeneous networks [30], where distributed Q-learning algorithm and heterogeneous multi-objective distributed strategies based on RL respectively are proposed.

The power control and interference coordination problem was considered in [31] for an ultra-dense small cell scenario. Both RL- and DRL-based methods were proposed, where the aim was to maximize the throughput of the target cell while minimizing the transmit power of this cell to limit interference with neighboring cells. Power control for multi-user cellular networks was considered in [32], where both inter- and intra-cell interference were taken into consideration. DRL methods, such as deep Q-network and deep deterministic policy gradient, were used to solve the problem.

19.3 New Design Expectations and Potential Research Directions

6G air interface designs should provide a brand new framework to support all emerging 6G radio access technologies in an efficient way. The following are the main desired features for such a design:

- More intelligent and greener with native AI and power saving capability;
- Allow more flexible spectrum utilization up to THz;
- Support efficient integration of communications and sensing;
- Facilitate tighter integration of terrestrial and non-terrestrial communications;
- Provide a simpler protocol and signaling mechanism with low overhead and complexity.

In this section, we will discuss the following: our expectations for the 6G intelligent air interface framework, which have a native personalized design for each individual user application and service with agile and lower overheads. This personalized design consists of AI-enabled intelligent PHY, intelligent MAC controller and intelligent protocol/signaling, followed by the potential new E2E AI-based link design.

19.3.1   AI-Enabled Personalized Air Interface

Intelligent protocol and signaling mechanisms are an essential part of an AI-enabled personalized air interface which will natively support intelligent PHY/MAC. This will be the main distinction between NR flexible air interface and 6G intelligent air interface.

- **Intelligent PHY:** AI/ML techniques have the capability to handle massive sampling data, solve non-linear mapping problems, and design self-evolvable transmission mechanisms. It is possible that AI/ML can provide a one-fits-many type of optimization module for different PHY functions to make PHY more adaptive and flexible. Meanwhile AI/ML can further exploit potential gain to improve the wireless link performance. In this context, an intelligent mechanism is needed to reform the PHY modules by fully facilitating this capability and supporting the fast and massive data processing-related use cases like sensing, localization, and extremely immersive experience.
- **Intelligent MAC controller:** AI/ML techniques have the capability to provide predictions and strategy. From the wireless data that have been collected, much can be learned. Based on this, we can imagine a situation in which intelligent MAC controllers are adopted into the system and the online adjustment of air interface algorithms and parameters is facilitated. By a cumulative learning method, this intelligent controller would autonomously become smart enough to memorize experiences and make decisions properly. Through joint configurations conducted by the MAC controller of parameters across different modules of the transceiver and cooperation among many network entities, enormous performance gains can be expected.
- **Intelligent protocol and signaling:** New protocol and signaling are needed for enabling intelligent PHY and intelligent MAC as well as integrated intelligent PHY and MAC. It is expected that 6G will design an intelligent mechanism to ensure the efficient operation of intelligent PHY and MAC.

In the following sections, we will provide some use cases in which AI can be used in PHY layer design. We will also provide more detailed discussion of some of the problems in the MAC layer that can benefit from AI/ML technologies.

AI-Enabled Intelligent PHY

Recent breakthroughs in AI/ML have encouraged researchers in the field of wireless communications to leverage AI/ML methods to design the physical layer modules for the next-generation wireless communication systems. Some AI/ML-based design can often adapt to the non-linear factors in some PHY modules; this is the case for example in channel coding, modulation, and waveform design, which is traditionally based on mathematical models. Moreover, it is expected that numerous channel and environment data collected by sensing in the 6G era can be utilized to facilitate AI/ML-based PHY layer designs, such as CSI for MIMO system. Some examples of the major PHY modules may benefit from using the AI/ML method, as follows.

- **Channel coding and decoding:** Channel coding is used for reliable data transmission over noisy channels. A good channel code can approximately approach the Shannon limit. Although Shannon information theory provides a goal or an evaluation criterion for a channel code, it does not directly provide the exact coding design. The channel coding design within the information theory framework is mainly based on the assumption of an AWGN channel. However, in practice, most of the channels are fading channels. In this case, the design of channel coding is lacking even theoretical guidelines and AI/ML can assist in this regard. The decoding is also difficult because it usually involves high computational complexity. Simplifying assumptions sometimes must be made to decode codes with affordable complexity, which sacrifices performance in exchange. In this context, AI/ML can also be used in channel decoders, where the decoding process can be modeled as a classification task.

- **Modulation and demodulation:** The main goal of a modulation module is mapping multiple bits into a transmitted symbol to achieve higher spectral efficiency given limited bandwidth. Classical modulation schemes such as M-QAM are widely adopted in wireless communication systems. Such square-shaped constellations guarantee low complexity for demodulation at the receiver. It is found that there exist some other constellation designs with additional geometric features, such as non-euclidean distance, and probabilistic shaping gains. Hence, AI/ML methods can be adopted to exploit the shaping gains and design suitable constellations for specific application scenarios.

- **MIMO and receiver:** MIMO is a very attractive technology which improves wireless communication in terms of robustness or throughput. With the growing numbers of antenna elements, MIMO systems have more gains to reap and consequently have become more complicated. It would thus be very helpful to employ AI-driven techniques to properly design the MIMO-related modules – such techniques as CSI feedback scheme, antenna selection, precoding, and channel estimation and detection. Most AI/ML algorithms can be deployed in an offline-training/online-inference way, which may fix the issue of the large training overhead caused by AI/ML methods.

- **Waveform and multiple access:** Waveform generation is responsible for mapping information symbols into signals suitable for electromagnetic propagation. A conventional module can be replaced by deep learning unit for waveform generation. For example, without using an explicit DFT module, learning-based methods have the capability of designing advanced waveforms. To take this a step further, it may be possible to directly design a new waveform to replace completely the standard OFDM by setting some particular requirements, for example, a PAPR constraint or a low level of out-of-band emission. This will support asynchronous transmission and so avoid the large overhead of synchronization signaling caused by massive terminals. It will also entail good localization property in the time domain to provide low-latency services and to support small-packet transmission efficiently.

Revolutionary changes can be foreseen in this AI-empowered system. Potential research directions in the future are as follows.

- **PHY element parameter optimization and update:** Optimized parameters (such as coding, modulation, MIMO parameters) for each module have great impact on the performance of communication systems. Optimized parameters may dynamically change owing to the fast time-varying channel characteristics of the physical layer in the real environment. By AI/ML methods, optimized parameters can be obtained by neural networks with much lower complexity than by traditional schemes. In addition, traditional parameter optimization is per building block, such as in the bit-interleaved coded modulation (BICM) model [33], while joint optimization of multiple blocks may provide additional performance gains from an AI neural network, for example, joint source and channel optimization. Furthermore, to adapt to fast time-varying channel status, self-learning of optimized parameters by AI/ML could be utilized to further improve performance.
- **Channel acquisition:** As a distinguishing property of wireless communications, acquiring information on wireless channel and transmission environment has always been the fundamental aspect of system design. Historic channel data and sensing data are stored as datasets, based on which a radio environment map can be drawn using AI/ML methods. From the radio map, channel information can be obtained not only through common measurement, but also by inference with other information, such as location.
- **Beamforming and tracking:** As the carrier frequency goes up to millimeter wave or even THz range, beam-centric design, such as beam-based transmission, beam alignment, and beam tracking, has been extensively applied in wireless communication. In this context, efficient beamforming and tracking algorithms become important. Relying on prediction capability, AI/ML methods can be used to optimize the antenna selection, beamforming, and precoding procedures jointly.
- **Sensing and positioning:** Data with high quality are fundamental to AI/ML technologies. In 6G systems, it is expected that both measured channel data and sensing and positioning data can be obtained thanks to the large bandwidth, new spectrum, dense network, and more LOS links. From the data, a radio environmental map can be drawn using AI/ML methods, where channel information is linked to its corresponding positioning or environmental information; therefore, the physical layer design can be enhanced.

AI-Enabled Intelligent MAC Controller

The controller in the MAC layer plays an essential role for the smooth operation of a radio access network. It makes many key decisions during the lifecycle of the communication system, such as TRP layout, beaming and beam management, spectrum utilization, channel resource allocation, MCS adaption, hybrid automatic repeat request (HARQ) management, Tx/Rx mode adaption, power control, and interference management. Wireless communication environments are highly dynamic due to the varying channel conditions, traffic conditions, loading, interference, and so on.

In general, the system performance is improved if transmission parameters can adapt to the fast-changing environment. However, traditional methods mainly rely on optimization theory to solve the above-mentioned problems, which are usually NP-hard and too complicated to implement. In this context, AI/ML could be a powerful tool to build an intelligent controller for air transmission optimization in the MAC layer.

When we consider the intelligent MAC controller design natively using AI/ML two problems should be carefully addressed:

- **Single agent or multi-agent:** In contrast with the DRL models used in other fields, where one agent may be enough for most applications, a multi-agent DRL framework is needed for wireless communication systems. Even though each controller for a base station can make individual decisions, the system always performs better with the joint decision of multiple base stations. Fortunately, nearly all DRL algorithms have their multi-agent counterparts. However, the training for a multi-agent DRL is much more difficult than for a single-agent one. The choice of single agent or multiple agents is fundamentally a trade-off between performance and training complexity.
- **Joint optimization or individual optimization:** Being limited by computational power, traditional algorithms are usually adopted in a small scope of the whole system. For example, many decisions are made within a cell in cellular networks, which leads to only local optima. With the help of multiple agents, joint optimization can be achieved in a large scope. The interaction among agents is an efficient way of building an intelligent MAC controller.

Following the discussion about intelligent controller at the beginning of this chapter, we believe that intelligent functionalities of the MAC layer could be key components for assembling an intelligent controller in the future. Moreover, joint optimization among these components is expected to provide even better performance. These key components are elaborated as follows.

- **Intelligent transmit/receive point (TRP) management:** Single TRP and multi-TRP joint transmission, for example, macro cells, small cells, pico-cells, femto-cells, remote radio heads, relay nodes, and so on, have been investigated in 5G. It has always been a challenge to design an efficient TRP management scheme while considering trade-offs between performance and complexity. The typical problems, including TRP selection, TRP turning on/off, power control, and resource allocation, are generally difficult to solve. This is especially the case with a large-scale network. Instead of using a complicated mathematical optimization method, AI/ML provides a better solution, which has less complexity and which can adapt to network conditions. For example, the policy network in DRL/multi-agent DRL can be designed and deployed to support intelligent TRP management for the integration of terrestrial and non-terrestrial networks.
- **Intelligent beam management:** Multiple antennas (or a phase shift antenna array) can dynamically form one or more beams, on the basis of the channel conditions, for directional transmissions to one or more users. The receiver also needs

to accurately tune the receiver antenna panel to the direction of the arrival beam. AI/ML can be used to learn environment changes and perform beam steering more accurately within a very short period of time. Rules can be generated to guide the operation of phase shifts of radio frequency devices, i.e., the antenna elements, which then can work in a smarter way by learning different policies under different situations.

- **Intelligent MCS:** Adaptive modulation and coding (AMC) is an important mechanism helping a system to fit with the dynamics of a wireless channel. Traditional AMC algorithms rely on the feedback from the receiver to make a decision reactively. Fast-varying channels, together with scheduling delays, often render the feedback out-of-date. To address this issue, AI/ML can be employed to decide the MCS settings. Through learning by experience and interaction with other agents, a smart agent is more likely to make a better decision and to make it proactively.

- **Intelligent HARQ strategy:** Besides the combining algorithms for multiple redundancy versions in the physical layer, the operation of the HARQ procedure also has impacts on performance, such as on the finite transmission opportunities and on the resources required to be allocated between new transmissions and retransmissions. To achieve a global optimization, it is necessary to consider the problem from a cross-layer point of view, where AI/ML is suitable for involvement owing to the large amount of information available from various sources.

- **Intelligent Tx/Rx mode adaption:** In a network with multiple communicating participants, coordination among them is key to efficiency. In real systems, both the system conditions, such as the wireless channel and buffer status, and the behavior of other players, are highly dynamic. Being so, they are extremely difficult, if not impossible, to predict with traditional methods. Although efforts have been made to deal with this issue, the solutions which currently exist may not be sufficient because of the increasingly huge size of problems, due to for example the larger number of players in the future. In this context, AI/ML can help by learning and prediction to provide more accurate reduction in the Tx/Rx mode adaption overhead and to improve the overall system performance.

- **Intelligent interference management:** Managing interference has been a key task for cellular networks. Interference changes dynamically and, without real-time communication, it is difficult to measure it accurately. AI/ML can be expected to learn the interference situation on the base station and UEs individually and jointly. The global optimal strategy can then be configured automatically by AI/ML in order to bring interference under control, hence achieving the greatest spectrum and power efficiency.

- **Intelligent channel resource allocation:** The scheduler for channel resource allocation can be viewed as the "brain" of a cellular network because it decides the allocation of transmission opportunities, determining system performance. Besides transmission opportunities, other radio resources such as spectrum, antenna port, and spreading codes can also be managed by smart agents together with intelligent TRP management. The coordination of radio resources among multiple base stations can be improved for higher global performance.

- **Intelligent power control:** The attenuation of radio signals and broadcasting characteristics of wireless channels mean that it is necessary to control power in wireless communications. Coverage should be guaranteed so that cell-edge users still can receive their information. At the same time, interference to other users should be kept as low as possible. Power control and interference coordination are therefore usually jointly optimized. Instead of solving a complicated optimization problem which must be repeated when the environment changes, AI/ML is expected to provide an alternative solution.

AI-Enabled Intelligent Protocol and Signaling

Intelligent PHY and intelligent MAC are desirable to support tailored air interface frameworks and so accommodate diverse services and devices. In order to support intelligent PHY and intelligent MAC natively, a new protocol and signaling mechanism is needed. This will allow the corresponding air interface to be personalized with customized parameters in order to meet particular requirements while minimizing signaling overheads and maximizing the whole system spectrum efficiency by personalized artificial intelligence technologies. Some examples of intelligent protocol and signaling are provided as follows:

- **Super-flexible frame structure and agile signaling:** For example, a super-flexible frame structure in a personalized air interface framework could be designed with more flexible waveform parameters and transmission duration. These can be tailored to adapt diverse requirements from a wide range of scenarios, such as for 0.1 ms extremely low latency. As a result, there are many options for each parameter in the system. The control signaling framework should aim to be a simplified and agile mechanism, requiring only few control signaling formats, while the control information can have a flexible size. It can also be detected with simplified procedures, minimized overheads, and UE capability. It can also be forward compatible, with no need to introduce a new format for future 6G developments.
- **Intelligent spectrum utilization:** As demonstrated in Chapter 12 the potential spectrum for 6G can be low-band, mid-band, mmWave bands, THz bands, and even the visible-light band. The spectrum range for 6G is thus much wider than that for 5G, and designing a high-efficiency system to support such a wide spectrum range is challenging.

  In the 4G and 5G networks, both CA and DC schemes are adopted to utilize jointly multiple pieces of wide spectrum. There are multiple DC schemes adopted in 5G to provide flexible usage of spectrum. With more combinations of frequency carriers for 6G, a new intelligent air interface with simplified and efficient operation is desired to support the whole range of spectrum operations.

  The current spectrum assignment and frame structure are usually associated with the duplex mode, either FDD or TDD, which may place restrictions on the efficient usage of spectrum. It is expected that full duplexing may get mature in the 6G era and that emerging wireless networks may contain more and more nodes (end user and access points) with different frequency bands.

Examples of communication that is not limited to the uplink and downlink directions are D2D communication, IAB communication, non-terrestrial communication, and so on. As a result, the intelligent air interface framework should provide a duplexing agnostic technology with adequate configurability to accommodate different communication nodes and communication types. In this case, a single frame structure can be designed to support all duplex modes and communication nodes, and resource allocation schemes in the intelligent air interface will be able to perform effective transmissions in multiple air-links.

- **Native intelligent power saving:**   For the design of a green network in 6G, power saving is a clear and essential requirement. A built-in power saving perspective should be considered fundamentally from conception, to minimize the power consumption both for devices and network nodes.

  The 6G air interface supports intelligent MIMO and beam management, intelligent spectrum utilization, and accurate positioning. These can dramatically reduce the power consumption both of devices and network nodes compared with traditional technologies, especially for data. The 6G air interface is thus a framework which provides greater power saving capability.

  It is expected that the data transmission duration can be significantly shortened by intelligent technologies in 6G. As a result, a device may be able to stay longer in an operating mode when it is not actively accessing or interacting with the network. This would make it feasible for operating a system with native power saving, which is especially important for energy-efficient devices and environmentally friendly networks.

  Effective transmission channels can be designed in such a way that control signaling can be optimized and the number of state transitions or power mode changes can be minimized in order to achieve maximal power saving for devices and network nodes. Moreover, as the 6G network has to support super-low-latency applications, such as enhanced URLLC (or URLLC plus) upon traffic arrival, these schemes or mechanisms in support of native power saving are also expected to provide flexible functionalities. They may provide ultra-fast access to networks and super-high data transmissions; an example is an optimized RRC state design with smart power mode management and operation.

  As the air interface is personalized for each device, different types of devices will have different requirements for power consumption. It will then be straightforward for power saving solutions to be personalized for different types of devices while meeting requirements for communication.

19.3.2   E2E AI-Based Link Design and Open Problems

Most of the recent attempts to introduce AI/ML to radio networks have tended to replace one or more conventional PHY components in an add-on manner. In the 6G era, it is envisioned that the basic network framework will be closely coupled with

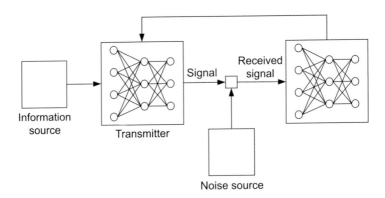

Figure 19.1 E2E intelligent communication link.

AI technology. This will distinguish it from both 5G and from previous wireless communication systems in which AI/ML techniques are not considered natively. As a result, a radio link procedure and AI technology will be jointly designed to realize extremely efficient information processing, transmission environment sensing, and intelligent air interface capabilities.

Figure 19.1 shows a totally different path for applying AI/ML, that is, E2E design. In this case, the transmission chain and modules may be reorganized. Some functionalities discussed previously may disappear. During the transmission and reception of information, the transmitter will intelligently match the real-time changes of the information source through the neural networks. The receiver will also extract effective information from the received data according to specific tasks. This is no longer a one-size-fits-all data processing approach based on a unique measure (delivering and restoring messages without errors) as described in Shannon's communication framework [34].

There are many theories used to understand artificial intelligence. Among them, the information bottleneck theory [35] is the most suitable for communication problems in the physical layer, because it uses the idea of mutual information, from information theory, as a metric of optimization. Based on the information bottleneck theory, an auto-encoder (AE) architecture is one of the key tools for end-to-end AI-based link design. More details about the information bottleneck theory and AE architectures can also be found in Part III of this book.

The above-mentioned intelligent communication link is able to introduce new possibilities and has the potential to become a basic communication framework in the future. However, AI/ML methods are data-hungry. In order to involve AI/ML in wireless communications, more and more data will need to be collected, stored, and exchanged. Unlike data with simple features, such as those in the field of computer vision or natural language processing, the characteristics of wireless data expand large ranges in multiple dimensions, for example, from sub-6-GHz millimeter to THz carrier frequency, from space to outdoor to indoor scenarios, and from text to voice to video. These data are so different that it will be hard to cover their collection, processing, and usage in a unified framework.

References

[1] Y. Sun, M. Peng, Y. Zhou, Y. Huang, and S. Mao, "Application of machine learning in wireless networks: Key techniques and open issues," *IEEE Communications Surveys & Tutorials*, vol. 21, no. 4, pp. 3072–3108, 2019.

[2] Q. Mao, F. Hu, and Q. Hao, "Deep learning for intelligent wireless networks: A comprehensive survey," *IEEE Communications Surveys & Tutorials*, vol. 20, no. 4, pp. 2595–2621, 2018.

[3] C. Zhang, P. Patras, and H. Haddadi, "Deep learning in mobile and wireless networking: A survey," *IEEE Communications Surveys & Tutorials*, vol. 21, no. 3, pp. 2224–2287, 2019.

[4] Y. Huangfu, J. Wang, R. Li, C. Xu, X. Wang, H. Zhang, and J. Wang, "Predicting the mumble of wireless channels with sequence-to-sequence models," in *Proc. 2019 IEEE 30th Annual International Symposium on Personal, Indoor, and Mobile Radio Communications (PIMRC)*. IEEE, 2019, pp. 1–7.

[5] Y. Huangfu, J. Wang, C. Xu, R. Li, Y. Ge, X. Wang, H. Zhang, and J. Wang, "Realistic channel models pre-training," in *Proc. 2019 IEEE Globecom Workshops*. IEEE, 2019, pp. 1–6.

[6] L. Huang, H. Zhang, R. Li, Y. Ge, and J. Wang, "AI coding: Learning to construct error correction codes," *IEEE Transactions on Communications*, vol. 68, no. 1, pp. 26–39, 2019.

[7] T. Gruber, S. Cammerer, J. Hoydis, and S. ten Brink, "On deep learning-based channel decoding," in *Proc. 2017 51st Annual Conference on Information Sciences and Systems (CISS)*. IEEE, 2017, pp. 1–6.

[8] Y. He, J. Zhang, C.-K. Wen, and S. Jin, "TurboNet: A model-driven DNN decoder based on max-log-MAP algorithm for turbo code," in *Proc. 2019 IEEE VTS Asia-Pacific Wireless Communications Symposium (APWCS)*. IEEE, 2019, pp. 1–5.

[9] N. Farsad, M. Rao, and A. Goldsmith, "Deep learning for joint source-channel coding of text," in *Proc. 2018 IEEE International Conference on Acoustics, Speech, and Signal Processing (ICASSP)*. IEEE, 2018, pp. 2326–2330.

[10] T. J. OShea, J. Corgan, and T. C. Clancy, "Convolutional radio modulation recognition networks," in *Proc. International Conference on Engineering Applications of Neural Networks*. Springer, 2016, pp. 213–226.

[11] O. Shental and J. Hoydis, "Machine learning: Learning to softly demodulate," in *Proc. 2019 IEEE Globecom Workshops*. IEEE, 2019, pp. 1–7.

[12] M. Stark, F. A. Aoudia, and J. Hoydis, "Joint learning of geometric and probabilistic constellation shaping," in *Proc. 2019 IEEE Globecom Workshops*. IEEE, 2019, pp. 1–6.

[13] M. Alrabeiah and A. Alkhateeb, "Deep learning for TDD and FDD massive MIMO: Mapping channels in space and frequency," in *Proc. 2019 53rd Asilomar Conference on Signals, Systems, and Computers*. IEEE, 2019, pp. 1465–1470.

[14] C.-K. Wen, W.-T. Shih, and S. Jin, "Deep learning for massive MIMO CSI feedback," *IEEE Wireless Communications Letters*, vol. 7, no. 5, pp. 748–751, 2018.

[15] T. Wang, C.-K. Wen, S. Jin, and G. Y. Li, "Deep learning-based CSI feedback approach for time-varying massive MIMO channels," *IEEE Wireless Communications Letters*, vol. 8, no. 2, pp. 416–419, 2018.

[16] F. Sohrabi, H. V. Cheng, and W. Yu, "Robust symbol-level precoding via autoencoder-based deep learning," in *Proc. 2020 IEEE International Conference on Acoustics, Speech and Signal Processing (ICASSP)*. IEEE, 2020, pp. 8951–8955.

[17] E. Balevi, A. Doshi, and J. G. Andrews, "Massive MIMO channel estimation with an untrained deep neural network," *IEEE Transactions on Wireless Communications*, vol. 19, no. 3, pp. 2079–2090, 2020.

[18] H. He, C.-K. Wen, S. Jin, and G. Y. Li, "A model-driven deep learning network for MIMO detection," in *Proc. 2018 IEEE Global Conference on Signal and Information Processing (GlobalSIP)*. IEEE, 2018, pp. 584–588.

[19] Z. Zhao, M. C. Vuran, F. Guo, and S. Scott, "Deep-waveform: A learned OFDM receiver based on deep complex convolutional networks," *arXiv preprint arXiv:1810.07181*, 2018.

[20] M. Kim, W. Lee, and D.-H. Cho, "A novel PAPR reduction scheme for OFDM system based on deep learning," *IEEE Communications Letters*, vol. 22, no. 3, pp. 510–513, 2017.

[21] B. Wang, Q. Si, and M. Jin, "A novel tone reservation scheme based on deep learning for PAPR reduction in OFDM systems," *IEEE Communications Letters*, vol. 24, no. 6, June 2020.

[22] W. Cui, K. Shen, and W. Yu, "Spatial deep learning for wireless scheduling," *IEEE Journal on Selected Areas in Communications*, vol. 37, no. 6, pp. 1248–1261, 2019.

[23] N. C. Luong, D. T. Hoang, S. Gong, D. Niyato, P. Wang, Y.-C. Liang, and D. I. Kim, "Applications of deep reinforcement learning in communications and networking: A survey," *IEEE Communications Surveys & Tutorials*, vol. 21, no. 4, pp. 3133–3174, 2019.

[24] M. P. Mota, D. C. Araujo, F. H. C. Neto, A. L. de Almeida, and F. R. Cavalcanti, "Adaptive modulation and coding based on reinforcement learning for 5G networks," in *Proc. 2019 IEEE Globecom Workshops*. IEEE, 2019, pp. 1–6.

[25] A. Destounis, D. Tsilimantos, M. Debbah, and G. S. Paschos, "Learn2MAC: Online learning multiple access for URLLC applications," *arXiv preprint arXiv:1904.00665*, 2019.

[26] J. Wang, C. Xu, Y. Huangfu, R. Li, Y. Ge, and J. Wang, "Deep reinforcement learning for scheduling in cellular networks," in *Proc. 2019 11th International Conference on Wireless Communications and Signal Processing (WCSP)*. IEEE, 2019, pp. 1–6.

[27] C. Xu, J. Wang, T. Yu, C. Kong, Y. Huangfu, R. Li, Y. Ge, and J. Wang, "Buffer-aware wireless scheduling based on deep reinforcement learning," in *Proc. 2020 IEEE Wireless Communications and Networking Conference (WCNC)*. IEEE, 2020, pp. 1–6.

[28] L. Liang, H. Ye, and G. Y. Li, "Spectrum sharing in vehicular networks based on multi-agent reinforcement learning," *IEEE Journal on Selected Areas in Communications*, vol. 37, no. 10, pp. 2282–2292, 2019.

[29] C. Fan, B. Li, C. Zhao, W. Guo, and Y.-C. Liang, "Learning-based spectrum sharing and spatial reuse in mm-wave ultradense networks," *IEEE Transactions on Vehicular Technology*, vol. 67, no. 6, pp. 4954–4968, 2017.

[30] G. Alnwaimi, S. Vahid, and K. Moessner, "Dynamic heterogeneous learning games for opportunistic access in LTE-based macro/femtocell deployments," *IEEE Transactions on Wireless Communications*, vol. 14, no. 4, pp. 2294–2308, 2014.

[31] L. Xiao, H. Zhang, Y. Xiao, X. Wan, S. Liu, L.-C. Wang, and H. V. Poor, "Reinforcement learning-based downlink interference control for ultra-dense small cells," *IEEE Transactions on Wireless Communications*, vol. 19, no. 1, pp. 423–434, 2019.

[32] F. Meng, P. Chen, L. Wu, and J. Cheng, "Power allocation in multi-user cellular networks: Deep reinforcement learning approaches," *IEEE Transactions on Wireless Communications*, 2020.

[33] G. Caire, G. Taricco, and E. Biglieri, "Bit-interleaved coded modulation," *IEEE Transactions on Information Theory*, vol. 44, no. 3, pp. 927–946, 1998.

[34] C. E. Shannon and W. Weaver, *The mathematical theory of communication*. University of Illinois Press, 1964.

[35] N. Tishby, F. C. Pereira, and W. Bialek, "The information bottleneck method," *arXiv preprint physics/0004057*, 2000.

20 Integrated Terrestrial and Non-Terrestrial Communication

20.1 Background and Motivations

The integration of non-terrestrial communication (NTN) into the terrestrial cellular system will play an important role in achieving truly global coverage, ensuring a high degree of availability and robustness for 6G networks, even in the presence of natural disasters. Such integration is essential to bridging the coverage gaps of underserved areas by extending the coverage of cellular networks through non-terrestrial nodes, which will be key to ensuring "anywhere, anytime" access and providing mobile broadband services to unserved or underserved regions in environments such as oceans, mountains, forests, or other remote areas, where it is difficult to deploy terrestrial access points or base stations. By exploiting the wide coverage of non-terrestrial components, services and applications can be enhanced for multicasting the same content to a large number of users, located in either urban or remote areas [1].

In addition to the prospective service enhancement benefits, the integration of terrestrial and non-terrestrial networks can enable a new range of services and applications, such as ubiquitous connection, remote sensing, passive sensing and positioning, navigation, tracking, and autonomous delivery. This calls for a single network design, whereby non-terrestrial network nodes (such as the satellite constellations with inter-satellite communications links), network nodes on board UAVs and HAPSs, and the terrestrial network nodes can be similarly treated as base stations in terms of their functionalities. User terminals can seamlessly access the terrestrial and non-terrestrial base stations. Alternatively, non-terrestrial network nodes, such as satellite constellations with inter-satellite communications links, can be treated as integrated access and backhaul (IAB) nodes with systematic deep integration, centralized coordination, communication capability, caching, or mobile edge computing (MEC).

For satellites or HAPS network nodes, the infrastructure of transmission nodes is always available, so on-demand services can be enabled by illuminating spot beams from satellites, HAPSs, or UAVs to a given region, thereby enhancing the reliability and elasticity of the overall network. Such a deployment is based on a service-aware mechanism and calls for efficient coordination of the transmission nodes from both the terrestrial and non-terrestrial networks.

For on-demand and temporary HAPSs, enhanced IAB can offer a demand-driven network densification solution. UAVs and low-altitude HAPSs, which may occasionally operate as relays using IAB technology, can provide short-term backhaul without the need for dense terrestrial infrastructure. IAB technology is also critical for enabling aerial vehicles such as airliners and cargo UAVs to operate as both aerial users and aerial base stations, thereby providing access to remote users or to users and base stations in congested parts of the terrestrial network. It is evident that avoiding congestion in both the terrestrial and non-terrestrial networks, while managing interference in the presence of IAB nodes, requires tight integration across different subsystems.

Accurate positioning is one of the key requirements for 6G networks, and this can be facilitated within the framework of the integrated terrestrial and non-terrestrial network. For example, the wide coverage and LOS features of the non-terrestrial network may result in better reception of the positioning reference signals. In such an environment, users may receive LOS signals from many base stations for better positioning performance. In addition, it is possible to measure the reflections of satellite-broadcasted signals for "passive sensing and positioning" of nearby objects.

To further enhance the link budget for accessing the non-terrestrial network, a set of enablers, such as strong on-board processing abilities, large antenna arrays for generating narrow beams, advanced air interface solutions, and on-demand deployment of dense non-terrestrial TRPs, needs to be combined to combat the high path loss and mobility in the context of satellite communications.

20.2   Overview of Existing Solutions

3GPP has studied different scenarios and enabling technologies for integrating non-terrestrial components into 5G new radio (NR) [1]. Such studies suggest different use cases for each of the 5G usage scenarios, i.e., eMBB, mMTC, and URLLC. Depending on the underlying scenario, the benefits of each prospective use case can be best achieved by integrating a different set of non-terrestrial components, such as UAVs, HAPSs, and satellites in different orbits.

Table 20.1 summarizes the main characteristics (altitude and propagation latency) and prospective use cases of various non-terrestrial network components [1].

While conventional geostationary-earth-orbit (GEO) satellites might be best utilized to broadcast public and popular content (such as media content, safety messages, and/or updates for connected cars) to a local server, they may not be of much use for latency-sensitive applications. Three GEO satellites, deployed 35,786 km above the equator and kept relatively stationary with respect to the earth's surface, are sufficient to provide global coverage (excluding the polar regions).

Low-earth-orbit (LEO) satellites may establish a better balance between wide coverage and propagation latency/path loss. Covering a large area, GEO and LEO satellites can be used to ensure service continuity for moving cells on ground, maritime, or aeronautical platforms, without the need for handover to a different cell or beam within hundreds of kilometers. They can also serve as the backhaul for fixed cells, especially those in remote areas. Though LEO satellites may not be directly accessible to ground users due to excessive path loss, it is envisioned that as the LEO satellite antenna technology continues to advance, user equipment will ultimately gain direct

Table 20.1 Characteristics and prospective use cases for different non-terrestrial components.

NTN node	Altitude	Propagation latency	Prospective use cases
GEO satellite	35,786 km	120 ms	Wide coverage (eMBB, mMTC) Media broadcast (eMBB) Public content broadcast (eMBB) Fixed/mobile cell backhaul connectivity Communication with remote/urban users
LEO satellite	400–1,600 km	1.3–5 ms	Fixed/mobile cell backhaul connectivity Communication with remote/urban users Media multicast services Wide-area IoT services
VLEO satellite	100–400 km	0.33–1.3 m	Fixed/mobile cell backhaul connectivity Communication with remote/urban users Media multicast services Wide-area IoT services Broadband Internet
HAPS	15–25 km	50–83 µs	Backhaul connectivity for aerial/ground base stations Communication with remote/urban users Media multicast service Local-area IoT service
UAV	0.1–10 km	0.33–33 µs	Backhaul connectivity for ground base stations Aerial relay/TRP Hotspots on-demand Regional emergency services

access to the 6G non-terrestrial network. Alternatively, airborne components such as UAVs or HAPSs may serve as a hub to provide non-terrestrial access for remote users who lack a satellite connection using IAB technology.

With the recent developments in space technology, it has become possible to stabilize satellites in VLEO (between 100–400 km), which is far below conventional LEO satellites. Combined with other enabling technologies that substantially lower the costs in launching satellites to lower orbits, such developments enable us to create large constellations of VLEO satellites for mobile broadband applications. The lower altitudes of these satellites would result in a better link budget, as well as shorter communication latency for ground users. In addition, lower altitudes make it easier to shrink the beam footprints, thereby enhancing frequency reuse over a fixed coverage. For example, SpaceX plans to launch a constellation of approximately 12,000 LEO and VLEO satellites in order to provide global Internet access by 2027.

Intermediary airborne components such as HAPSs are advantageous in that they provide air interface with lower path loss for users who have a limited power budget, while leveraging the HAPS LOS feature and other advantages (such as large antenna arrays) to effectively communicate with satellites or other ground stations in the back-haul. HAPSs can deliver an excellent trade-off between propagation latency/path loss and coverage for base stations in remote areas, and may also serve as the backbone (providing communication and computing capabilities) for remote base stations, or those in urban areas where the terrestrial network is overloaded during peak hours [2, 3]. In [4], the authors proposed the use of HAPSs as super macro-base stations to complement the terrestrial network coverage in urban regions, while also enabling other applications. Potential applications include IoT, intelligent transportation systems, high-stake cargo UAVs, providing on-demand service during temporary unpredictable events, and computation offloading, which are enabled by providing HAPSs with storage and computing capabilities in addition to communication infrastructure. Particularly, some researchers propose an architecture whereby a HAPS operates as an intermediary node between aerial and terrestrial nodes and satellite mega-constellation nodes in an integrated vertical heterogeneous network. They also propose the integration of reconfigurable smart surfaces into the payload of the HAPS, thereby providing power-efficient backhaul for aerial and terrestrial base stations.

Lower-altitude platforms such as UAVs could still be used to achieve dense deployment, since their coverage can be limited to a local area. The flexibility to adjust UAV's trajectory also provides an additional DoF for deployment [5]. Their ease of deployment makes them an effective solution to establish on-demand hotspots (e.g., to cover a certain event or sports game), or to develop an ad-hoc network for emergency responders.

Given the progress of on-board antennas, high-power amplifiers, large antenna arrays, and signal processing capabilities, air/space-borne non-terrestrial platforms have significantly improved, especially in terms of their ability to provide high-throughput services.

While the throughput of non-terrestrial transmission nodes can be comparable to, and in some instances even surpass, the design target of 5G base stations, the area capacity is still quite limited compared with the achievable area capacity of the terrestrial network. The link-level spectrum efficiency of the non-terrestrial network remains low due to a limited link budget.

The following technical limitations have been identified as critical for improving the service quality of the non-terrestrial network.

- **Limited spectrum efficiency:** Owing to the progress of on-board processing capabilities, co-channel interference mitigation technology can be employed in satellite communication. The currently deployed multi-color frequency reuse schemes, typically employing frequency reuse factors 3 to 4, are not suitable for achieving high overall spectrum efficiency. Advanced transmission schemes such as joint transmission involving multiple satellites can lead to improved user throughput; however, such advanced transmission schemes need further study and consideration and are not ready for practical deployment.
- **Lack of on-demand adaptive coverage:** Due to the absence of scheduling coordination among satellites and on-board antenna cost limitations, the radio coverage of non-terrestrial platforms is typically either fixed or pre-planned, irrespective of

the service requirements. As such, the benefits gained by providing adaptive on-demand services via satellite communications cannot be fully leveraged.

- **Mobility and beam management overhead:** The high-layer signaling and power consumption during mobility procedures due to the fast motion of satellites would lead to an overwhelming signaling burden, especially considering the potentially huge number of mobile users in 6G networks. In beam-based communications, the beam management overhead may become prohibitive in the presence of VLEO satellites due to their orbital movement, which makes beam sweeping and beam failure recovery procedures – as well as the tracking of UE movement – all the more challenging. In addition, such procedures are costly in terms of latency. Consequently, optimizing mobility and beam management procedures is an important research area for 6G networks.

- **Lack of sensing support:** Sensing support is not currently available in NR networks; however, sensing is expected to play a more prominent role in 6G networks, where non-terrestrial and terrestrial network-based sensing could be used together to enhance UE experience.

- **Inherent latency:** Most existing non-terrestrial platforms are not equipped with on-board base stations. The deployment of low-altitude platforms such as VLEO satellites, HAPSs, and UAVs with full on-board signal processing capabilities and stable inter-satellite communications links should provide the basis for enabling low-latency transmission.

- **Lack of tight integration between terrestrial and non-terrestrial networks:** The non-terrestrial network integration in NR R16 is an add-on feature to NR R15. The terrestrial-oriented NR design philosophy is not inherently optimized for on-board RF hardware and non-terrestrial network platform mobility. It is widely recognized that improvements in the non-terrestrial network are still needed in order to integrate non-terrestrial components into the terrestrial network, thereby unleashing a broader range of services and applications based on a fully integrated terrestrial and non-terrestrial network. However, it is clear that those services and applications may not be effectively implemented by either the terrestrial or the non-terrestrial network individually. As such, a tight integration approach for jointly operating terrestrial and non-terrestrial networks is required. Note that this type of integration is different from the integrated non-terrestrial network approach considered in 3GPP in the context of 5G NR Release 16 and later releases [1], where the integration only aims to enhance NR air interface to accommodate non-terrestrial networks as an extension to terrestrial networks rather than achieving tight integration between terrestrial and non-terrestrial networks.

20.3 New Design Expectations and Potential Research Directions

20.3.1 Integrated Multi-Layer Network

The result of integrating non-terrestrial components into the terrestrial communication system is a heterogeneous network comprising multiple layers, as illustrated in

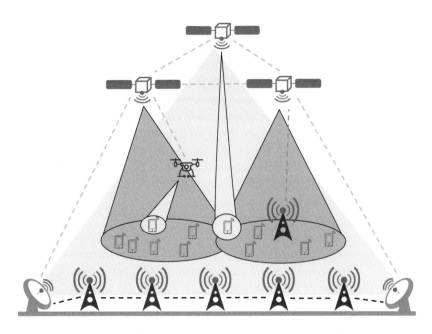

Figure 20.1 Integrated terrestrial and non-terrestrial network (gray dashed lines represent wireless backhaul connections, and black dashed lines represent fiber connections).

Figure 20.1. The main design target for integrating terrestrial networks and non-terrestrial networks is to achieve better overall performance through efficient multi-link joint operation, more flexible functionality sharing, and faster physical layer link switching between terrestrial and non-terrestrial networks.

The heterogeneity and diversity in such an environment can potentially enhance both network reliability and coverage by providing users with multiple connections across different layers of the integrated network (including satellites in different orbits, HAPSs, and other airborne or terrestrial access points). However, it is worth pointing out that each layer may not individually achieve the desired performance. For instance, while a particular UE or IoT device can access multiple terrestrial base stations, VLEO satellites, or even HAPSs, each candidate access link may not be reliable all the time. As a result, fast selection and switching to the best available access link are critical, and, in order to enhance reliability, it is necessary to opportunistically exploit different paths, which requires tight integration of different layers. In 5G NR, the terrestrial and non-terrestrial systems are designed to operate separately. However, in the 6G networks of the future, the functions and operations, as well as the resource and mobility management of both systems, are expected to be integrated. In the following sections we discuss different integration approaches and the expected improvements in each case.

Different Integration Approaches

Application-layer integration is the simplest form of loose integration. In order to allow different subsystems to co-exist with each other, each subsystem implements all layer functionalities and application-layer interface in order to let the user exchange information to and from the application server over multiple connections across different subsystems. This approach is easy to implement, but may not be efficient as there is no coordination across different subsystems [6].

Core network-based integration is another loose-integration approach that is typically adopted for interworking between 3GPP cellular networks (such as LTE) and non-3GPP networks (such as WLAN). In this approach, the radio access part of both networks operates individually, but there is some interworking across the core networks. Users are provided with some network-assisted information and configurations related to operators' policies [6]. Users then select one of the RANs for forwarding packets based on network configurations, radio link qualities, user preferences, etc. In this way, the users establish connections through the best RAN. The user decision, however, is based on local signaling or measurements, resulting in a suboptimal selection. In addition, radio access resources may not be efficiently utilized, as each part of the resources is separately managed by a different RAN.

Another alternative solution is *RAN-based tight integration*, wherein different subsystems converge at the RAN level, possibly sharing the same resources [6]. An example of RAN-based integration is the 3GPP solution to establish dual-/multi-connectivity across different radio access technologies (such as LTE and NR), which may share bandwidth from the same carrier. Depending on backhaul connections, the convergence here can range from partial coordination, such as LTE/NR multi-connectivity solutions, to fully fledged integration that corresponds to a single radio access technology (RAT) solution using a unified air interface between terrestrial and non-terrestrial networks. In partial coordination, each subsystem provides some assistance to coordinate interference across different TRPs or subsystems, whereas in the single-RAT solution, the TRPs of terrestrial and non-terrestrial networks are coordinated by the same centralized control unit (CU) in each region, and the regional CUs are interconnected through high-capacity backhaul interface. Exploiting a unified control plane, the network jointly manages the radio resources for terrestrial and non-terrestrial networks, adapting the physical layer parameters of the unified air interface on the basis of the instantaneous channel conditions, thereby efficiently utilizing resources while enhancing the reliability and QoS for different UEs.

5G terrestrial and non-terrestrial networks use essentially one design: due to the forward compatibility of 5G NR, non-terrestrial network integration is carried out after the specifications of 5G terrestrial networks are completed.

In contrast, we envision a single E2E design for 6G, as terrestrial and non-terrestrial networks will have a common design from day one. Terrestrial nodes and terrestrial networks, and non-terrestrial nodes and non-terrestrial networks, are merely different implementations. Consequently, after deployment of LEO constellations, the 6G network will be capable of full earth coverage. Terrestrial networks will provide higher data rates for end users, and non-terrestrial networks will provide basic data rates in locations where non-terrestrial networks are absent. In this 6G architecture, multiple airborne or space-borne layers can be utilized to complement the terrestrial network coverage by providing on-demand supplementary coverage in urban regions.

In particular, it may not be cost-effective to cover the sporadic (in both time and location) peak user demands through costly deployment of densified terrestrial infrastructure. Exploiting the wide coverage of non-terrestrial networks, it is possible to opportunistically provision non-terrestrial resources to parts of the networks that are in desperate need of additional coverage. Non-terrestrial resources should be jointly coordinated through a tight integration of different layers in the integrated multi-layer 6G network, wherein the resources of different transceiver points (at different layers) are orchestrated by the same control plane entity.

Challenges and Potential Research Directions

Over the past few decades, wireless networks have predominantly consisted of static terrestrial access points. However, considering the prevalence of UAVs, HAPSs, and VLEO satellites and the desire to integrate satellite communications into cellular networks, future-proof systems will no longer be horizontal and two-dimensional. The emerging 3D vertical networks consist of many moving and high-altitude access points (other than geostationary satellites), such as UAVs, HAPSs, and VLEO satellites, as illustrated in Figure 20.2.

The new challenge for 6G is to support a diverse and heterogeneous range of access points, which requires self-organization to seamlessly integrate a new UAV, or a passing low-orbit satellite, into the network without needing to reconfigure users. Owing to their relative proximity to the ground, UAVs, HAPSs, and VLEO satellites can carry out functions similar to terrestrial base stations, and can thus be seen as a new type of base station in the 6G era, albeit bringing a new set of challenges that would need to be overcome. While they can utilize similar air interface and frequency bands in terrestrial communication systems, a new approach is required for cell planning, cell acquisition, and handover among non-terrestrial access nodes or between terrestrial and non-terrestrial access nodes. Moreover, similarly to their terrestrial counterparts, non-terrestrial nodes and their clients need adaptive and dynamic wireless backhaul to

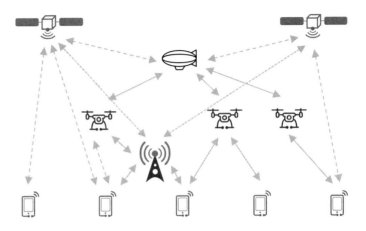

Figure 20.2 UAVs, HAPSs, and VLEO satellites as new in the integrated 6G network.

maintain their connectivity. Supporting such diverse and heterogeneous access points requires self-organization to seamlessly incorporate a new UAV into the network without the need for high overhead reconfiguration. Such solutions (those based on a virtualized air interface) should simplify cell and TRP acquisition as well as data and control routing to efficiently and seamlessly integrate those aerial nodes with the underlying terrestrial network. Consequently, the addition and deletion of aerial access points should be largely transparent to end users, beyond the physical-layer operations such as UL/DL synchronization, beamforming, measurement, and feedback associated with vertical access points.

The 6G terrestrial and non-terrestrial networks aim to share a unified PHY and MAC layer design, so that the same modem chip equipped with an integrated protocol stack can support both terrestrial and non-terrestrial communications. Although a single chipset makes sense from a cost perspective, it is quite challenging to achieve owing to the different design requirements for terrestrial and non-terrestrial networks, which may impact such factors as physical-layer signal design, waveform, and AMC. For example, satellite communication systems may have a stringent PAPR requirement. Although NR numerology has been optimized for low-latency communications, satellite communications should be able to accommodate long transmission latency. While a more detailed discussion on the potential enhancements to non-terrestrial communications will be provided in the next section, it is expected that the unified PHY/MAC design framework can be flexibly dimensioned and tailored via several parameters to accommodate different deployment scenarios, with native support for airborne or space-borne non-terrestrial communications.

20.3.2   Enhanced Non-Terrestrial Communications

Although non-terrestrial components such as UAVs, HAPSs, and VLEO satellites will be part of the 6G network infrastructure, mostly providing functionalities similar to terrestrial base stations, there is still room for improvements in terms of non-terrestrial node design to overcome the stringent link budget requirements, especially for high-altitude platforms. Indeed, based on the expected progress of on-board RF modules and processing capabilities, there will be ample room for accommodating potential breakthroughs in 6G air interface design. Two general design principles should be considered:

- Overcoming the challenges posed by non-terrestrial communications.
- Taking advantage of properties that are specific to non-terrestrial nodes.

For the limitations in the current non-terrestrial communications highlighted in Section 20.2, the corresponding solutions and potential research directions to improve the efficiency of non-terrestrial networks are discussed below.

High Spectrum Efficiency Transmission Technology
The overall spectrum efficiency of existing satellite systems is far lower than that of cellular networks, partly due to the inefficient link budget caused by long transmission

distances, and partly due to the strong interference in satellite communications. The latter impairment will persist despite the progress of on-board processing and/or user reception capabilities. In current deployment, multi-color frequency reuse is employed to mitigate co-channel interference from adjacent beams at the cost of low spectrum efficiency. Multi-beam precoding is a very mature and effective technique used in cellular networks for mitigating co-channel interference and achieving full frequency reuse in the context of satellite communications, enhancing the overall user experience in an integrated terrestrial and non-terrestrial network. Low-overhead channel feedback schemes are required to support efficient channel information acquisition. A precoding strategy which takes RF constraints and other deployment restrictions into account needs to be further investigated.

Polarization multiplexing is another promising satellite-specific technology which can enhance spectrum efficiency, given that satellites are typically equipped with circularly polarized antennas. Thanks to LOS channel properties, right-hand circular polarization (RHCP) and left-hand circular polarization (LHCP) signals can be immune to cross-polarization notwithstanding the long transmission distance. Polarization provides a new orthogonal dimension, on top of traditional dimensions such as time and frequency, and should be further investigated in non-terrestrial network system design. By taking advantage of the good isolation of different polarizations in the context of satellite communications, adjacent spot beams can employ the same frequency band with different polarizations to avoid co-channel interference. Other advanced transmission schemes such as MIMO [7] and space modulation [8] can also be applied over the two isolated polarization channels to further improve transmission efficiency.

Multi-satellite joint transmission is another promising, albeit challenging, solution to improve transmission efficiency. The actual transmission rate is expected to increase if a user can receive signals from multiple satellites simultaneously, or if multiple satellites can jointly detect uplink signals from one user, leveraging distributed MIMO techniques. The latter option will be extremely helpful in addressing the link budget bottleneck due to limited user transmission power in the uplink. Processing gains can be achieved when joint transmission and reception is implemented at the physical layer. The channel capacity of such a distributed MIMO scheme should be optimized in terms of the topology and link budget of the coordinated satellites. For LEO constellations, additional efforts are needed to address the synchronization and receiver issues to/from different satellites.

Smart On-Demand Coverage

A constellation-based network should be smart in order to provide flexible and on-demand coverage, as shown in Figure 20.3. Assuming we have ideal backhaul connections between non-terrestrial network components, the infrastructure deployed in the sky can be viewed as a resource pool, and all resources contained within it can be scheduled simultaneously to improve the overall resource efficiency. The network should be aware of user demands and dynamically coordinate available resources to meet these demands by managing the allocation of time/frequency resources and

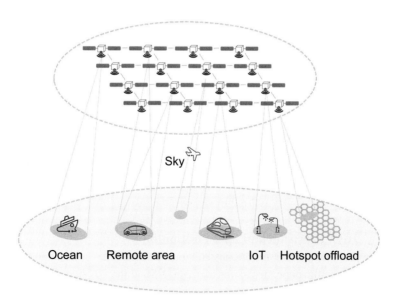

Figure 20.3 On-demand satellite coverage.

deploying highly directional spot beams. In the event of a sudden demand arising from a certain hotspot area, all the resources in the visible sky should be pooled to this area in order to boost the regional capacity density. Similarly, whenever a satellite node fails, other satellites should be able to immediately fill the coverage gap brought by the failed node to mitigate sudden capacity loss.

The footprints of satellite beams should be dynamically adjusted according to changing interference conditions as the satellites move along their orbits. As the service requirements within a satellite footprint at a given time may vary significantly from area to area, and the perceived requirement pattern at the satellite side will change from time to time due to the satellite's motion, the resource scheduler should adapt accordingly so that sufficient time, frequency, and power resources can be conserved from lightly loaded areas in order to better serve hotspots. A basic response is still maintained for users in the lightly loaded areas, and the resource allocation pattern can be updated in real time when the satellite footprint moves away to a new area. The scheduling algorithm should take into account any restrictions regarding on-board capabilities and geolocation-related service requirements.

Efficient Mobility Management

The satellite network should enable efficient mobility management to cope with the frequent spot beam switching caused by the motion of non-GEO satellites, thereby reducing signaling overhead, interruption time, and power consumption.

The widely applied power-based reselection and handover triggering strategy for terrestrial networks will not work in the context of satellite communications, as the "near–far effect" cannot be observed. As a result, new metrics related to the relative positions of users and satellite spot beams are required. The handover and reselection procedures should also be optimized with regard to the features of specific switching scenarios, including intra-satellite, inter-satellite, and satellite-cellular switching. As the satellite's motion dominates the relative position changes between the satellite and users, we can mitigate the signaling overhead caused by switching by predicting this motion.

Satellite spot beam switching may result in a connection failure in cases where resources from the new spot beam are unavailable. To ensure service continuity, a resource reservation mechanism capable of guaranteeing users the required resources after switching should be employed. As such, mobility procedures in non-terrestrial networks can be assisted by terrestrial networks, and vice versa.

High Precision and Fast Positioning

Low-earth-orbit mega-constellations can provide built-in positioning capability to reduce the dependence on external GNSS at the UE side, while also ensuring an even better user experience. The geometric dilution of precision (GDOP) values of a mega-constellation are much improved when compared with those of the existing GNSS [9]. As it provides similar positioning accuracy, positioning based on mega-constellations can reduce the accuracy demands of on-board oscillators, and can tolerate increased orbit errors due to perturbations. When combined with the communication system, fast and accurate time of arrival (TOA) and/or frequency of arrival (FOA) measurement can be achieved using advanced positioning schemes. For instance, scheduling positioning beams in dedicated slots can avoid interference during data transmission. However, scheduling the positioning beams and reference signals to achieve this target while also lowering the overheads is an issue still to be resolved.

References

[1] 3GPP, "Study on new radio (NR) to support non-terrestrial networks," 3rd Generation Partnership Project (3GPP), Technical Report (TR) 38.811, 10 2019, version 15.2.0. [Online]. Available: https://portal.3gpp.org/desktopmodules/Specifications/SpecificationDetails.aspx?specificationId=3234

[2] O. Kodheli, E. Lagunas, N. Maturo, S. K. Sharma, B. Shankar, J. Montoya, J. Duncan, D. Spano, S. Chatzinotas, S. Kisseleff *et al.*, "Satellite communications in the new space era: A survey and future challenges," *arXiv preprint arXiv:2002.08811*, 2020.

[3] P. Wang, J. Zhang, X. Zhang, Z. Yan, B. G. Evans, and W. Wang, "Convergence of satellite and terrestrial networks: A comprehensive survey," *IEEE Access*, vol. 8, pp. 5550–5588, 2019.

[4] G. Kurt, M. G. Khoshkholgh, S. Alfattani, A. Ibrahim, T. S. Darwish, M. S. Alam, H. Yanikomeroglu, and A. Yongacoglu, "A vision and framework for the high altitude platform station (HAPS) networks of the future," *arXiv preprint arXiv:2007.15088*, 2020.

[5] H. Wang, H. Zhao, W. Wu, J. Xiong, D. Ma, and J. Wei, "Deployment algorithms of flying base stations: 5G and beyond with UAVs," *IEEE Internet of Things Journal*, vol. 6, no. 6, pp. 10 009–10 027, 2019.

[6] S. Andreev, M. Gerasimenko, O. Galinina, Y. Koucheryavy, N. Himayat, S.-P. Yeh, and S. Talwar, "Intelligent access network selection in converged multi-radio heterogeneous networks," *IEEE Wireless Communications*, vol. 21, no. 6, pp. 86–96, 2014.

[7] A. Byman, A. Hulkkonen, P.-D. Arapoglou, M. Bertinelli, and R. De Gaudenzi, "MIMO for mobile satellite digital broadcasting: From theory to practice," *IEEE Transactions on Vehicular Technology*, vol. 65, no. 7, pp. 4839–4853, 2015.

[8] P. Henarejos and A. I. Pérez-Neira, "Dual polarized modulation and reception for next generation mobile satellite communications," *IEEE Transactions on Communications*, vol. 63, no. 10, pp. 3803–3812, 2015.

[9] T. G. Reid, B. Chan, A. Goel, K. Gunning, B. Manning, J. Martin, A. Neish, A. Perkins, and P. Tarantino, "Satellite navigation for the age of autonomy," in *Proc. 2020 IEEE/ION Position, Location and Navigation Symposium (PLANS)*. IEEE, 2020, pp. 342–352.

21 Integrated Sensing and Communication

21.1 Background and Motivations

Cellular networks were originally designed for wireless communication, and the rapidly increasing demand for location-based applications has drawn a considerable amount of attention to positioning research in cellular networks. As described in Chapter 3, some of the more intriguing 6G applications involve sensing environments through high-precision positioning, mapping and reconstruction, and gesture/activity recognition. Sensing will be a new 6G service, and it can be described as the act of obtaining information about a surrounding environment. It can be realized through a variety of activities and operations, and classified into the following categories.

- **RF sensing:** This essentially involves sending an RF signal and learning the environment by receiving as well as processing the reflected signals.
- **Non-RF sensing:** This involves exploiting pictures and videos obtained from the surrounding environment (e.g., via camera).

By sending an electromagnetic wave and receiving echoes, RF sensing is able to extract information about the objects in an environment, such as existence, texture, distance, speed, shape, and orientation. In current systems, RF sensing is limited to radar, which is used to localize, detect, and track passive objects, i.e., objects that are not registered to the network. Existing RF sensing systems have two main limitations:

- They are stand-alone and application-driven, meaning they do not interact with other RF systems.
- They only target passive objects and cannot exploit the distinct features of active objects, i.e., objects registered to the network.

Today, the main objective of designing a wireless network is to optimize communication performance, and this includes improving spectral efficiency and reliability while minimizing latency and power consumption. As such, more cognizant, efficient, and agile communication systems that improve service quality and coverage by embracing diverse use cases are needed now more than ever, as are sensing systems that provide the required knowledge. The conventional practice involves two different subsystems that exchange a limited amount of information to facilitate a certain level of cognition, but this approach has many shortcomings, such as large overhead, large power footage, inefficiency, and bulkiness of the subsystems. An alternative approach relies on the fact that future wireless systems are expected to shift more towards the higher frequencies (such as mmWave and even THz) with plenty of spectrum available. This will enable communication systems to have similar capabilities to those of radar systems. Therefore, to reduce both power footage and form factor, some hardware components can be shared between the two types of systems, such as antennas, power amplifiers, and oscillators. In addition, the systems can leverage shared resources (time and spectrum), instead of dedicated resources, to further boost their respective performance.

As new techniques emerge, improvements will continue to be made in terms of high data rates, low latency, and massive connection capabilities of wireless communication systems. This will also bring other system capabilities such as sensing/imaging and localization, thereby introducing a plethora of innovative applications and boosting the future wireless system's performance. In the integrated sensing and communication (ISAC) system, sensing and communication will be two mutually beneficial functions. Following this trend, the positioning in 5G (which is limited to active devices) will be expanded into more sensing services. This requires new metrics (such as sensing accuracy and sensing resolution) to serve as the new KPIs in 6G, replacing the positioning accuracy KPI in 5G. As described in Chapter 3, KPIs are proposed that are based on different application scenarios, which can be as tight as approximately 1 cm to 10 cm for accuracy, and up to 1 mm in resolution. These goals depend on the enabling technologies that will be discussed in this chapter.

21.2   Overview of Existing Solutions

The 3rd Generation Partnership Project (3GPP) has been working on the integration of communication and positioning since the era of GSM [1], and "positioning enhancement to address high accuracy and latency requirements in indoor industrial situations" was recently planned as a feature in 5G NR Release 17 [2]. However, like the global navigation satellite system (GNSS), the positioning capability in 5G NR is designed for active objects (device-based), which means devices send or receive signals to or from multiple network nodes and then estimate the position information locally on the device or remotely through a network entity. However, there are many scenarios where detection and localization of passive objects (device-free) is required, such as environment sensing, gesture recognition, and the surveillance of restricted zones.

Passive object detection falls within the scope of traditional radar research. Over the years, radar has been used in many applications, including airport and harbor traffic control, earth remote sensing, high-precision detection of small surface deformation, deforestation measurements, as well as volcano and earthquake monitoring. More recently, the trend has been shifted to include vehicle cruise control and collision avoidance, and health status monitoring of heartbeat, respiration, and vocal cord movement. As a ubiquitous sensor, the scenarios in which radar is being applied have exceeded all expectations from when it was first developed and designed, e.g.,

Google Soli [3]. Although 5G NR does not natively support radar sensing from a system design perspective, a significant amount of research is being conducted on the use of existing cellular signals for monostatic or bistatic target detection [4, 5]. For example, [6] uses downlink waveforms in the LTE and NR as radar illumination signals for sensing purposes and investigates the full-duplex problem in a monostatic configuration.

Typically, there are limitations involved with directly applying the physical signal in the traditional cellular system (including 5G NR) for sensing. First, although the reference signals used for channel estimation and phase tracking in 5G NR can potentially be reused for sensing, they do not have sufficient spectrum, time, or space resources to perform this with high accuracy and resolution [7]. This is because the spectrum is a very scarce resource from a communication perspective, and reference signals are carefully designed to minimize overhead. On top of that, communication does not require very accurate estimation of the surrounding environment. Second, passive bistatic or multistatic radar systems that use cellular signals have limitations from a system design perspective. The reason for this is mainly that the passive coherent radar receiver requires a separate directional antenna (direct channel) to receive LOS signals as a timing reference for coherent detection [8]. That said, in reality, no such LOS reference signal exists in many cases, owing to noise, interference, and fading effects. This problem is aggravated further by the fact that different signals are emitted from the base station side via beamforming (digital and/or analog) to users in different directions, and as such, the target reflection signal may not be a replica of the delayed and attenuated version of the LOS reference signal. As a result, coherent detection is impractical and passive radar performs unpredictably. In addition, although the LOS signal is normally useful in communication, it is imperative that we suppress the strong LOS signal leaked into the surveillance channel in the passive radar system to extract weak echoes from targets. This makes the dynamic range requirements for passive radar receivers much more stringent than those for communication receivers. Last but not least, there is no cooperation between Tx and Rx nodes in a passive multistatic configuration, making it considerably more difficult to mitigate interference and conduct coherent processing.

As a sensing service, high-resolution imaging using RF signals can obtain more object information and provide basic data for object classification and recognition [9]. Imaging can essentially be modeled as an inverse electromagnetic scattering problem [10]. The main principle of imaging involves using electromagnetic waves to irradiate the target and reconstructing the target information by collecting the scattered echo signals. The research on imaging can be divided into two categories: 2D imaging and 3D tomography. Two-dimensional imaging is used for perfect conductor targets and 3D diffraction tomography is used for dielectric body targets, as shown in Figure 21.1. In traditional imaging systems, the distance between the transmitter and imaging target is usually large; therefore, the 2D imaging model can be simplified as a linear inverse problem in the far field [11]. In 3D tomography, non-linear inverse problems need to be solved to obtain quantitative information, such as the dielectric constant and conductivity, which in turn need to be transformed into the

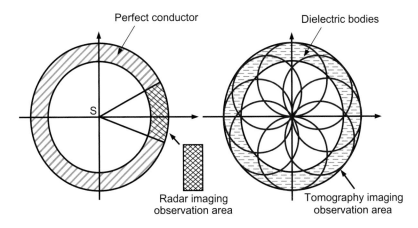

Figure 21.1 Observation of radar imaging and tomography.

imaging result. Compared with 2D imaging, tomography is more complicated since the multi-scattering effect results in a non-linear imaging problem [12]. In recent years, researchers proposed a high-resolution imaging method based on multiple observations relying on the spatio-temporal random radiation field [13]. In 2013, a coherent imaging model based on the compression perception theory of bone tuberculosis and the Green's function correlation calculation was proposed [14], in which the positioning of the target is achieved by using a specially designed metamaterial antenna. With the development of metamaterials and tunable materials, the performance of this imaging system can be improved to achieve higher resolution.

In 6G networks, sensing may be considered as a parallel service to communication, since it can independently generate revenue. This means that the system should be designed to shift from communication only to interaction between sensing and communication, thereby satisfying the KPIs for both services.

The interaction level between communication and sensing systems can be classified into the following categories [15].

- **Co-existence:** This refers to the case when two systems view each other as interference and therefore do not share any information between themselves.
- **Cooperation:** This refers to the case when two systems are designed separately; however, they share information between themselves with the goal of reducing inter-system interference.
- **Joint/integrated design:** This refers to the case when the two systems are designed to behave as a single unified system.

Most research regarding integrated design mainly focuses on the joint waveform, which is elaborated in Section 22.2.4. The main challenge of this design arises from the fact that communication and sensing have contradicting KPIs. More specifically, the former mainly aims to maximize spectral efficiency, whereas the optimum waveform design for sensing aims for accurate estimations and resolution. Put differently, when only range estimation is required, the optimum sensing waveform is the delta-like time-domain auto-correlation function, along with the high-processing gain for enabling parameter estimation when strong noise or interference exists.

Although waveform design is an important aspect of ISAC, it does not cover the whole picture. As such, some research work proposes integrating the sensing and communication design beyond just the waveform level. To give a few examples, the superposition of radar and communication signals in the power domain was proposed in [16] while a joint power- and spatial-domain method was proposed in [17]. However, almost all the existing literature focuses on the link-level aspects of the integrated design, excluding the system-level perspective. In our view, a system-wise architecture design is required to ensure that full integration of the communication and sensing systems can provide the expected benefits for the various corresponding services and applications.

21.3 New Design Expectations and Potential Research Directions

Looking to the future, 6G is expected to operate over higher-frequency ranges with wider bandwidths (e.g., THz) and ultra-massive antenna arrays will become more available. This will provide us with a unique opportunity to widen the scope of cellular networks applications from pure communication to dual communication and sensing functionalities. In our view, 6G should naturally lead to an integrated solution for communication and sensing, as opposed to the existing solution in which the main components (including 5G NR, GNSS, and radar systems) operate independently, as shown in Figure 21.2. In addition, as new enabling technologies evolve in tandem with 6G advancements (including the introduction of metamaterials, reconfigurable intelligent antennas, and AI), this integrated solution will be even more justifiable, especially as these technologies are key driving forces for both services.

Like any other new technology, ISAC comes with both opportunities and challenges. The main way to go about unleashing the immense potential of this technology involves first understanding and addressing the following challenges:

- How can communication and sensing modules co-exist in the integrated solution at the system level, and how can system-level design ensure more effective co-existence?
- How will the integrated solution benefit communication and sensing services? More specifically, which challenges in the current communication and sensing systems will integration address, and what new challenges will it bring?

The answers to these questions help us define the new potential research directions discussed in subsequent sections.

21.3.1 System Design Aspects for Integrated Sensing and Communications

With Figure 21.3, a simple description of communication and sensing can be explained. The figure shows an RF source S transmitting a signal $S(t)$, which is

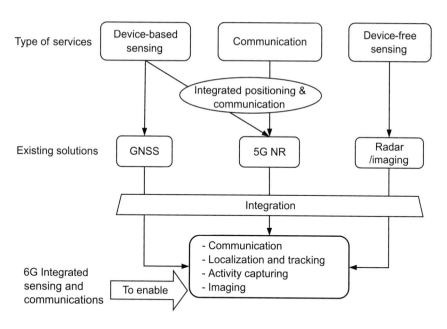

Figure 21.2 Overview of the proposed ISAC solutions.

propagated through the communication medium $\mathcal{M}$ and received or processed by one or many receivers. In communication, the RF signal $S(t)$ is manipulated by the RF source to embed the information (d_S) in it. The theoretical bounds on the transmission rate are well understood and can be expressed as the Shannon limit $I(S(t); Y(t)|\mathcal{M}(t))$, where $Y(t)$ denotes the received signal at the intended receiver and $I(x; y|z)$ denotes the information theoretic mutual information between the variables x and y given variable z. In the preceding formulation, all variables are assumed to be functions of time (t). As for the sensing operation, the RF signal is manipulated by the communication medium, but not by the RF source. In other words, we can view sensing as a special communication method in which the information $(d_{\mathcal{M}})$, which is embedded in the communication medium, is conveyed to the receiver(s) through an RF signal that can be transmitted from a third party. As in communication, we can define a sensing rate that is bounded by $I(Y(t); \mathcal{M}(t)|S(t))$ [18]. Put differently, the difference from traditional communication and sensing is whether the information source and RF source are co-located or not.

This suggests that sensing and communication can both be represented under a general communication umbrella in which the transmitted RF source can be manipulated by both the source data (d_S) and the medium information $(d_{\mathcal{M}})$. This is the starting point for defining a framework to design an ISAC network in which the nodes are capable of both functionalities with the following benefits.

- **More efficient resource usage:** The integrated design allows for efficient resource sharing between the two, rather than conservative resource splitting (e.g., FDMA and TDMA).

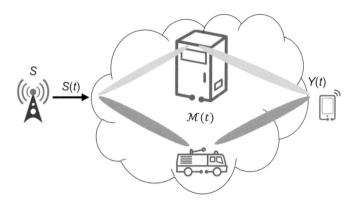

Figure 21.3 Communication and sensing under the same platform.

- **Interference mitigation between sensing and communication signals:** It goes without saying that the main benefit of the integrated design will be the avoidance or mitigation of interference between the two systems when they share resources in the time, frequency, or spatial domains.
- **Better communication performance:** The integrated design allows for better communication performance in two aspects. The first involves medium-aware communication through more efficient beamforming and interference mitigation techniques thanks to sensing information. The second aspect involves more opportunistic and predictive communication that occurs in optimal situations using knowledge of the current and future status of the communication channel. This is based on the environment characterization and prediction of the changes within it. In fact, by knowing the communication environment, we can be aware of the root cause of changes in the channel state information without needing to observe the latter.
- **More efficient sensing:** The ISAC network enables more efficient, on-demand sensing. In other words, sensing is no longer only an application-driven service, but can now be triggered by a demand from another network node.
- **Less power footprint:** Provided that a big portion of the electrical power in RF-based systems is dissipated in an analog front-end (e.g., power amplifiers and ADC/DAC), power can be used much more efficiently if the systems are integrated to a certain degree.

A major challenge in the analysis of an ISAC system is obtaining the performance bounds. One of the first attempts in this line of research was [19] which proposed an ISAC framework in a single-objective system. One important observation in this work is that, as the total energy of the transmitted signal from the signal source is fixed, a trade-off must be made between the echo and the absorbed waves. The proposed ISAC framework is a very specific example in the sense of only considering mono-static sensing, i.e., the transmitter and receiver of the sensing signal are collocated. Another well-known example of ISAC in the coherent communications paradigm is the "channel estimation" process in which some "known" signals, which are referred to as reference signals or pilots, are transmitted along with the information-carrying signal and the receiver of the communication signal and sensing signals is a single device. This way, the transmitted RF signal is manipulated both by the source information and the propagation channel. However, this is an example of ISAC in which the sensing information (propagation channel) is not valuable on its own and it is only used for the purpose of helping the receiver to decode the source data. In a general framework of ISAC, sensing is considered as a separate service which is not necessarily used to facilitate communications. Even if sensing is to assist communications, the utility of sensing may expand beyond the narrow step of detection and equalization.

The future 6G system will continue to use advanced technologies to further propel the performance of the mobile communication system. Some of these technologies, which are key enablers for the 6G system and include terminals and network infrastructures, also lead to the following endogenous super-sensing capabilities:

- More and higher spectrum with larger bandwidth
- Evolved antenna design with extremely large arrays and metasurface
- Larger scale of collaboration between base stations and UEs
- Advanced techniques for interference cancellation
- Integrated advanced signal processing and AI

There are five potential levels of the ISAC design. The first two levels address how to share the available resources (including spectrum), hardware, and RF chains. The third level discusses how to share the digital signal processing resources and capabilities such as processing algorithms and modules in the physical layer. Further integration at the next level may require the protocol interface design to enable cross-layer, cross-module, and cross-node information sharing. Finally, the most efficient and ideal level involves sharing all the available resources and information to enhance the performance of both services. As such, different integration levels might co-exist in the 6G network according to specific use cases, KPIs, and implementation prices. For instance, vehicles with extreme requirements on communication delay and the probability of object recognition or detection might need the five levels of design for ISAC.

Although this design may seem straightforward at first glance due to the inherent commonalities between communication and sensing, the following major challenges need to be addressed to implement such an integrated system.

- **Diverse node capabilities:** Compared to 5G, future wireless networks will have even more wireless nodes with a diverse range of capabilities in terms of processing, memory, bandwidth, and RF. This will make integration challenging compared to communication, where the node capability has a much higher impact on sensing performance.
- **Half-duplex nodes:** Although full-duplex transceivers have already been proto-typed in current systems with acceptable performance, and are expected to be com-mercialized in the near future, we still have a long way to go before networks feature

full-duplex nodes only. As future networks will feature quite a number of half-duplex devices, the half-duplex capability of nodes will make sensing challenging especially in the mono-static mode, where transmitters and receivers are collocated.

- **Limited sensing coverage:** One of the main challenges of sensing in cellular networks is the fact that the supported sensing range increases as the fourth root of the transmit power, as opposed to second root, which is the case for communication.

The following research directions will take us closer to integrating communication and sensing:

- **Performance analysis of general ISAC networks:** On the basis of observations of the fundamental performance bounds obtained in the simple framework presented in [19], it would be desirable to extend the results to more general setups of multi-node multi-target networks with multistatic sensing configurations and find the answer to the following questions: (1) what is the best trade-off between the sensing performance of sensing and the communication capacity; (2) how to design practical schemes approaching such a trade-off, e.g., waveforms, codebooks, frameworks, and protocols. One attempt in this line of research is presented in [20], in which the sensing targets are considered as virtual energy receivers, and then the ISAC is modeled as resource allocation problem of information and energy over the wireless channel. From the perspective of information theory, the sensing targets are relays which receive the probing waveform and forward it back to the transmitter with their own parameter information embedded in the echo wave. More approaches and insights are needed to get a good perspective on the performance bounds of ISAC networks in various scenarios.

- **Radio access network design for ISAC:** The most important problem to address involves how this integration affects the radio access network design for different layers. From the physical-layer perspective, we need the following:

  - a design to enable flexible and healthy co-existence between communication and sensing signals as well as the related configurations, thereby ensuring that the performances of the communication and sensing systems are not compromised;
  - system-wide solutions to collaboratively exploit the sensing capabilities of different nodes, including network nodes and user devices;
  - signaling mechanisms that offer support between the network entities to enable the design and configure the related parameters.

- **Sensing-assisted communication:** Although sensing will be introduced as a separate service in the future, it might still be beneficial to look at how the information obtained through sensing can be used in communication. The most trivial benefit of sensing will be environment characterization, which enables medium-aware communication owing to more deterministic and predictable propagation channels. Some examples of sensing-assisted communication are shown in Figure 21.4, which demonstrates how the environmental knowledge gained through sensing can improve communication. The first example illustrates how environmental knowledge is used to optimize the beamforming to the UE (medium-aware beamforming),

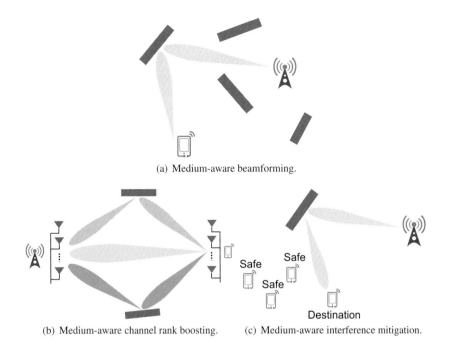

(a) Medium-aware beamforming.

(b) Medium-aware channel rank boosting.   (c) Medium-aware interference mitigation.

Figure 21.4 Examples of sensing-assisted communication.

while the second example shows how this knowledge can be used to exploit all potential DoF in the propagation channel (medium-aware channel rank boosting). The last example shows how medium awareness can reduce or mitigate inter-UE interference.The degree to which sensing benefits communication should not be limited to throughput improvement and interference mitigation. One interesting line of research could be devoted to extending the scope of sensing-assisted communication and studying how functionalities that are usually performed by the communication system can be performed by the sensing submodule instead. This will lead to huge savings in terms of both overhead and latency.

- **Sensing-enabled communication:** Another paradigm shift in the ISAC design involves using an alternative approach for communication through sensing. As previously mentioned, communication and sensing can be viewed as a general platform in which the RF signal from a source can be manipulated by data in another location and received by a third party. With this platform, there is no need to be constrained by a collocated data source and RF source, and this would be particularly beneficial in scenarios where devices with limited processing capabilities (most IoT devices in future systems) are deployed to collect data. Instead of transmitting collected data, devices can manipulate the RF signal transmitted by a different source in the RF domain (as opposed to baseband) to save a significant amount of power. This type of sensing-enabled communication is referred to as backscatter communication [21]. Another example is media-based communication [22], in which the communication medium is deliberately changed to convey information.

- **Communication-assisted sensing:** The communication platform allows us to achieve more efficient and smarter sensing by connecting the sensing nodes. In a network of connected users, on-demand sensing can be realized, i.e., sensing can be performed on the basis of a different node's request or delegated to another node. Furthermore, it enables collaborative sensing in which multiple sensing nodes obtain environmental information. All these advanced features require a carefully designed RAN in order to accommodate communication between the sensing nodes through DL, UL, and SL channels with minimum overhead and maximum sensing efficiency, which is an interesting research topic.
- **Sensing-assisted positioning:** Active localization, also referred to as positioning, involves localizing UEs through the transmission or reception of signals to or from the UEs. Its main advantage is simple operation. Even though accurate knowledge of UE locations is extremely valuable, it is difficult to obtain due to many factors including multipaths, imperfect time/frequency synchronization, limited user sampling/processing capabilities, and the limited dynamic range of UEs. On the other hand, passive localization involves obtaining the location information of active or passive objects by processing the echoes of a transmitted signal at one or multiple locations. Compared to active localization, passive localization through sensing features the following distinct advantages:

  - It helps in identifying LOS links and mitigating the residual NLOS bias.
  - It is much less impacted by synchronization errors between the UEs and the network.
  - It can improve positioning resolution and accuracy for cases where the localization bandwidth is constrained by the target UEs.

In light of this, a potential research direction could focus on how passive localization through sensing improves the shortcomings of active localization. Put differently, how can we enjoy the best of both worlds? As for passive localization, the matching problem is worth mentioning. This is due to the fact that received echoes do not have a unique signature to unambiguously associate them with the objects (and their latent location variables) from which they are reflected. This is in complete contrast to active- or beacon-based localization where the signature recorded from the beacon or landmarks uniquely identifies their associated objects. We therefore need advanced solutions to associate the sensing observations with the locations of active devices, substantially improving the active localization accuracy and resolution.

21.3.2 RF Sensing Design and Algorithms

In this section, we first delve into the technical challenges and potential research directions of general RF sensing, then we address the specific technical challenges involved with imaging-based sensing.

Compared to 5G, the use of much larger bandwidth and antenna array resources in the 6G ISAC system presents us with both opportunities and challenges. The most evident selling point from a sensing perspective is the improved resolution and accuracy, and the main challenges can be summarized as follows.

- **Resource fragmentation:** As a 6G subsystem, RF sensing cannot use the full bandwidth, time slot, and antenna elements simultaneously, and resources allocated to RF sensing will be discontinuous in certain corresponding domains. Given this, one of the main challenges involves efficiently designing the signal and coordinating resource allocation in the time, frequency, and spatial domains without degrading the spectral efficiency and sensing performance.
- **High implementation complexity:** In 6G RF sensing, as for data communication, large bandwidth is accompanied by very high sampling rates, instantaneous large-bandwidth waveforms, and massive processing channels. This brings forth major challenges in the form of computational complexity and high power consumption.
- **Reduced system coverage:** In 6G, as the operating frequency tends to the higher bands, the system coverage decreases under the assumption that the number of BSs and the power consumption remain constant. In addition, the NLOS nature of the propagation channel worsens the coverage problem due to harsh environments (e.g., indoors and urban canyons). Therefore, even though high frequency can potentially bring high resolution and high-accuracy RF sensing performance, the system may not have a sufficient number of nodes to provide high-quality service consistently.
- **High-resolution requirement:** In 6G, sensing can be used in scenarios where high resolution is required, which will enable a series of applications, as mentioned in the use cases chapters. At the same time, this requirement brings challenges such as near-field sensing with large-array antennas, RF signaling design to reduce sensing time, sensing of objects with complicated internal structure, etc.

To address these challenges, we discuss several potential research directions as follows.

- **Joint signal design for RF sensing:** The communication and sensing services in a 6G system will share time-, frequency-, and spatial-domain resources. As such, it is essential that we coordinate these resources so that both services can meet their respective design requirements. In fact, RF sensing may not need to occupy all the bandwidth, time slots, and antenna array elements, meaning we can reduce the resource allocation for the sensing service through sparsification. Considering that the sensing target or environmental parameters usually do not change abruptly owing to physical law constraints (i.e., the channel does not suddenly change in the coherence time), the preceding attribute can be exploited to partition the time–frequency subspace. As a result, each time slot's instantaneous bandwidth will only be a discrete segment of the full bandwidth, and the composite post-processing resolution is the same as the full-bandwidth performance. Looking ahead to 6G, the need for valuable transmitting resources can be reduced by sparse design of the sensing signal, while also ensuring ultra-high-accuracy sensing services. We can therefore use the saved resources to guarantee the quality of communication services, and, as such, it is worthwhile to look into the sparsification of sensing signals in the joint signal space (the frequency–time space) without degrading the sensing performance.

• **Compressed sensing algorithm for RF sensing:** The compressed sensing (CS) and sampling theory provides us with a mathematical framework for acquiring and processing a wide class of analog signals at a rate lower than the Nyquist rate. At its core, CS involves the recovery of sparse high-dimensional vectors from a fairly small amount of measured data, and it can easily be extended to sparse signals on any appropriate basis.

On top of that, CS can effectively reduce computational burden. Although the conventional RF sensing techniques reliably estimate target parameters, they require the signal to be sampled at no less than the Nyquist rate. For instance, they use matched filtering (MF) or pulse compression to maximize the signal-to-noise ratio (SNR) in the presence of additive white Gaussian noise. In some applications, other filters can also be used to optimize different metrics, such as peak to sidelobe ratio (PSLR) and integrated sidelobe ratio (ISLR) [23–25]. Sensing resolution is inversely proportional to the support of the ambiguity function, thereby limiting the ability to perform finer resolution on closely spaced targets. As a result, to achieve fine range resolution, many modern RF sensing systems use wide bandwidths typically ranging from a few hundred MHz to even tens of GHz. This requires high-speed ADCs and leads to a significant amount of processing power overhead. In order to increase the feasibility of high-resolution algorithms with wide bandwidth, a proposal has been put forward to sample the signal and estimate the parameters of the target at a sub-Nyquist rate [26]. In addition to the frequency domain, similar subsampling in the Doppler domain has also been proposed [27]. Radio frequency sensing devices with antenna arrays face comparable spatial-domain sampling problems, and, as such, it would be worth exploring how Nyquist processing with spatial subsampling can be used in MIMO array devices, using fewer antenna elements without reducing angular resolution.

• **Near-field high-resolution sensing:** Due to the channel characteristics, a large number of antennas can fully leverage the propagation path for transmission, thereby increasing the number of ranks and capacity in the communication channel. In addition, because of the development of high-frequency applications and chips, large-array antennas of the centimeter level may be widely integrated and deployed at high frequency in the future, when the sub-THz or THz spectrum is available. Following this, high-resolution sensing that relies on large-array antennas will be viewed as a native function of the ISAC system. Further, considering the limitations with regard to transmit power and path loss at higher frequencies (e.g., THz), it would be more natural to place the ISAC function within a short range of the handheld terminal. For instance, the UE or gloves mentioned in Chapter 3 will sense within dozens of or even several meters, meaning the system will work in the near-field assumption.

Conventional sensing systems mainly function by radiating an electromagnetic signal in free space and receiving echo signals through scatterers. The signal used

Table 21.1 Near-field distance for various antennas.

	Near-field distance calculated by $2D^2$		
Frequency	256 Elements (16×16)	1024 Elements (32×32)	2048 Elements (45×45)
3.5 GHz	11 m	44 m	87 m
6 GHz	6 m	26 m	51 m
10 GHz	4 m	15 m	30 m
39 GHz	0.98 m	3.9 m	7.8 m
73 GHz	0.5 m	2.1 m	4.2 m
140 GHz	0.3 m	1.1 m	2.2 m

for imaging can radiate in a near or far field depending on the antenna aperture size D and the transmitter's operating wavelength. Table 21.1 shows the relationship of the near-field distance to the antenna array size and frequency.

In far-field imaging, electromagnetic waves are incident as plane waves. At the same time, under the far-field assumption, a small difference in distance has little effect on the amplitude of echo signals, and, as such, we can take each point on the target as being approximately equal to the radar distance. By using the simplified radar echo formula, the target image can be reconstructed through a Fourier transform, which is fast. However, in the near-field assumption, the phase difference between the center point and edge of the target is relatively large; therefore, the spherical wave hypothesis must be used. Thus, in the imaging formula, the nonlinear phase term appears and approximate processing cannot be used in distance compression to reduce the calculation burden.

- **Concurrent active and passive localization (CAPL):** The localization problem might seem to be daunting when we do not know the position of navigating users and the environment has not been given. Indeed, one of the fundamental findings in the robotics field involves the derivation of solutions that simultaneously localize a sensing user with an unknown location and iteratively construct a map of the environment. This paradigm was named simultaneous location and mapping (SLAM), and is based on the premise that mapping and positioning are not separate problems. In other words, suboptimal answers will be yielded if we treat them as independent problems.

   The entangled nature of the preceding problem was also mathematically proven by the pioneers in robotics during the 1990s [28, 29]. There is a logical explanation to the map/localization entanglement problem as well: an inaccurate map directly impacts the sensing device's positioning accuracy, and an inaccurate location directly translates into an inaccurate map. Therefore, if the sensing device was mounted on a static map clutter instead of on the mobile navigating device, the measurements would likely be the same. This also means that regardless of where the sensing module is mounted (on a navigating device or at a fixed point), the only problem involves finding the objects that the sensing signal interacts with.

This interaction can be reflection, reception, or transmission. All in all, not only is there no conceptual difference between passive and active localization, but the two problems are indeed entangled.

All this suggests that concurrent passive and active localization (CAPL) with only handheld devices as sensors faces severe challenges. Fortunately, the emergence of numerous use cases in 6G, such as UAVs, vehicles, and IoT devices, could facilitate CAPL. With UAVs, vehicles, IoT devices, and UEs assisting networks in localizing themselves and building a map of the environment, cellular systems can build a virtual environment in cyber space.

- **Cooperative multi-nodes RF sensing:** RF sensing through cooperation refers to sensing nodes that share their observations with each other and attempt to reach a common consensus on the surrounding environment, and this has been shown to significantly improve localization performance [30]. The specific process involves the cooperative nodes forming a dynamic reference grid through distributed transmission and processing. The cooperation reduces measurement uncertainty and provides greater coverage as well as higher sensing accuracy and resolution through data fusion.

References

[1] R. S. Campos, "Evolution of positioning techniques in cellular networks, from 2G to 4G," *Wireless Communications and Mobile Computing*, vol. 2017, Article ID 2315036, 17 pages, 2017. [Online]. Available: https://doi.org/10.1155/2017/2315036

[2] B. Bertenyi, "5G in Release 17 – strong radio evolution," Technical Report, 2019. [Online]. Available: https://www.3gpp.org/news-events/2098-5g-in-release-17-%E2%80%93-strong-radio-evolution

[3] J. Lien, "Soli radar-based perception and interaction in pixel 4," Technical Report. [Online]. Available: https://ai.googleblog.com/2020/03/soli-radar-based-perception-and.html

[4] M. Bica, K.-W. Huang, V. Koivunen, and U. Mitra, "Mutual information based radar waveform design for joint radar and cellular communication systems," in *Proc. 2016 IEEE International Conference on Acoustics, Speech and Signal Processing (ICASSP)*. IEEE, 2016, pp. 3671–3675.

[5] M. Schmidhammer, S. Sand, M. Soliman, and F. de Ponte Muller, "5G signal design for road surveillance," in *Proc. 2017 14th Workshop on Positioning, Navigation and Communications (WPNC)*. IEEE, 2017, pp. 1–6.

[6] C. B. Barneto, T. Riihonen, M. Turunen, L. Anttila, M. Fleischer, K. Stadius, J. Ryynänen, and M. Valkama, "Full-duplex OFDM radar with LTE and 5G NR waveforms: Challenges, solutions, and measurements," *IEEE Transactions on Microwave Theory and Techniques*, vol. 67, no. 10, pp. 4042–4054, 2019.

[7] R. M. Rao, V. Marojevic, and J. H. Reed, "Probability of pilot interference in pulsed radar-cellular coexistence: Fundamental insights on demodulation and limited CSI feedback," *IEEE Communications Letters*, vol. 24, no. 8, pp. 1678–1682, Aug. 2020.

[8] D. E. Hack, L. K. Patton, B. Himed, and M. A. Saville, "Detection in passive MIMO radar networks," *IEEE Transactions on Signal Processing*, vol. 62, no. 11, pp. 2999–3012, 2014.

[9] J. Yan, X. Feng, and P. Huang, "High resolution range profile statistical property analysis of radar target," in *Proc. 6th International Conference on Signal Processing, 2002*, vol. 2. IEEE, 2002, pp. 1469–1472.

[10] M. Bertero and P. Boccacci, *Introduction to inverse problems in imaging*. CRC Press, 1998.

[11] D. M. Sheen, D. L. McMakin, and T. E. Hall, "Three-dimensional millimeter-wave imaging for concealed weapon detection," *IEEE Transactions on Microwave Theory and Techniques*, vol. 49, no. 9, pp. 1581–1592, 2001.

[12] J. P. Guillet, B. Recur, L. Frederique, B. Bousquet, L. Canioni, I. Manek-Hönninger, P. Desbarats, and P. Mounaix, "Review of terahertz tomography techniques," *Journal of Infrared, Millimeter, and Terahertz Waves*, vol. 35, no. 4, pp. 382–411, 2014.

[13] Y. Guo, D. Wang, X. He, and B. Liu, "Super-resolution staring imaging radar based on stochastic radiation fields," in *Proc. 2012 IEEE MTT-S International Microwave Workshop Series on Millimeter Wave Wireless Technology and Applications*. IEEE, 2012, pp. 1–4.

[14] J. Hunt, T. Driscoll, A. Mrozack, G. Lipworth, M. Reynolds, D. Brady, and D. R. Smith, "Metamaterial apertures for computational imaging," *Science*, vol. 339, no. 6117, pp. 310–313, 2013.

[15] B. Paul, A. R. Chiriyath, and D. W. Bliss, "Survey of RF communications and sensing convergence research," *IEEE Access*, vol. 5, pp. 252–270, 2016.

[16] X. Zheng, T. Jiang, and W. Xue, "A composite method for improving the resolution of passive radar target recognition based on WiFi signals," *EURASIP Journal on Wireless Communications and Networking*, vol. 2018, no. 1, p. 215, 2018.

[17] J. A. Zhang, X. Huang, Y. J. Guo, J. Yuan, and R. W. Heath, "Multibeam for joint communication and radar sensing using steerable analog antenna arrays," *IEEE Transactions on Vehicular Technology*, vol. 68, no. 1, pp. 671–685, 2018.

[18] M. R. Bell, "Information theory and radar waveform design," *IEEE Transactions on Information Theory*, vol. 39, no. 5, pp. 1578–1597, 1993.

[19] M. Kobayashi, G. Caire, and G. Kramer, "Joint state sensing and communication: Optimal tradeoff for a memoryless case," in *Proc. 2018 IEEE International Symposium on Information Theory (ISIT)*. IEEE, 2018, pp. 111–115.

[20] F. Liu, C. Masouros, A. Petropulu, H. Griffiths, and L. Hanzo, "Joint radar and communication design: Applications, state-of-the-art, and the road ahead," *IEEE Transactions on Communications*, vol. 68, no. 6, pp. 3834–3862, 2020.

[21] H. Stockman, "Communication by means of reflected power," *Proceedings of the IRE*, vol. 36, no. 10, pp. 1196–1204, 1948.

[22] A. K. Khandani, "Media-based modulation: A new approach to wireless transmission," in *Proc. 2013 IEEE International Symposium on Information Theory*. IEEE, 2013, pp. 3050–3054.

[23] N. Levanon, *Radar principles (First Edition)*. John Wiley & Sons, 1988.

[24] J. E. Cilliers and J. C. Smit, "Pulse compression sidelobe reduction by minimization of l/sub p/-norms," *IEEE Transactions on Aerospace and Electronic Systems*, vol. 43, no. 3, pp. 1238–1247, 2007.

[25] J. George, K. Mishra, C. Nguyen, and V. Chandrasekar, "Implementation of blind zone and range-velocity ambiguity mitigation for solid-state weather radar," in *Proc. 2010 IEEE Radar Conference*. IEEE, 2010, pp. 1434–1438.

[26] Y. C. Eldar, *Sampling theory: Beyond bandlimited systems*. Cambridge University Press, 2015.

[27] J. Akhtar, B. Torvik, and K. E. Olsen, "Compressed sensing with interleaving slow-time pulses and hybrid sparse image reconstruction," in *Proc. 2017 IEEE Radar Conference (RadarConf)*. IEEE, 2017, pp. 0006–0010.

[28] S. Thrun, W. Burgard, and D. Fox, "A probabilistic approach to concurrent mapping and localization for mobile robots," *Autonomous Robots*, vol. 5, no. 3-4, pp. 253–271, 1998.

[29] H. Durrant-Whyte and T. Bailey, "Simultaneous localization and mapping: Part I," *IEEE Robotics & Automation Magazine*, vol. 13, no. 2, pp. 99–110, 2006.

[30] M. Z. Win, Y. Shen, and W. Dai, "A theoretical foundation of network localization and navigation," *Proceedings of the IEEE*, vol. 106, no. 7, pp. 1136–1165, 2018.

22 New Waveforms and Modulation Schemes

22.1 Background and Motivation

The current 4G and 5G communication systems at sub-6 GHz frequency bands face extensive multipath fading. Both LTE and NR standards use OFDM as the major waveform to tackle multipath fading. OFDM is also compatible with MIMO, which is one of the most important techniques for achieving high spectrum efficiency. However, compared with single-carrier waveforms, OFDM has a higher peak-to-average power ratio (PAPR). In scenarios where uplink transmission is coverage limited, low-PAPR discrete-Fourier-transform-spread OFDM (DFT-s-OFDM) waveform is also supported.

In terms of modulation, LTE and NR both utilize regular QAM constellations with gray labeling. QAM facilitates system design by achieving simple demodulation with real–imaginary separation but at the expense of some loss of shaping gain. In addition, $\pi/2$ binary phase shift keying (BPSK) modulation is introduced to work with DFT-s-OFDM waveforms to further reduce the PAPR for extended coverage scenarios with very low spectrum efficiency.

While multi-carrier waveforms, particularly OFDM, and regular QAM may continue to play a central role in wireless access systems in the future, new waveforms and modulation schemes may be introduced, as new use cases, devices, and spectrum emerge these are discussed below specifically.

First, a low PAPR is critical for a wide range of applications. For example, new spectrum at very high frequencies (e.g., THz bands) will play an essential role in addressing the ever-increasing demand for higher data rates and new types of traffic. Low PAPR will be vital for any waveform and modulation scheme at these high frequencies due to the challenges associated with designing efficient high-frequency wideband PAs, coping with the experienced path loss, and handling a relatively flat channel with very sparse scattering in the temporal–spatial domain [1]. To address such challenges, we can introduce new waveforms and modulation schemes. This is also true in other scenarios where the transmitter and receiver are close together and communication is performed over a predominantly LOS channel that remains flat. In such scenarios, the benefits of using cyclic prefix OFDM (CP-OFDM) to combat multipath fading disappear but the cyclic prefix (CP) overhead remains. Because high capacity is a key requirement for short-range communications, it is not desirable to achieve a low PAPR by sacrificing capacity. In satellite communications, due to the power limitations of satellites and the non-linearity of PAs, a new low-PAPR waveform is needed. Low PAPR is also a key requirement to achieve low power consumption for low-cost devices with inexpensive PAs, low computational capabilities, and limited power supply capabilities.

Second, for low-cost devices, complexity is another key issue in addition to low PAPR. The waveform and modulation scheme should have low processing complexity with good tolerance to phase noise, carrier frequency offset, timing offset, non-linearity, and more. This is because low-cost hardware will introduce RF distortion at certain levels.

Third, in high-speed scenarios, the Doppler effect leading to time selectivity in wireless channels cannot be ignored. If the channels are LOS-dominant, the Doppler effect can be compensated for at the transmitter or receiver side by pilot-aided estimation or a priori knowledge. However, if the channels are doubly selective, especially in MIMO systems, we should consider advanced schemes that have low overheads and complexity while also meeting use case-specific requirements.

Fourth, for URLLC usage scenario, low latency and high reliability are the key requirements. This means that the desired waveform and modulation scheme should have short time durations and sufficient decoding performance without compromising other performance metrics, such as the spectrum efficiency.

Last but not the least, ISAC further impacts the waveform design for wireless systems in the future. Using the same waveform for communications and sensing is desirable but means that additional characteristics unique to sensing must be considered during the waveform design. Specifically, a waveform suitable for communications typically has high spectrum efficiency and low out-of-band emission (OOBE), whereas the optimal waveform design for sensing targets has different objectives, such as estimation accuracy and precision. In particular, when only range estimation is required, this reduces to delta-like time-domain auto-correlation for better evaluation of the time delay. Furthermore, the waveform used for sensing requires a sufficient processing gain to enable parameter estimation in scenarios with strong noise and interference, multipath fading, and hardware imperfections (e.g., lack of time/frequency synchronization, phase noise, and PA non-linearity). Although such errors – within reason – are acceptable in communication systems, they may severely compromise the accuracy of sensing tasks.

22.2   Overview of Existing Solutions

The waveform and modulation design for cellular communications mainly considers the following requirements:

- Wide range of use cases, including eMBB, mMTC, and URLLC
- High spectrum efficiency to meet the exponentially growing data traffic for eMBB use cases
- Sufficiently wide coverage

- Unified design for downlink, uplink, and sidelink with good MIMO compatibility for high spectrum efficiency
- Unified design for low and high frequencies
- Low complexity with simple implementation and good energy efficiency

Every waveform has its own advantages and disadvantages, meaning that a single waveform will not outperform all other waveforms in meeting all the preceding requirements. The following sections discuss the different waveforms.

22.2.1 Multi-Carrier Waveforms

Multi-carrier waveforms utilize the spectrum very effectively, but they do so at the cost of a very high PAPR due to the random addition of signals from different subcarriers. We discuss several variations of multi-carrier waveforms below, highlighting their advantages and disadvantages. For detailed performance evaluations and comparisons of the different waveforms, please refer to [2].

- **CP-OFDM with spectral confinement:** CP-OFDM used in LTE downlink (DL) is very flexible in frequency-domain processing, including user and channel multiplexing, resource allocation, and carrier aggregation. It has very low complexity at the transmitter and receiver sides, and it achieves a good block error rate (BLER) performance. Furthermore, it is compatible with MIMO, which is key to achieve high spectrum efficiency and/or reliability. Yet despite these advantages, it has a very high out-of-band emission (OOBE). To address this, filtering- and windowing-based spectral confinement techniques are proposed, which are illustrated in Figure 22.1.

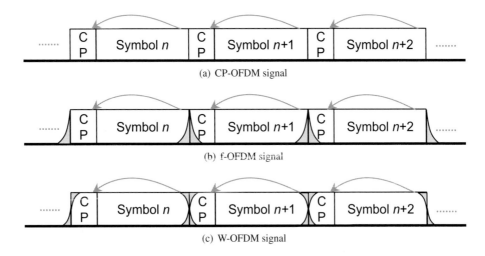

Figure 22.1 Time-domain illustrations of filtering- and windowing-based spectral confinement techniques.

- **Filtered OFDM (f-OFDM):** Subband-based filtering is applied to OFDM signals to suppress inter-subband interference if they use different numerologies or have asynchronous transmission between each other [3, 4]. The time-domain orthogonality between consecutive OFDM symbols in each subband is broken intentionally in order to achieve a lower OOBE with only a negligible performance loss in other aspects. Compared with conventional OFDM, f-OFDM supports mixed numerologies and asynchronous transmission across subbands, eliminating the need for global synchronization. In addition, f-OFDM consumes significantly fewer guard band resources, utilizing spectrum more efficiently (due to good OOBE and BLER performance) if the computation complexity is affordable.
- **Windowed OFDM (W-OFDM):** A time-domain non-rectangular window is applied to smooth the transition between consecutive OFDM symbols, which reduces processing complexity [5]. However, part of the CP length is used for windowing, limiting the performance of W-OFDM owing to the shortened effective CP length.

Figures 22.2 and 22.3 compare the BLER performance of f-OFDM, W-OFDM, and OFDM in DL [6] and UL [7] mixed-numerology cases, respectively, under 0 and 12 guard tone numbers. The comparisons are based on an NLOS tapped-delay line channel (TDL-C) with 1000 ns delay spread at the 4 GHz band and with ideal channel estimation. In Figure 22.2, the subcarrier spaces for the target subband and interfering subband are 15 kHz and 30 kHz, respectively. In Figure 22.3, the subcarrier spaces for the target UE and two interfering UEs are 15 kHz and 30 kHz, respectively, and the interfering UE has power 5 dB larger than the target UE. The figures show that f-OFDM provides much better performance. However, as long as the transmit signal matches the RF requirements, different schemes may be selected for different scenarios to achieve a balance between complexity and performance. Furthermore,

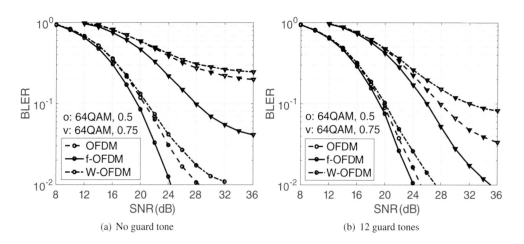

(a) No guard tone   (b) 12 guard tones

Figure 22.2 BLER comparisons of f-OFDM, W-OFDM, and OFDM in the DL mixed-numerology case.

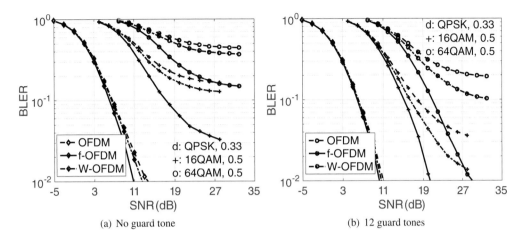

Figure 22.3 BLER comparisons of f-OFDM, W-OFDM, and OFDM in UL mixed-numerology case.

the receiver can implement filtering or windowing to suppress the interference from adjacent bands. In 5G, CP-OFDM is used without explicit restriction on filtering- or windowing-based spectral confinement, depending on the practical implementations.

- **Universal filtered multi-carrier (UFMC):** UFMC applies subband filtering to each OFDM symbol and, due to the linear convolution of the filtering, CPs are replaced by symbol extensions to form guard intervals between symbols [8]. The guard interval length limits the UFMC filter length and consequently limits the OOBE performance. Because UFMC lacks CPs, it is more complex in terms of demodulation and more sensitive to timing synchronization compared with CP-OFDM.

- **Generalized frequency-division multiplexing (GFDM):** GFDM applies subcarrier filtering in order to achieve good OOBE performance. High-order subcarrier-wise circular filtering and tail biting are needed to suppress the inter-carrier interference (ICI) because the subcarriers are close together and not mutually orthogonal [9]. However, advanced receivers are still needed to alleviate the ICI that remains after filtering, especially for high-order modulation, making the pilot design and MIMO transmission complex. Furthermore, GFDM is processed in a block-wise manner; although the CP overhead can be reduced, doing so increases the processing latency and is therefore not suitable for low-latency transmission.

- **Spectrally precoded OFDM (SP-OFDM):** SP-OFDM applies a precoder to data symbols before OFDM modulation in order to reduce OOBE [10]. This reduction is limited, especially if the bandwidth is small. Although precoding creates inter-carrier interference, this can be mitigated (e.g., by multiplying by the inverse of the precoder) at the receiver if the receiver has the precoder information. Consequently, the decoding complexity is higher and may need additional signaling. Furthermore, precoding may cause in-band ripple, which may degrade detection performance.

- **Filter bank multi-carrier with offset QAM (FBMC-OQAM):** In FBMC-OQAM, each subcarrier is filtered individually, achieving very good OOBE performance [11]. Compared with other prefix- or padding-based systems (e.g., CP-OFDM), FBMC-OQAM does not need a guard interval, with timing overhead saving. However, due to subcarrier filtering, the filter length is typically very long (e.g., more than three times the symbol duration). As a result, FBMC-OQAM is not suitable for short symbol transmission as needed for low-latency applications. Furthermore, it employs a complex channel estimation scheme and cannot be easily extended to MIMO due to the non-orthogonality in complex domains, limiting its application scope.

- **Filter bank multi-carrier QAM (FBMC-QAM):** This is designed to mitigate the complication of the FBMC-OQAM waveform where even subcarriers and odd subcarriers use different filters [12]. FBMC-QAM optimizes the two types of filters to minimize the intrinsic ICI caused by the non-orthogonality in complex domains. It ensures low OOBE at the cost of using a long filter due to the nature of subcarrier-based filtering. However, FBMC-QAM cannot eliminate the intrinsic ICI, which will cause performance loss even with an advanced non-linear receiver. Furthermore, the non-orthogonality complicates the application of FBMC-QAM in MIMO and non-orthogonal multiple access (NOMA) scenarios.

- **Weighted circular convolution FBMC-OQAM (WCC-FBMC-OQAM):** To remove the tails caused by linear convolution of the subcarrier filter in the FBMC-OQAM scheme, a weighted circular convolution is used [13]. This eliminates the timing overhead without causing any ICI or inter-symbol interference (ISI) to signals. In addition, weighted time-domain windowing is used to provide smooth transitions on the edges of the signal, thereby solving the issue of sharp signal edges in the time domain caused by circular convolution. WCC-FBMC-OQAM achieves exceptional OOBE performance with only limited time-domain window overhead. Although it removes signal tails through filtering, it possesses the other disadvantages of FBMC-OQAM (e.g., it is not compatible with MIMO, which is key to achieving high spectrum efficiency).

- **Flexible configured OFDM (FC-OFDM):** FC-OFDM is a structure enabling FBMC-OQAM and OFDM co-existence in one fast Fourier transform (FFT), where guard subcarriers are reserved to separate interference from FBMC-OQAM to OFDM [14]. In other words, it enables a unified transmit structure to support these two waveforms. However, in order to suppress the OFDM OOBE, additional filtering or windowing is required. Furthermore, the FBMC-OQAM overlapping factor (filter length divided by FFT size) is limited to 1 for low-complexity processing, which loses the OOBE advantage of the general FBMC-OQAM, which uses a large overlapping factor (e.g., 4).

- **Orthogonal time–frequency space (OTFS):** OTFS uses a two-dimensional Fourier transform to transform time–frequency-domain data to the delay-Doppler domain, and this transformed data can be transmitted by the traditional OFDM modulator [15]. OTFS assumes that channels are sparse and unchanged in the delay-Doppler domain, where data are convoluted with these channels. It may achieve potential gains in high-speed scenarios, but this comes at the cost of long symbol duration and very high equalization complexity, not to mention the extension to MIMO applications. Note that CP-OFDM with additional demodulation reference signals (DMRSs) for high-mobility users would have very good performance especially under LOS-dominant channels.

- **Spectrally efficient frequency-division multiplexing (SEFDM) and overlapped frequency-domain multiplexing (OVFDM):** Compared with OFDM, SEFDM/OVFDM uses a subcarrier spacing that is smaller than the subcarrier width [16, 17]. Consequently, signals are compressed in the frequency domain at the cost of introducing ICI. Higher compression means that less bandwidth is consumed, but this increases the severity of ICI. To suppress ICI, an advanced receiver can be used, for example, by using the sphere detection method. However, this approach is typically complex, especially for high modulation and large numbers of subcarriers. The gain is a type of coding gain, meaning that if the system employs powerful coding (e.g., Polar code), additional capacity gains may not be possible. Because high modulation is sensitive to noise and interference, capacity loss may occur due to the introduction of ICI.

- **Vector OFDM (V-OFDM):** In vector OFDM [18], N-subcarrier data of each OFDM are divided into K sub-vectors of length N/K. An N/K inverse fast Fourier transform (IFFT) is applied to each sub-vector, and the resulting K sub-vectors are interleaved to generate an interleaved vector of length N, which is transmitted together with a CP. Compared with OFDM, which uses N subcarriers, V-OFDM reduces the FFT size to simplify processing on the transmitter and reduce PAPR. But because signals are interleaved, the spectrum is not localized. Furthermore, ISI exists among the K sub-vectors, resulting in a higher equalization complexity in V-OFDM than in CP-OFDM.

- **Non-orthogonal frequency-division multiplexing (NOFDM)/Pulse-shaped OFDM (P-OFDM)/Filter bank OFDM (FB-OFDM):** Instead of using a rectangular pulse shape (as used in CP-OFDM), NOFDM/P-OFDM/FB-OFDM considers the pulse shape as a DoF to meet different design requirements [19–21]. For example, the pulse shape can be designed to have a sharp frequency-domain decay in order to enhance the frequency localization compared with OFDM. As a result of the uniform DFT filter bank, the transceiver can be efficiently realized by a polyphase network after inverse discrete Fourier transform (IDFT) at the transmitter side and before DFT at the receiver side. Various multi-carrier waveforms, such as CP-OFDM, W-OFDM, GFDM, and FBMC-OQAM, can share similar processing for transmission, but they use different pulse shape functions and symbol intervals for different waveforms. This means that the advantages and disadvantages differ depending on which pulse functions and symbol intervals are used.

- **Wavelet-OFDM:** Instead of using an FFT (as used in OFDM), wavelet-OFDM uses the discrete wavelet transform (DWT) [22]. Specifically, inverse discrete wavelet transform (IDWT) and DWT are applied at the transmitter and receiver sides, respectively, using filter banks for implementation. Similar to the subcarrier filtering schemes, wavelet-OFDM achieves good OOBE performance due to symbol overlapping (time-domain tails of long filters), and does not need a guard interval for time overhead saving. Due to the long time-domain tails, it is not suitable for low-latency transmission. Furthermore, its low-complexity extension to MIMO needs further in-depth research.

- **Lagrange Vandermonde division multiplexing (LVDM):** LVDM generalizes the zero padding OFDM (ZP-OFDM) where the transmitter modulates the transmit symbol vector using a Lagrange matrix, while the receiver demodulates the received symbol vector using a Vandermonde matrix [23]. The signature roots to construct the Lagrange and Vandermonde matrices are optimized according to fading channels, requiring knowledge about the channels. If the signature roots have been spread over a unit circle with unknown channels, LVDM reduces to ZP-OFDM. More research is needed to determine the performance gain (e.g., BLER comparison under coded systems).

- **Chirp transform-based waveform:** As a development of the Fourier transform used in OFDM, a fractional Fourier-transform-based waveform was proposed in [24] and an affine Fourier-transform-based waveform was proposed in [25]. Their purpose is to handle time-varying channels caused by the Doppler effect. Introducing freedom to the chirp parameter makes the system more robust to ICI compared with OFDM in cases where multipath channels with a linear relationship between delays and Doppler shifts are experienced. For general doubly selective channels, more research is needed compared with the OFDM system.

- **Slepian-basis-based non-orthogonal multi-carrier (SNMC):** In SNMC [26], one subcarrier is multiplexed by multiple orthogonal Slepian basis functions, which carry data information (e.g., QAM symbols). Because the energy concentration of the Slepian basis is optimal in both time and frequency domains, SNMC achieves very good OOBE performance. Although the Slepian basis is orthogonal in one subcarrier, ICI exists due to the non-orthogonality between subcarriers. Consequently, an advanced receiver is needed to suppress the ICI. The ICI issue will also complicate the pilot design and channel estimation schemes. In MIMO scenarios, the use of SNMC is expected to be much more complex compared with OFDM.

22.2.2   Single-Carrier Waveforms

Single-carrier waveforms (e.g., DFT-s-OFDM, adopted in LTE and NR UL) are more advantageous in terms of PAPR compared with multi-carrier waveforms. Single-carrier waveforms are also more robust to phase noise with relatively simple time-domain estimation and compensation. However, single-carrier waveforms have some common drawbacks, which limit their application scenarios. Specifically, supporting frequency-division multiplexing (FDM) operations (e.g., FDM users, FDM reference signal and data, and carrier aggregation) will deteriorate the PAPR performance. In addition, the frequency resource allocation and data mapping for users served by single-carrier waveforms are highly restricted in order to maintain low PAPR and reasonable processing complexity. With multipath frequency-selective wireless channels, applying simple frequency-domain equalization (as applied in

OFDM) will cause performance loss. The performance loss is much higher in MIMO scenarios, where OFDM can implement frequency-domain precoding (even on a per subcarrier basis) without affecting the PAPR. Conversely, single-carrier waveforms can implement only full-band precoding to maintain low PAPR. The following are some typical single-carrier waveforms.

- **DFT spreading-based single carrier:** Most of the multi-carrier waveforms discussed earlier can be transformed to corresponding single-carrier versions. A typical example of this is DFT-s-OFDM. In DFT-s-OFDM, DFT precoding transforms the input signals into the frequency domain. Frequency-domain spectral shaping (FDSS) can be applied to further reduce the PAPR, which will typically consume more bandwidth (i.e., at the cost of reducing the spectrum efficiency). The FDSS filter cannot be transparent for high-order modulation, since it will significantly reduce the decoding performance, potentially requiring additional signaling overhead. Frequency-domain signals are mapped to the desired bandwidth and then transformed back to the time domain after IFFT. CPs are added to enable single-tap frequency-domain equalization at the receiver side. Although the resource allocation is compatible with OFDM in terms of frequency-domain processing, the allocation is restricted to limited options in order to simplify the DFT operation.
- **SC-QAM/SC-FDE:** SC-QAM/SC-FDE is a traditional single-carrier waveform, where all processing is performed in the time domain [27]. Similarly to DFT-s-OFDM, SC-QAM/SC-FDE adds CPs, zeros, or unique words (UWs) to enable single-tap frequency-domain equalization at the receiver side. Compared with DFT-s-OFDM, SC-QAM/SC-FDE has lower complexity (because neither DFT nor IFFT is used at the transmitter side) and is less flexible in terms of frequency-domain operations. Specifically, SC-QAM/SC-FDE only supports full bandwidth transmission/reception without considering FDM user, channels, or reference signals.
- **ZT/UW-DFT-s-OFDM:** ZT/UW-DFT-s-OFDM is a variant of DFT-s-OFDM where zeros or unique words (UWs) are added before DFT to replace CPs [28]. In DFT-s-OFDM, the CP overhead is fixed and common to all users, whereas the length of zeros or UWs can be user-specific to dynamically adjust the overhead based on the spread delay and propagation delay in ZT/UW-DFT-s-OFDM. This could potentially reduce some overhead for low-delay-spread users. However, owing to the lack of CPs, the symbol duration is different from all the CP-based waveforms described earlier. As such, ZT/UW-DFT-s-OFDM cannot multiplex with those users within the same band. The channel delay spread should be known at the base station and UE sides, and additional signaling overhead will be required. In addition, higher-order modulation performance will be limited due to the lack of channel circular-convolution characteristics.
- **Faster than Nyquist (FTN)/overlapped time-domain multiplexing (OVTDM):** Unlike SEFDM/OVFDM, FTN/OVTDM packs symbols in the time domain for transmission at an FTN rate [17, 29]. The drawbacks in FTN/OVTDM are the same as those in SEFDM/OVFDM; namely, a complex receiver with potentially limited or no capacity gain compared with systems that employ a powerful coding

scheme. Note that OVFDM and OVTDM are jointly called overlapped X domain multiplexing (OVXDM) [17].

22.2.3   Modulation Schemes

Compared with regular QAM, there exist other modulation schemes which can provide better shaping gains, lower PAPR, and better robustness to RF distortion. Some of the modulation schemes are discussed below.

- **Rotated QAM:** In rotated QAM, phase rotation is applied to regular QAM. For example, $\pi/4$ rotated quadrature phase shift keying (QPSK) is applied in narrowband Internet of Things (NB-IoT) UL, where the phase rotation is equal to $\pi/4 \times (n \mod 2)$ and n is the symbol index. The additional phase rotation can reduce the phase offset between adjacent modulated symbols, which in turn can reduce PAPR by using single-carrier waveforms. However, the PAPR gain is extremely limited for higher-order modulation above QPSK. Furthermore, rotated QAM is typically used together with non-transparent FDSS for DFT-s-OFDM to provide promising PAPR and BLER performance.

- **Non-regular QAM:** Several non-regular QAM schemes have been proposed in 5G, including 1D/2D optimized constellation non-uniform QAM, probabilistically shaped coded modulation [30], and amplitude phase shift keying (APSK). These schemes have the potential to provide higher shaping gains, PAPR gains, and robustness to phase noise. However, such gains come at the cost of higher demodulation complexity.

- **Constellation interpolation:** To reduce the phase offset between adjacent modulated symbols, constellation interpolation was proposed in [31]. Specifically, the modulated constellation is first interpolated along a smooth, constant-envelope trajectory (for QPSK) or near-constant-envelope trajectory (for QAM) before being input to the DFT of DFT-s-OFDM. The interpolation is achieved by using a simple Q state trellis coder, where Q is the original constellation size. A larger interpolation ratio implies a more constrained constellation trajectory, leading to a smaller PAPR at the cost of higher interpolation complexity. FDSS can adjust the trade-off between spectrum efficiency and PAPR by using different spectral shaping filters. A standard single-tap frequency-domain equalizer can be used for channel equalization. The equalized symbol is then fed into a Q state trellis decoder to produce the log likelihood ratio (LLR) for the forward error correction (FEC) decoder. Constellation interpolation can achieve extremely good PAPR performance, but would do so at the cost of transmission and reception complexity. Currently, further research is needed to investigate the BLER performance at different spectrum efficiency and PAPR trade-offs.

- **Multi-dimensional modulation:** Multi-dimensional modulation exploits more freedom for performance improvements at the expense of complexity. It can improve the shaping gain compared with the traditional QAM constellations. Furthermore, it can provide new freedom for massive user access, such as the sparse code multiple access (SCMA) scheme with provision of the coding gain [32, 33]. In addition, Grassmannian constellations can be applied for pilot-free non-coherent detection, which have the potential to significantly reduce the pilot overhead for certain applications (e.g., mMTC and massive MIMO systems) at the expense of increased detection complexity [34].

- **Index modulation:** In index modulation, the indices of building blocks in communication systems are used to convey additional information. Two applications of index modulation are spatial modulation and OFDM with index modulation [35]. Spatial modulation transmits information by using the indices of the transmit antennas in addition to a traditional symbol modulation scheme (e.g., QAM). On the other hand, OFDM with index modulation transmits information by using the indices of the transmit subcarriers. Index modulation can achieve good energy efficiency, but the spectrum efficiency will be lower compared with traditional QAM communication systems.

22.2.4   Sensing Waveforms

Sensing – and how to reuse communication waveforms – has attracted a great deal of interest in academic circles. Some of the typical sensing waveforms are discussed below.

- **Multi-carrier waveforms:** Multi-carrier waveforms and in particular CP-OFDM have been dominant options for communications due to their high spectrum efficiency, scalability, and flexibility, thanks to the introduction of CP. Hence, many researchers have considered using these waveforms for sensing as well. Given the fact that CPs may deteriorate the time-domain auto-correlation [36], a novel frequency-domain processing approach is adopted, which can efficiently estimate parameters for CP-OFDM with the maximal processing gain [37]. In addition, CP-OFDM has been proven free of range-Doppler coupling issues, meaning that range estimation and Doppler estimation can be treated as independent issues in CP-OFDM [38]. Furthermore, CP-OFDM parameters, such as carrier distance, guard interval length, frame length, and pilot design, can be adopted to optimize the sensing detection and data communications performance and robustness [39]. However, such advantages rely on perfect synchronization (time and frequency) between the transmitter and receiver – perfect synchronization may not be possible, though, especially for bi-static sensing where the transmitter and receiver of the sensing signal are not co-located. In this case, CP may not provide any performance benefits for sensing, and multi-carrier waveforms without CP can also be considered. The main drawback of removing CP is the complexity involved in data detection (due to the introduction of ISI), which needs to be addressed. Nevertheless, the large PAPR of multi-carrier waveforms (with or without CP) is another major issue for sensing applications where power efficiency is extremely important.
- **Frequency modulated continuous wave (FMCW):** This waveform is traditionally used for radar. It has features suitable for sensing, including good auto-correlation

properties, robustness to hardware imperfections, and good PAPR. However, it is not adaptable to data embedding, which is crucial for ISAC. To make it more suitable for communications, some researchers have proposed the modification of FMCW. For example, Saddik *et al.* in 2007 proposed to use up-chirp for communications and down-chirp for radar [40], and Han and Wu in 2010 proposed a trapezoidal frequency modulation continuous wave (TFMCW) [41], which is a novel scheme that multiplexes radar cycles and communication cycles in the time domain. Although these schemes enable efficient multiplexing of communications data and sensing signals, they still suffer from low spectrum efficiency due to the chirp sensing signals.

- **Single-carrier waveforms:** Single-carrier waveforms are based on code-domain spreading of joint radar and communications signals [42–47], in which the radar performance is affected by the auto-correlation of sequences. However, long spreading code, which results in good auto-correlation, reduces the spectrum efficiency for communications. In this case, Doppler estimation is not trivial, requiring algorithms that are more complex.

22.3   New Design Expectations and Potential Research Directions

The requirements for use cases, devices, and spectrum which are relevant to waveform and modulation design are summarized below. It is expected that 6G waveforms and modulation schemes should meet these requirements, which may necessitate more than one design.

- **Very high frequency:** The key requirements are low PAPR for coverage, high robustness to RF distortion (e.g., phase noise), low complexity for wideband operations, and compatibility with MIMO. The channel characteristic is high path loss with sparse scattering.
- **Satellite communications:** The key requirements are low PAPR for coverage and low complexity for power saving. The channel characteristics are high path loss due to long distances and high speed due to satellite movement.
- **Short-range communications:** The key requirements are high capacity (e.g., compatible with MIMO) with robustness to RF distortions. The channels are mostly LOS-dominant with a flat frequency response.
- **Low-cost devices:** The key requirements are low PAPR, low complexity for power saving, and high robustness to low-cost hardware. The channel characteristics depend on the application scenarios. For example, the channels are relatively flat in wide-coverage narrowband scenarios.
- **High mobility:** The key requirements are high robustness to the Doppler effect and compatibility with MIMO. The channels are time selective or doubly selective due to the Doppler effect.

- **URLLC:** The key requirements are ultra-low latency and high reliability (e.g., compatibility with MIMO). The channel characteristics depend on the application scenarios.
- **ISAC:** In additional to the communication requirements discussed earlier, the key requirements for sensing are estimation accuracy and precision, especially under RF distortions (e.g., time/frequency synchronization error, phase noise, and non-linearity). The channel characteristics are high path loss for round-trip sensing and ability to share multipath wireless channels with communications.

On the basis of these requirements, 6G system designs should consider the following factors – some of which have already been covered in the technical report [48] for NR beyond 52.6 GHz.

- **PA efficiency:** PA efficiency for high frequencies is expected to degrade, and some use cases (e.g., satellite communications) are sensitive to power consumption. It is therefore necessary to consider low-PAPR waveforms to minimize PA back-off and maximize efficiency. Furthermore, some high-electron mobility transistors for THz are not capable of sustained power transmission, meaning that a waveform with intermittent pulses may be required in the future [49].
- **Dynamic ranges of analog-to-digital converter (ADC) and digital-to-analog converter (DAC):** A wider bandwidth leads to increased difficulty in ensuring a large effective number of bits (ENOB) in ADC/DAC at a given power consumption. But, in order to accommodate higher-PAPR baseband signals, a higher ENOB is required in the transmit DAC. Furthermore, low-cost devices may use low-ENOB DACs/ADCs. These factors need to be considered, as they are all affected by waveforms and modulation schemes.
- **Modulated signal accuracy and OOBE:** The entire RF chain is designed and adjusted to meet RF requirements, such as spectrum emission mask (SEM), adjacent channel leakage ratio (ACLR), in-band emission (IBE), OOBE, and error vector magnitude (EVM). Waveforms and modulation schemes should meet these requirements to offer appropriate in-band signal quality characteristics and minimize adjacent channel interference and impact to signals in adjacent channels. The occupied signal bandwidth and guard band for a given channel bandwidth determine the frequency utilization, which is critical to achieving high throughput (e.g., for short-range communications).
- **Complexity and performance:** Given a high data rate and high sampling rate during high-frequency low-cost device operations or power-sensitive satellite communications, the complexity and performance trade-offs for waveform generation/modulation and reception/demodulation should be considered in the design.
- **Spectrum flexibility:** Use cases and frequency allocations by different government bodies may require various bandwidths to be supported. Therefore, spectrum flexibility should be considered in the designs.
- **Robustness to time offset, frequency offset, and phase noise:** The carrier frequency offset and phase noise are much higher with imperfect PAs and crystal oscillators, and are more critical for high frequencies. In addition, the Doppler shift and

spread increases with the carrier frequency and relative movement speed. For low-cost devices, in addition to the frequency offset and phase noise (depending on the operating frequency), time offsetting should also be considered because only coarse timing synchronization may be available. The accumulation of signals is required to achieve high-resolution sensing, which relies on the coherency of echoes. Phase noise is therefore a major factor affecting system coherence.

- **Compatibility with MIMO:** MIMO is an efficient way to enhance the spectrum efficiency. Millimeter wave and THz communications can support at least 2×2 MIMO with two polarizations, even for pure LOS channels. Therefore, new waveforms should be easily extended to MIMO operations with reasonable complexity.
- **Sensing:** The biggest challenge is contradicting KPIs for communications and sensing. For communications, the main objectives are to maximize the spectrum efficiency and OOBE. For range and Doppler sensing, however, the optimal waveform design aims at maximizing the estimation accuracy and precision.

Given the requirements and considerations outlined above, we discuss some potential research directions below.

- **Low PAPR:** Multi-carrier waveforms with advanced PAPR reduction techniques and single-carrier waveforms with advanced modulation techniques are two promising research directions, which can reduce PAPR without sacrificing the spectrum efficiency. Moreover, AI-optimized constellations and demodulators may also help to reduce PAPR.
- **Low complexity:** For low-cost devices, waveforms are designed for narrowband scenarios where the processing complexity is low to enable power saving. It is also desirable to ensure robustness to the restrictions posed by low-cost hardware (e.g., time and frequency offsets). An example of this is on–off modulation/demodulation, which is simple to implement if the target data rate and access rate meet scenario-specific requirements. On the other hand, for mmWave/THz bands, waveforms are designed for very wide bandwidths. In particular, a simple equalization-based waveform is desirable to avoid high power consumption.
- **Time–frequency localization:** An enhanced time–frequency localization waveform will improve the frequency utilization and spectrum efficiency for short-range communications with high data rate requirements. It will also relax the time synchronization requirements for low-cost device UL access. In addition to filtering and windowing, other techniques (e.g., FTN transmission) may warrant investigation. Furthermore, URLLC requires good time localization–order to achieve relatively short symbol durations and good decoding performance.
- **Spectrum efficiency:** Spectrum efficiency can be improved by using good time–frequency localization, and further improvements in spectrum efficiency can be realized by using AI-optimized high-order constellations and demodulators, which may help to improve the shaping gain with reasonable demodulation complexity. Pilot-free non-coherence modulation and demodulation could also be considered for pilot overhead-sensitive applications.

- **High mobility:** For high-speed use cases with doubly selective channels, advanced schemes that meet overhead, complexity, and other requirements (e.g., low latency for latency-sensitive applications) should be considered. OFDM with additional DMRSs for channel tracking in NR systems can be considered a baseline.
- **Robustness to RF distortion:** Traditional signal-processing-based techniques, such as phase noise tracking and digital pre-distortion for PA non-linearity, have attracted a great deal of interest in academic circles. However, a significant gap exists between real and ideal performances. AI-based techniques can be considered to further reduce the gap in order to achieve near-ideal performance.
- **ISAC:** Although a significant amount of research is targeted at waveform design for ISAC systems, there is still plenty of room for a waveform design that strikes a balance between good communications and sensing performance, owing to their contradictory performance requirements. In particular, the waveform design should consider whether ISAC signals should be used for dedicated sensing purposes, data transmission purposes with general sensing, or both.

References

[1] C. Lin and G. Y. L. Li, "Terahertz communications: An array-of-subarrays solution," *IEEE Communications Magazine*, vol. 54, no. 12, pp. 124–131, 2016.

[2] X. Zhang, L. Chen, J. Qiu, and J. Abdoli, "On the waveform for 5G," *IEEE Communications Magazine*, vol. 54, no. 11, pp. 74–80, 2016.

[3] J. Abdoli, M. Jia, and J. Ma, "Filtered ofdm: A new waveform for future wireless systems," in *Proc. 2015 IEEE 16th International Workshop on Signal Processing Advances in Wireless Communications (SPAWC)*. IEEE, 2015, pp. 66–70.

[4] X. Zhang, M. Jia, L. Chen, J. Ma, and J. Qiu, "Filtered-ofdm-enabler for flexible waveform in the 5th generation cellular networks," in *Proc. 2015 IEEE Global Communications Conference (GLOBECOM)*. IEEE, 2015, pp. 1–6.

[5] R. Zayani, Y. Medjahdi, H. Shaiek, and D. Roviras, "WOLA-OFDM: A potential candidate for asynchronous 5G," in *Proc. 2016 IEEE Globecom Workshops*. IEEE, 2016, pp. 1–5.

[6] Huawei and HiSilicon, "Waveform evaluation updates for case 2," 3rd Generation Partnership Project (3GPP), RAN1 (R1) 166120, Aug. 2016. [Online]. Available: https://www.3gpp.org/ftp/tsg_ran/wg1_rL1/TSGR1_86/Docs/

[7] Huawei and HiSilicon, "Waveform evaluation updates for case 4," 3rd Generation Partnership Project (3GPP), RAN1 (R1) 166091, Aug. 2016. [Online]. Available: https://www.3gpp.org/ftp/tsg_ran/wg1_rL1/TSGR1_86/Docs/

[8] F. Schaich and T. Wild, "Waveform contenders for 5G OFDM vs. FBMC vs. UFMC," in *Proc. 2014 6th International Symposium on Communications, Control and Signal Processing (ISCCSP)*. IEEE, 2014, pp. 457–460.

[9] N. Michailow, M. Matthé, I. S. Gaspar, A. N. Caldevilla, L. L. Mendes, A. Festag, and G. Fettweis, "Generalized frequency division multiplexing for 5th generation cellular networks," *IEEE Transactions on Communications*, vol. 62, no. 9, pp. 3045–3061, 2014.

[10] X. Huang, J. A. Zhang, and Y. J. Guo, "Out-of-band emission reduction and a unified framework for precoded OFDM," *IEEE Communications Magazine*, vol. 53, no. 6, pp. 151–159, 2015.

[11] F. Schaich, "Filterbank based multi carrier transmission (FBMC) evolving OFDM: FBMC in the context of WiMAX," in *Proc. 2010 European Wireless Conference (EW)*. IEEE, 2010, pp. 1051–1058.

[12] C. Kim, K. Kim, Y. H. Yun, Z. Ho, B. Lee, and J.-Y. Seol, "QAM-FBMC: A new multi-carrier system for post-OFDM wireless communications," in *Proc. 2015 IEEE Global Communications Conference (GLOBECOM)*. IEEE, 2015, pp. 1–6.

[13] M. J. Abdoli, M. Jia, and J. Ma, "Weighted circularly convolved filtering in OFDM/OQAM," in *Proc. 2013 IEEE 24th Annual International Symposium on Personal, Indoor, and Mobile Radio Communications (PIMRC)*. IEEE, 2013, pp. 657–661.

[14] H. Lin, "Flexible configured OFDM for 5G air interface," *IEEE Access*, vol. 3, pp. 1861–1870, 2015.

[15] R. Hadani, S. Rakib, M. Tsatsanis, A. Monk, A. J. Goldsmith, A. F. Molisch, and R. Calderbank, "Orthogonal time frequency space modulation," in *Proc. 2017 IEEE Wireless Communications and Networking Conference (WCNC)*. IEEE, 2017, pp. 1–6.

[16] X. Liu, T. Xu, and I. Darwazeh, "Coexistence of orthogonal and non-orthogonal multi-carrier signals in beyond 5G scenarios," in *Proc. 2020 2nd 6G Wireless Summit (6G SUMMIT)*. IEEE, 2020, pp. 1–5.

[17] D. Li, "Overlapped multiplexing principle and an improved capacity on additive white gaussian noise channel," *IEEE Access*, vol. 6, pp. 6840–6848, 2017.

[18] X.-G. Xia, "Precoded and vector OFDM robust to channel spectral nulls and with reduced cyclic prefix length in single transmit antenna systems," *IEEE Transactions on Communications*, vol. 49, no. 8, pp. 1363–1374, 2001.

[19] W. Kozek and A. F. Molisch, "Nonorthogonal pulseshapes for multicarrier communications in doubly dispersive channels," *IEEE Journal on Selected Areas in Communications*, vol. 16, no. 8, pp. 1579–1589, 1998.

[20] Z. Zhao, M. Schellmann, Q. Wang, X. Gong, R. Boehnke, and W. Xu, "Pulse shaped OFDM for asynchronous uplink access," in *Proc. 2015 49th Asilomar Conference on Signals, Systems and Computers*. IEEE, 2015, pp. 3–7.

[21] X. Yu, Y. Guanghui, Y. Xiao, Y. Zhen, X. Jun, and G. Bo, "FB-OFDM: A novel multicarrier scheme for 5G," in *Proc. 2016 European Conference on Networks and Communications (EuCNC)*. IEEE, 2016, pp. 271–276.

[22] S. Galli, H. Koga, and N. Kodama, "Advanced signal processing for PLCs: Wavelet-OFDM," in *Proc. 2008 IEEE International Symposium on Power Line Communications and Its Applications*. IEEE, 2008, pp. 187–192.

[23] K. Tourki, R. Zakaria, and M. Debbah, "Lagrange Vandermonde division multiplexing," in *Proc. 2020 IEEE International Conference on Communications (ICC)*. IEEE, 2020, pp. 1–6.

[24] M. Martone, "A multicarrier system based on the fractional Fourier transform for time-frequency-selective channels," *IEEE Transactions on Communications*, vol. 49, no. 6, pp. 1011–1020, 2001.

[25] T. Erseghe, N. Laurenti, and V. Cellini, "A multicarrier architecture based upon the affine Fourier transform," *IEEE Transactions on Communications*, vol. 53, no. 5, pp. 853–862, 2005.

[26] X. Yang, X. Wang, and J. Zhang, "A new waveform based on Slepian basis for 5G system," in *Proc. 2016 Wireless Days Conference (WD)*. IEEE, 2016, pp. 1–4.

[27] F. Pancaldi, G. M. Vitetta, R. Kalbasi, N. Al-Dhahir, M. Uysal, and H. Mheidat, "Single-carrier frequency domain equalization," *IEEE Signal Processing Magazine*, vol. 25, no. 5, pp. 37–56, 2008.

[28] G. Berardinelli, F. M. Tavares, T. B. Sørensen, P. Mogensen, and K. Pajukoski, "Zero-tail DFT-spread-OFDM signals," in *Proc. 2013 IEEE Globecom Workshops*. IEEE, 2013, pp. 229–234.

[29] J. B. Anderson, F. Rusek, and V. Öwall, "Faster-than-Nyquist signaling," *Proceedings of the IEEE*, vol. 101, no. 8, pp. 1817–1830, 2013.

[30] O. İşcan, R. Böhnke, and W. Xu, "Probabilistic shaping using 5G new radio polar codes," *IEEE Access*, vol. 7, pp. 22 579–22 587, 2019.

[31] MediaTek Inc., "A new DFTS-OFDM compatible low PAPR technique for NR uplink waveforms," 3rd Generation Partnership Project (3GPP), RAN1 (R1) 1609378, Oct. 2016. [Online]. Available: https://www.3gpp.org/ftp/TSG_RAN/WG1_RL1/TSGR1_86b/Docs/

[32] H. Nikopour and H. Baligh, "Sparse code multiple access," in *Proc. 2013 IEEE 24th Annual International Symposium on Personal, Indoor, and Mobile Radio Communications (PIMRC)*. IEEE, 2013, pp. 332–336.

[33] M. Taherzadeh, H. Nikopour, A. Bayesteh, and H. Baligh, "Scma codebook design," in *Proc. 2014 IEEE 80th Vehicular Technology Conference (VTC2014-Fall)*. IEEE, 2014, pp. 1–5.

[34] R. H. Gohary and H. Yanikomeroglu, "Noncoherent MIMO signaling for block-fading channels: Approaches and challenges," *IEEE Vehicular Technology Magazine*, vol. 14, no. 1, pp. 80–88, 2019.

[35] E. Basar, "Index modulation techniques for 5G wireless networks," *IEEE Communications Magazine*, vol. 54, no. 7, pp. 168–175, 2016.

[36] B. Paul, A. R. Chiriyath, and D. W. Bliss, "Survey of RF communications and sensing convergence research," *IEEE Access*, vol. 5, pp. 252–270, 2016.

[37] C. Sturm, E. Pancera, T. Zwick, and W. Wiesbeck, "A novel approach to OFDM radar processing," in *Proc. 2009 IEEE Radar Conference*. IEEE, 2009, pp. 1–4.

[38] M. Braun, C. Sturm, and F. K. Jondral, "Maximum likelihood speed and distance estimation for OFDM radar," in *Proc. 2010 IEEE Radar Conference*. IEEE, 2010, pp. 256–261.

[39] M. Braun, C. Sturm, A. Niethammer, and F. K. Jondral, "Parametrization of joint OFDM-based radar and communication systems for vehicular applications," in *Proc. 2009 IEEE 20th International Symposium on Personal, Indoor and Mobile Radio Communications*. IEEE, 2009, pp. 3020–3024.

[40] G. N. Saddik, R. S. Singh, and E. R. Brown, "Ultra-wideband multifunctional communications/radar system," *IEEE Transactions on Microwave Theory and Techniques*, vol. 55, no. 7, pp. 1431–1437, 2007.

[41] L. Han and K. Wu, "Radar and radio data fusion platform for future intelligent transportation system," in *Proc. 7th European Radar Conference*. IEEE, 2010, pp. 65–68.

[42] K. Mizui, M. Uchida, and M. Nakagawa, "Vehicle-to-vehicle 2-way communication and ranging system using spread spectrum technique: Proposal of double boomerang transmission system," in *Proc. Vehicle Navigation and Information Systems Conference*. IEEE, 1994, pp. 153–158.

[43] S. Lindenmeier, K. Boehm, and J. F. Luy, "A wireless data link for mobile applications," *IEEE Microwave and Wireless Components Letters*, vol. 13, no. 8, pp. 326–328, 2003.

[44] S. Xu, Y. Chen, and P. Zhang, "Integrated radar and communication based on DS-UWB," in *Proc. 2006 3rd International Conference on Ultrawideband and Ultrashort Impulse Signals*. IEEE, 2006, pp. 142–144.

[45] Z. Lin and P. Wei, "Pulse amplitude modulation direct sequence ultra wideband sharing signal for communication and radar systems," in *Proc. 2006 7th International Symposium on Antennas, Propagation & EM Theory*. IEEE, 2006, pp. 1–5.

[46] Z. Lin and P. Wei, "Pulse position modulation time hopping ultra wideband sharing signal for radar and communication system," in *Proc. 2006 CIE International Conference on Radar*. IEEE, 2006, pp. 1–4.

[47] M. Bocquet, C. Loyez, C. Lethien, N. Deparis, M. Heddebaut, A. Rivenq, and N. Rolland, "A multifunctional 60-GHz system for automotive applications with communication and positioning abilities based on time reversal," in *Proc. 7th European Radar Conference*. IEEE, 2010, pp. 61–64.

[48] 3GPP, "Study on requirements for NR beyond 52.6 GHz," 3rd Generation Partnership Project (3GPP), Technical Report (TR) 38.807, Jan. 2020, version 16.0.0. [Online]. Available: https://portal.3gpp.org/desktopmodules/Specifications/Specification Details.aspx?specificationId=3522

[49] J. M. Jornet and I. F. Akyildiz, "Femtosecond-long pulse-based modulation for terahertz band communication in nanonetworks," *IEEE Transactions on Communications*, vol. 62, no. 5, pp. 1742–1754, 2014.

23 New Coding

23.1 Background and Motivations

Channel coding is a fundamental component in wireless communication. From 2G to 5G, wireless systems have always adopted state-of-the-art channel coding technologies. Innovations in channel coding design – for example, convolutional codes for 2G, turbo codes for 3G and 4G, as well as polar and low-density parity-check (LDPC) codes for 5G – have accelerated the development of coding technologies. Following Moore's law, high performance encoders and decoders with higher performance and lower power consumption continue to emerge at lower costs, enabling a wide range of advanced channel coding technologies. As of the year 2020, the channel coding performance has almost reached the theoretical Shannon limit for an additive white Gaussian noise (AWGN) channel, with a reasonable implementation cost.

Channel coding is used to address hostile and dynamic channel environments, ensuring reliable communication and QoS. Because 6G will cover many diverse use cases, some with extreme requirements such as ultra-high data rates, ultra-low latency, and ultra-low power consumption, innovations in 6G channel coding are needed to deliver a powerful channel code solution that can provide the optimal code construction for a given channel condition and usage scenario. This is discussed in Section 23.2.

The foundational work of Shannon's theorem allows channel coding and information source coding to be separated as two layers while achieving the maximum channel capacity. This is true if the information block size is very large, but it increases the implementation complexity and latency for both the encoder and decoder. As such, using the current separate source coding and channel coding schemes to achieve ultra-high data rates with ultra-low latency is not optimal. Instead, 6G research will explore a cross-layer approach to jointly design source and channel coding. Section 23.3 discusses this in more detail. Such research will be important for performance optimization in 6G applications.

Beyond the classical Shannon's theorem, there is motivation to revisit a broader definition of intelligent communications. Warren Weaver, in his joint book with Shannon published in 1949 [1], proposed a three-level intelligent communications hierarchy: the first level is Shannon's theorem (which is technical), the second level is semantic, and the third level is effective. In terms of 6G, semantic communication has many applications in most 6G use cases, especially for machine-to-machine and human-to-machine communications. In the intelligent communications hierarchy, the wireless transmission channel uses inner channel coding, which is based on Shannon's theorem and uses an outer semantic channel. In this semantic channel, source codes extract semantics of interest other than simply compressing semantics towards information entropy. Integrating the wireless transmission channel and outer semantic channel is a new and fundamental area of research in 6G, attracting some exploration already [2]. Semantic extraction is usually performed with machine learning using a data-driven method. This is discussed later in Section 23.3.

Channel coding can be applied to P2P communications over AWGN channels, and multipoint-to-point (MP2P), point-to-multipoint (P2MP), and multipoint-to-multipoint (MP2MP) communications over non-AWGN or non-stationary channels. In 6G, with more machine-to-machine and human-to-machine communications, network coding can be applied in MP2P, P2MP, and MP2MP communication scenarios to help improve the overall communication spectrum efficiency. The basic idea of network coding is to combine several packets into one for better coding gains. Another critical aspect of using network coding is for interference cancellation, in either the digital or the analog domain. The coding gains can be achieved and obtained by all participating devices in the network. Network coding is typically performed at the application layer, but it may be performed at the physical layer in 6G due to latency and throughput requirements. Section 23.4 discusses this in more detail.

23.2   Channel Coding Schemes

23.2.1   Background

Since the introduction of Shannon's theorem in 1948, various channel coding schemes have been proposed and implemented in different communication systems. Early schemes used algebraic codes whose information bits and encoded bits satisfied linear algebraic relationships. Their designs aimed to maximize the code distance with linear algebraic properties in order to maximize error correction capabilities. Among the first channel coding schemes were Hamming codes [3], Golay codes [4], and Reed–Muller (RM) codes [5]. These code constructions were so restrictive that the code lengths and rates they supported were limited. The discovery of cyclic codes [6] led to code constructions that were much more flexible. Two prime examples of this are Bose–Chaudhuri–Hocquenghem (BCH) codes [7, 8] and Reed–Solomon codes [9]. These algebraic cyclic codes exhibit excellent code distance properties, offering good performance under soft decoders [10]. Powerful soft decoding schemes such as list decoding [11] and ordered statistics decoding [12] yield performance reaching the maximum likelihood for many short codes.

Channel coding has been a key driver in helping wireless systems approach the Shannon limit. Although algebraic codes made it possible to transmit digital signals over long distances, their coding gains are far from the Shannon limit owing to their relatively short-length applicability. Wireless communication systems have used a variety of coding schemes. For example, 2G used convolutional codes [13], whereas 3G and 4G used turbo codes [14]. Modern coding schemes such as turbo codes approach the theoretical channel capacity with a larger block size, significantly improving the spectrum efficiency of mobile communications. The success of the turbo iterative decoder led to the rediscovery of low-density parity-check (LDPC) codes. Irregular constructions for LDPC codes [15] were shown to approach the Shannon limit with a gap of only 0.0045 dB. The first codes proven to achieve the theoretical channel capacity and close the gap were polar codes [16].

Channel coding capable of achieving the theoretical capacity helps to improve user experience in many ways. For instance, a high coding gain extends network coverage and improves QoS, especially at cell edges. Furthermore, a highly parallelized decoder boosts peak data rates; and a simple, hardware-friendly design substantially reduces power consumption, thereby extending battery life. These benefits have been central to the success of previous wireless generations. 5G NR standards adopt polar codes and LDPC codes for control channels and data channels, respectively, and use rate-compatible codes with hardware-friendly features to deliver coding gains.

6G applications will require performance surpassing that available in 5G: faster data rates, higher reliability, lower complexity, and lower power consumption. They also require a more diverse range of KPIs that are not present in previous generations. The following sections discuss the key requirements and KPIs for 6G channel coding design, as well as some design principles.

23.2.2 Target KPIs of 6G Channel Coding

6G channel coding will continue to make performance improvements in existing 5G usage scenarios, namely, eMBB, URLCC, and mMTC. Such improvements will include raising the peak data rate to the Tbit/s level (the current eMBB data plane decoding rate is 10–20 Gbit/s), eliminating the block decoding error floor for URLLC, and improving the short block length decoding performance for mMTC towards the finite-length performance bound [17]. Other use cases, such as autonomous driving, industry verticals, and satellite communications, require a variety of target KPIs, including sub-millisecond latency, seven to ten nines reliability, high power efficiency, and long battery life. Figure 23.1 shows the KPIs achieved by 5G NR polar codes for control channels and 5G NR LDPC codes for data channels, with respect to 6G's diverse channel coding KPIs. Some of the typical target KPIs are described below.

- **Tbit/s throughput:** As 6G will become commercially available by 2030, VR/AR applications will become mainstream and many of them will leverage 6G's wider bandwidth for 1 Tbit/s of wireless throughput over short distances. As such, a peak data rate of 1 Tbit/s will be a basic KPI for 6G.
- **Performance of short-length codes:** A wide variety of machine-to-machine communication use cases will adopt short-length codes (i.e., shorter than 200 bits). The coding performance optimizations will improve coverage and benefit many relevant applications. At present, however, the channel coding schemes used at the

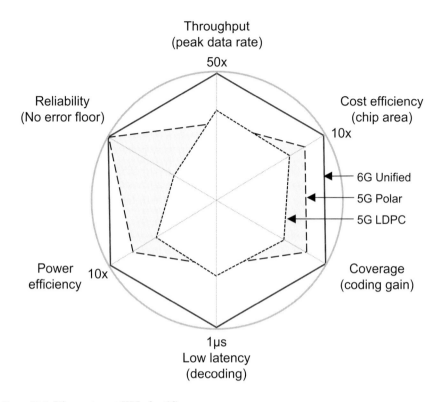

Figure 23.1 Diverse target KPIs for 6G.

5G data plane have a gap of 1–2 dB to the finite-length performance bound by a practical decoder with reasonable implementation complexity. 6G will not only narrow this gap, but also simplify decoding algorithms to improve the decoder's power efficiency.

- **Low encoding and decoding latency and long battery life:** Application-specific KPIs, such as high energy efficiency, long battery life, and low decoding complexity, depend on the code constructions. This should be considered in conjunction with air interface latency, transmission reliability, and coverage. Furthermore, the code constructions should be sufficiently flexible to adapt to different KPI combinations for different use cases. For example, high-reliability and low-latency applications may require coverage or mobility, whereas other applications may accept a moderate decoding latency in order to obtain a higher coding gain (e.g., to compensate for fading induced by high mobility).

As silicon technology continues to advance, more channel coding schemes will become feasible for 6G applications. And because complex decoding algorithms can be implemented at lower cost, trade-offs between the performance gain and implementation complexity will be explored further in 6G design.

23.2.3   6G Channel Coding Design Principles

Unified and Flexible Channel Coding

In 5G, it has been difficult to select a channel coding scheme that meets multiple requirements in multiple use cases – no one-size-fits-all channel coding scheme exists. This issue will remain in 6G, where diverse use cases will pose new challenges for 6G channel coding. Because customized, use-case-specific channel coding schemes will increase implementation complexity, there is strong motivation to either design a family of codes that reuse hardware and can be applied to diverse applications and specifications or devise a framework that efficiently adapts multiple codes to different applications.

Such a framework, comprising component codes or base codes, will need to be unified and flexible. It will also need to be adaptive in order to meet the requirements of diverse 6G use cases. Two examples of how to design this framework are briefly introduced below.

Example 1: Unified and flexible framework based on polar codes.

Polar codes demonstrate superior performance over a wide range of code lengths and rates. With an optimized code construction, polar codes achieve stable performance at one-bit granularity [18] in terms of information encoding block length. They also support multiple decoders [19], ranging from low-complexity successive cancellation (SC) decoding to higher-complexity SC-list (SCL) decoding with higher performance. A soft-cancellation decoder [20] or belief propagation (BP) decoder may also be utilized to support soft bit output in advanced iterative receivers. And in scenarios where high throughput is desirable, a bit-flipping (BF) decoder with parallel scheduling can be implemented. Although SC-based decoders are inherently serial, some parallelism can be introduced for URLLC scenarios [21]. In mMTC scenarios, where short block lengths are typical, polar codes can be optimized in terms of the code distance in order to achieve comparable performance with RM codes and extended BCH codes [22]. For ultra-high-throughput applications, novel constructions within the G_N-coset code framework (described later) support fully parallelized decoding [20]. Polar-based schemes either adapt the decoder or optimize code constructions to serve specific purposes. As such, the decoding core and processing elements can be reused in hardware, and the code description can be unified into one framework.

Example 2: Multi-code adaptation framework.

The introduction of multiple codes needs a code adaptation scheme. LDPC codes – especially quasi-cyclic raptor-like LDPC codes [23] – support fine-granularity code constructions and HARQ. These codes, which have been the subject of extensive research, perform well in high-rate and long-length scenarios. However, their performance and implementation efficiency are undermined by low-rate and short-length codes. In very short block length scenarios, polar codes and algebraic codes (such as the BCH reference signal) are more competitive than LDPC codes [24]. In very long block length scenarios, spatially coupled codes [25] with component codes such as LDPC codes [26], polar codes, and other generalized coupling structures have the potential to achieve an optimal balance between complexity and performance. Rate-matching [27] and HARQ schemes of spatially coupled codes require further study. To combine the strengths of these codes effectively, a code adaptation framework can associate each channel coding scheme with a specific use case or type of user device.

Tbit/s Coding Design

One of the most important target KPIs is high throughput. In order to deal with the higher complexity and larger power consumption, powerful channel encoders and decoders are required. For example, achieving 1 Tbit/s throughput requires about a 50-fold increase in throughput over 5G with limited resources. At present, neither LDPC codes nor polar codes can achieve an area efficiency greater than 1 Tbit/s per mm^2 and power efficiency less than 1 pJ/bit [28]. The LDPC decoder is limited due to silicon-based routing, whereas the polar decoder is limited due to its successive decoding architecture. As such, a novel decoding architecture that achieves high efficiency and fast implementation is a target for further research.

The design of a 1 Tbit/s coding scheme should prioritize low complexity and a high degree of decoding parallelism, paying special attention to the hardware implementation [29]. For example, codes implemented with simpler routing schemes will require a smaller die area and exhibit a higher degree of regularity in both encoding and decoding. Alternatively, some existing codes can be optimized for higher parallelism, higher regularity, and lower complexity, either on the decoding side [30] or on both code construction and decoding [31].

Parallelized polar (the G_N-coset code framework) includes polar and RM codes, and can support high-throughput decoding with both parallelism and regularity. G_N-coset codes can be linear block codes with the same generator matrix as polar codes but different information sets. A parallel decoding algorithm can be used on the factor graph of G_N-coset codes [31]. This algorithm exhibits both high parallelism and high regularity. It treats G_N-coset codes as concatenation codes and decodes the inner codes in parallel. To avoid decoding the outer codes in series, the algorithm either constructs an equivalent decoding graph by swapping the outer and inner codes in the previous factor graph, as shown in Figure 23.2, or decodes the inner codes on both decoding graphs.

Short-Length Coding

To close the 1 dB performance gap to the theoretical finite-length error correction bound, polar codes and short algebraic codes such as RM and BCH/RS codes may be considered [24]. Polar codes decoded by a successive cancellation list (SCL) decoder strike an optimal balance between performance and complexity in short-length scenarios, and are therefore adopted for 5G control channels. For the same reason, some 6G scenarios where short block lengths will be prevalent can also adopt polar codes for

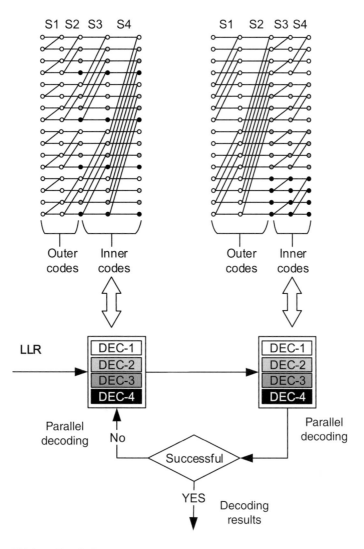

Figure 23.2 Tbit/s coding design.

data channels, with the additional benefit of reusing 5G legacy hardware. By increasing the list size, short polar codes approach the random coding union bound at the low error rate region [24]. In addition, AI techniques can be used to further optimize code constructions for list decoders [32, 33]. Some research has been conducted into fast and simplified SCL decoding implementations [19, 34], and the following areas are being explored:

• Finite-length performance can be improved by modifying the polarization kernels in order to increase the polarization rate [35]. Some examples of this include convolutional polar codes [36, 37], polar codes with large binary kernels [38], mixed-kernel polar codes [39], and non-binary polar codes [40].

- To further enhance the performance of polar codes, polarization-adjusted convolutional (PAC) codes [41] can be concatenated by polar transformations. With Fano decoding techniques, this construction has exhibited bound-achieving performance.
- Code constructions based on outer code concatenation can be unified under pre-transformed polar codes, where the minimum code distance is guaranteed to improve [22].

Algebraic codes were primarily used in optical communications and storage applications, but, for use in wireless communications, their code constructions need to be customized. The design principles are as follows:

- Code distance has a major impact on performance. Research is needed into how to construct efficient short channel codes with good minimum distance properties.
- Adaptability to varying packet sizes is required. Practical rate and length adaptation methods for high-performing short codes need to be studied.
- Novel decoding algorithms that strike a balance between computational complexity and implementation complexity need to be devised. In this regard, ordered statistics decoding [12] is a viable candidate, but it needs further optimization to minimize complexity. An impractical decoder cannot be implemented even if it achieves close-to-bound performance.

Coding for Mission-Critical Applications

In addition to conventional broadband applications, 6G will enable a multitude of mission-critical applications that provide extreme performance guarantees. Such guarantees may not be fulfilled by existing coding schemes, so application-specific codes must be devised for some cases.

- **Extremely high reliability,** from 10^{-7} to 10^{-10} without an error floor and with good performance. Although an error floor is inevitable with LDPC and other iteratively decoded codes [42], it is possible to lower this below 10^{-10}. Theoretically, polar codes and related code constructions have been proven to have no error floor, and are therefore suitable for these applications. Advanced features such as non-binary or convolutional constructions may be employed to further enhance reliability.
- **Extremely low latency,** from 1 ms to 0.1 ms to support autonomous driving and industrial automation. These applications typically require high reliability along with extremely low encoding and decoding latency. Algebraic codes may be suitable, but existing schemes require ordered statistics decoding that is too complex for a bounded-latency application. New code constructions and decoding schemes need to be devised in order to support this type of use case. Decoding latency for short polar codes has been demonstrated to be very low [21], and can be lowered further with segmented decoding [43] or parallelized decoding [44].

- **Extremely low power and cost,** to support battery-less and energy-harvesting implementations over a 10-year to infinite lifespan. Smart devices will modulate and reflect received RF signals, or they will harvest ambient radio energy for RF transmissions. Such devices typically use low-cost hardware, mandating extremely low power consumption. This may be delivered by a joint code-modulation design, where channel coding is co-designed with waveform and reference signals [45].
- **Extremely high density,** up to 100 million devices per km^2. For densely deployed IoT systems, numerous collisions are to be expected. The codes need to carry both UE IDs and data. A joint sequence coding design with an extremely large sequence/code space supporting non-coherent detection/decoding is desirable.

23.3 Joint Source and Channel Coding

23.3.1 Research Background

Joint source and channel coding (JSCC) has been the subject of extensive research for more than 40 years [46]. For finite-length transmissions, upper bounds for the channel coding rate were derived by introducing channel dispersion [17]. A couple of years later, lower bounds for the lossy source coding rate of finite block lengths were derived by employing source d-tilted information [47] . The results of these works were combined, proving that JSCC outperforms separate source and channel coding (SSCC) in terms of overall rate in finite-length regions [48]. Figure 23.3 highlights the differences between JSCC and SSCC.

These works prove that source coding in a finite length is often imperfect and its ensuing channel decoder can exploit the leftover redundancy. This has triggered a great deal of investigation into JSCC. For example, the high-level or a posteriori information of source codes (e.g., Huffman [49], JPEG [50], Arithmetic [51], and Lempel–Ziv [52]) can be employed to assist sparse-graph-based channel decoding. In [53], a joint decoding scheme based on polar codes and a language decoder was proposed, in which a dictionary is used to correct most early-stage decoding errors. To leverage the memory of sources, some researchers have built joint Markov source and channel decoding based on the assumption that some sources can be modeled as a Markov (or hidden Markov) process [54, 55]. To solve the problem of distributive compression of multi-terminal sources, various source models and joint channel

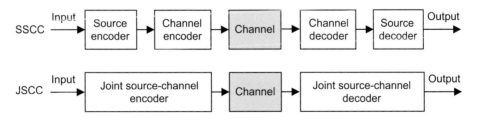

Figure 23.3 Illustration of differences between SSCC and JSCC.

decoding schemes have been proposed based on the well-known Slepian–Wolf theorem [56, 57]. Although these JSCC approaches outperform their SSCC counterparts, they depend heavily on source compression schemes and suffer from poor generalization. In other words, if a source compression scheme is changed, the joint decoder must be redesigned.

Another area of JSCC attracting significant research is wireless image/video transmission. For example, [58] proposed an analog transmission scheme called SoftCast. In this scheme, the original image pixels are linearly coded to analog coefficients and transmitted over OFDM, and the receiver decodes pixels with the instantaneous channel quality. To address the low spectrum efficiency and compression gain inherent in analog transmission, some hybrid digital–analog transmission schemes have been proposed. For example, [59] proposed transmitting a digital base layer via traditional digital methods and transmitting the enhancement layer via pseudo-analog and spatial multiplexing. In [60], the Shannon–Kotel'nikov mapping was adapted to further compress the analog coefficients. JSCC has also been discussed recently for deep-space image transmission. An example of this is [61], which proposed using a fixed-to-fixed-length linear coding approach based on Raptor codes to replace the traditional entropy coding, thereby enhancing robustness to decoding errors and avoiding extremely long retransmission time. As these works demonstrate, JSCC can significantly reduce the transmission latency and provide acceptable image/video quality even when the channel condition is poor. However, the proposed designs all target a specific scenario: image/video transmission. The joint design of source coding and channel coding is highly integrated, making it difficult for such a design to gain popularity.

Some researchers have explored combining ML techniques with JSCC. For example, [62] proposed using auto-encoder-based JSCC for wireless image transmission, where the noisy communication channel is represented as a middle layer in the encoder-decoder pair. In [63] and [64], joint transmission–recognition based on a DNN was examined for transmitting data wirelessly to a server for recognition tasks. In addition to images, ML-based JSCC can also be used to protect text transmissions over a noisy channel, as discussed in [65] and [66]. Similarly to previous works, these ML-enabled JSCC approaches can enhance channel adaptability and increase the optimization space compared with traditional JSCC; however, the corresponding design is task-specific and not a generalized solution. The articles [65] and [66] discussed how to preserve the semantic information of sentences instead of minimizing bit error rates in wireless transmissions. In the next section, we explore the key points involved in extracting and protecting semantics in machine type communications.

23.3.2   JSCC Based on ML

The development of AI and ML will affect JSCC. For 6G, Weaver's semantic communication has many applications in most 6G use cases related to machine-to-machine and human-to-machine communications. This architecture involves an inner channel based on classical channel coding and an outer semantic channel based on deep learning. Integrating these two channels is the subject of research in 6G, with some studies already published [2].

With the help of AI and ML, one important goal for 6G machine-to-machine communication is to create a new source coding paradigm for AI-to-AI communications. In advanced source coding for AI agents, deep-neural-network-based ML is a major driver for AI-to-AI communications. Specifically, a trained DNN partitions a high-dimensional space, with the training considered as an "interpolator" and its inference considered as an "extrapolator." Conversely, a DNN is also a dimension reducer that compresses or source-encodes information by removing irrelevant and less important components in a low-dimensional representation. Furthermore, a DNN can be reconstructed as a dimension expander, decompressor, or source decoder from its low-dimensional representation.

Typically, encoder and decoder DNNs are connected through AE architecture that reflects the rate-distortion theorem in general: an increasing dimension of the bottleneck layer decreases distortion, whereas decreasing dimension increases distortion. The theoretical foundations behind semantic communication are discussed in Chapter 8. Features of the JSCC for AI-to-AI communications may include the following.

- **JSCC beyond classic Shannon information theory:** The objective of classical source coding is to compress the information source bit rate. This is performed to minimize the transmission cost and maximize the network capacity. In most cases, source coding is optimized primarily for the fidelity of human perception. At the transmit side, the bit rate is maximally compressed while retaining the Shannon information and avoiding information loss. At the receive side, the receiver decompresses the information, ensuring no impact on human perception. In 6G, however, the objective of source coding is not to preserve the fidelity of human perception but rather to enable machine-to-machine communications. It may be possible to further compress the bit rate, even when source coding is applied to big data in machine learning (i.e., training DNNs).
- **A new metric beyond the rate distortion theory:** In practical applications, an optimal balance has been achieved between two classical metrics: rate and distortion. In human-to-human communications, however, such a balance requires an additional metric, i.e., human perception via scoring systems with human participation. In 6G research, we also need to explore a new metric for machine-to-machine communications, i.e., machine perception, which may be implemented by a DNN.
- **Multi-terminal JSCC:** Classical information theory addresses multi-terminal source code based on the Wolf theorem, which assumes that two input streams are independent and equally important. However, for many use cases in 6G, sensory data may not be independent or equally important. For example, two sensors may measure the same physical world but do so from different locations or using different sensing technologies. Furthermore, one sensor may capture more information than the other does, experience more channel interference, or be dependent on the other sensor's channel coefficient. Research into multi-terminal JSCC is gaining popularity, especially in the applications of DNN construction, where multi-sensory big data is compressed for training machine learning.

The DNN supports multi-terminal inputs. One such input must be set as the main input for the training of machine learning. During the training process, the other inputs will be autonomously weighed and fused by the neurons. In this case, the DNN will discard unsuitable information input from a sensor by assigning it a low weight.

- **Task-driven JSCC:** Traditional source coding focuses on the E2E distortion to meet human perception requirements. However, machine perception requirements may differ significantly in aspects such as object detection, classification, and reconstruction. Intuitively, we can compress more information for target detection or classification than for target reconstruction. Various ML tasks can be realized through DNN encoder and decoder training.

23.3.3   6G JSCC Design Principles

Compared with SSCC, JSCC is more efficient for the finite coding of block lengths and latency constraints. Many existing systems optimize the JSCC design based on the compression scheme (e.g., Huffman compression [49, 53]) or the type of data source (e.g., image [61, 62] or text [65, 66]). In 6G, however, there are three opportunities to apply novel JSCC. The first is in low-latency short block basic wireless link machine-to-machine communications. The second is in high-speed visual communications for the virtual world, where hybrid analog and digital JSCC can achieve low latency and low power consumption due to the simple compression scheme, especially for short-range communications. In ML, data-driven training can be compressed with auto-encoder techniques based on the information bottleneck. This is a new area for AI-to-AI communications in 6G.

To enhance the performance of JSCC and its applications in 6G, a generalized design for JSCC is needed. Additionally, ML techniques can be utilized to extract semantics of interest in machine type communications. Some design guidelines and further research directions are listed below.

- It is desirable to develop a generalized framework for JSCC. Such a framework should apply to various applications and not be highly dependent on the compression scheme of sources.
- Enhanced scalability is required to meet the requirements of various applications. For this purpose, research can be conducted into bit-wise and message-wise JSCC, for example, to define and protect against errors at the bit level and message level, respectively.
- A new design metric is necessary for JSCC especially in machine-to-machine communications. Such a metric should consider lossy transmission, information timeliness, attention control, memory of sources and channels, etc.
- For applications with massive connections, especially for machine type communications, the information shared between different users may be highly correlated. This correlation can be exploited to further enhance the performance of JSCC.

- ML-enabled source coding and communication offer a wider optimization space in JSCC compared with traditional approaches. The research in [62, 66] has provided a basis for new design dimensions, but further research is needed. And to protect user privacy, a design based on "black box" content and "white box" channels is desirable.

23.4   PHY Network Coding

23.4.1   Background

Shannon's theorem is built upon P2P communication for AWGN memoryless channels, whereas the network coding scheme deals with coding for multi-point networks. In this case, PHY network coding delivers performance enhancements via coding gain and diversity gain, especially for many 6G use cases and applications that have low-latency constraints. Due to the potential performance gain, it is of interest to research network coding for the PHY layer, and even for the analog signals (i.e., analog network coding). Network coding was first introduced by Ahlswede, Cai, Li, and Yeung in 2000 in their seminal paper [67]. A processing node encodes incoming information flows to improve spectrum efficiency. To do this network coding, the encoding node receives and combines multiple packets into a combined packet and transmits it out. The input flows can be from one or more nodes, called intra-session or inter-session respectively. A longer code block means a higher coding gain, and different transmission paths represent some diversity gain. The articles [68–70] discuss several types of PHY network coding, which are briefly described below.

- **Linear network coding:** An encoding node performs a linear combination of the inputs. It can be as simple as performing an exclusive OR (XOR) operation. Linear network coding is relatively simple and widely used for broadcast and relay. Random linear network coding is similar but uses random coefficient to tackle unknown topologies or distributive communication.
- **Analog network coding (ANC):** It operates directly on analog signals by super-imposing electromagnetic waves from S_1 to relay and S_2 to relay. ANC is typically used in two-way relay communications, as shown in Figure 23.4(c).

Different Application Scenarios for PHY Network Coding
- **Network coding for broadcast/groupcast/unicast:** 5G introduced groupcast transmission with HARQ for sidelink communication to improve reliability. However, HARQ efficiency and latency can be further improved, especially in broadcast and groupcast scenarios, as shown in Figure 23.4(a). Because one piece of information is transmitted to multiple UEs, the probability of information loss occurring is high, especially for UEs in poor channel conditions. This probability can be reduced by using network coding (e.g., with XOR for HARQ retransmission) to combine lost packets from different receivers or the same receiver [71, 72]. During unicast transmission, the information sent to one UE can be overheard by

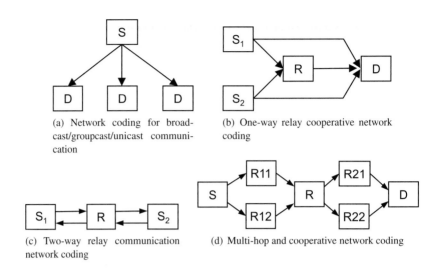

(a) Network coding for broadcast/groupcast/unicast communication

(b) One-way relay cooperative network coding

(c) Two-way relay communication network coding

(d) Multi-hop and cooperative network coding

Figure 23.4 Illustrations of application scenarios for network coding.

another UE, meaning that network coding-based HARQ retransmission can be enabled [73].

- **One-way relay cooperative network coding:** One-way relay cooperative communication will enhance reliability. In Figure 23.4(b), both S_1 and S_2 send information to D in orthogonal resource units in the first phase, and the relay node R forwards the information received from S_1 and S_2 in other orthogonal resource units in the second phase. PHY network coding can keep the same diversity order with similar spatial path and fewer resources compared to a scheme with orthogonal resource units in the second phase [74]. In the second phase, node R can perform an XOR operation between the received bits from S_1 and S_2. Whereas a traditional scheme requires four resource units, a PHY network coding scheme requires only three to achieve the same diversity order.

- **Two-way relay communication network coding:** Two-way relay communication will enable two sources, e.g., S_1 and S_2 in Figure 23.4(c), to communicate with each other via a relay. If digital network coding is used in the second phase, three resource units are required, whereas if ANC is used, only two resource units are required. One type of ANC is physical-layer network coding (PNC) [68, 69, 75]: in the first phase, the two sources transmit signals simultaneously to the relay, assuming adequate synchronization. Upon receipt, the relay will first make channel decoding of the superimposed signal, then will transform the superimposed information to network-coded information, and finally channel-encodes the information for broadcast to the two sources. Because the interference is unknown to each source, the source nodes perform XOR operations to obtain the expected signal from the broadcast signal. Differently from PNC, an ANC [70] relay simply amplifies and broadcasts the received network-coded superimposed signal to both sources. Each source node can then decode its expected signal based on the known signal.

- **Multi-hop and cooperative network coding:** Multi-hop cooperative communication is a type of mesh network transmission, as shown in Figure 23.4(d). However, the trade-offs between throughput, latency, and reliability for mesh networks are not fully understood. One example scheme is adaptive casual network coding [76, 77]. Network coding in multi-hop cooperative transmission has been considered. Both prior and posterior FEC network coding are used based on ACK/NACK feedback and the number of new packets. Between adjacent hops, rate matching is proposed to align the path rate and maximize the throughput [77].

23.4.2 6G PHY Network Coding Design Principles

Network coding offers two significant benefits: coding gain and diversity gain. From the perspective of information theory, side information will facilitate a decoder cancelling out known radio interference. As such, we can extend network coding from broadcast/groupcast to cooperative and relay transmission in 6G.

- **Instant decodable network coding:** To meet the ultra-low latency and cost requirements in 6G, instantly decodable network coding (IDNC) schemes [78, 79] have been proposed. Only simple binary XOR is used by network-coding encoders and decoders to reduce latency and memory consumption despite information loss. A progressive IDNC decoder is beneficial for latency-sensitive communication but requires enough feedback to obtain the decoding status from the receiver.
- **Joint channel and network coding:** Network coding applied at the PHY layer is considered as "outer code." To improve performance, channel and network coding can be designed together, sharing redundancy information between them. Part of the redundancy information is included in a channel code word, while the rest is included in a network code word and transmitted over another path via a relay node [80]. A decoding node can use successive cancellation to decode information from both code words. The two paths may not be strictly independent or memoryless, meaning that a suitable code rate partition between the two paths could benefit from these paths, resulting in diversity gain. In 6G, a high percentage of users are expected to be machines, which can work together to form adaptive network coding [81]. Whether the coding and diversity gains outweigh the redundancy benefits depends on the design of side information and partition among cooperative users.
- **Analog network coding (ANC):** ANC is expected to deliver lower latency and higher efficiency. A type-1 ANC scheme was proposed in 2006, and both theoretical work and prototype verification have been carried out [69, 82, 83]. Type-1 ANC requires that the two links to a relay be accurately synchronized in terms of amplitude, phase, time, and frequency; however, achieving this is challenging [83], and may involve higher overhead for calibration, handshaking, and the corresponding period. Type-2 ANC [70] relaxes the synchronization requirements while still achieving low latency, but amplifies both signals and noise. In high signal-to-noise ratio (SNR) scenarios, type-2 ANC is desirable, because a high SNR regime can be strengthened by beam scheduling.

References

[1] C. Shannon and W. Weaver, *The mathematical theory of communication*. University of Illinois Press, 1949.

[2] H. Xie, Z. Qin, G. Y. Li, and B.-H. Juang, "Deep learning enabled semantic communication systems," *arXiv preprint arXiv:2006.10685*, 2020.

[3] R. W. Hamming, "Error detecting and error correcting codes," *The Bell System Technical Journal*, vol. 29, no. 2, pp. 147–160, 1950.

[4] M. J. Golay, "Notes on digital coding," *Proc. IEEE*, vol. 37, p. 657, 1949.

[5] D. E. Muller, "Application of boolean algebra to switching circuit design and to error detection," *Transactions of the IRE Professional Group on Electronic Computers*, no. 3, pp. 6–12, 1954.

[6] E. Prange, *Cyclic error-correcting codes in two symbols*. Air Force Cambridge Research Centre, 1957.

[7] A. Hocquenghem, "Codes correcteurs derreurs," *Chiffres*, vol. 2, no. 2, pp. 147–56, 1959.

[8] R. C. Bose and D. K. Ray-Chaudhuri, "On a class of error correcting binary group codes," *Information and Control*, vol. 3, no. 1, pp. 68–79, 1960.

[9] I. S. Reed and G. Solomon, "Polynomial codes over certain finite fields," *Journal of the Society for Industrial and Applied Mathematics*, vol. 8, no. 2, pp. 300–304, 1960.

[10] R. Silverman and M. Balser, "Coding for constant-data-rate systems," *Transactions of the IRE Professional Group on Information Theory*, vol. 4, no. 4, pp. 50–63, 1954.

[11] M. Sudan, "Decoding of reed solomon codes beyond the error-correction bound," *Journal of Complexity*, vol. 13, no. 1, pp. 180–193, 1997.

[12] M. P. Fossorier and S. Lin, "Computationally efficient soft-decision decoding of linear block codes based on ordered statistics," *IEEE Transactions on Information Theory*, vol. 42, no. 3, pp. 738–750, 1996.

[13] P. Elias, "Coding for noisy channels," *IRE Convention Record*, vol. 3, pp. 37–46, 1955.

[14] C. Berrou, A. Glavieux, and P. Thitimajshima, "Near Shannon limit error-correcting coding and decoding: Turbo-codes," in *Proc. ICC'93 – IEEE International Conference on Communications*, vol. 2. IEEE, 1993, pp. 1064–1070.

[15] T. J. Richardson, M. A. Shokrollahi, and R. L. Urbanke, "Design of capacity-approaching irregular low-density parity-check codes," *IEEE Transactions on Information Theory*, vol. 47, no. 2, pp. 619–637, 2001.

[16] E. Arikan, "Channel polarization: A method for constructing capacity-achieving codes for symmetric binary-input memoryless channels," *IEEE Transactions on information Theory*, vol. 55, no. 7, pp. 3051–3073, 2009.

[17] Y. Polyanskiy, H. V. Poor, and S. Verdú, "Channel coding rate in the finite blocklength regime," *IEEE Transactions on Information Theory*, vol. 56, no. 5, pp. 2307–2359, 2010.

[18] H. Zhang, R. Li, J. Wang, S. Dai, G. Zhang, Y. Chen, H. Luo, and J. Wang, "Parity-check polar coding for 5G and beyond," in *Proc. 2018 IEEE International Conference on Communications (ICC)*. IEEE, 2018, pp. 1–7.

[19] X. Liu, Q. Zhang, P. Qiu, J. Tong, H. Zhang, C. Zhao, and J. Wang, "A 5.16 Gbps decoder ASIC for polar code in 16nm FinFET," in *Proc. 2018 15th International Symposium on Wireless Communication Systems (ISWCS)*. IEEE, 2018, pp. 1–5.

[20] J. Tong, H. Zhang, X. Wang, S. Dai, R. Li, and J. Wang, "A soft cancellation decoder for parity-check polar codes," *arXiv preprint arXiv:2003.08640*, 2020.

[21] H. Zhang, J. Tong, R. Li, P. Qiu, Y. Huangfu, C. Xu, X. Wang, and J. Wang, "A flip-syndrome-list polar decoder architecture for ultra-low-latency communications," *IEEE Access*, vol. 7, pp. 1149–1159, 2018.

[22] B. Li, H. Zhang, and J. Gu, "On pre-transformed polar codes," *arXiv preprint arXiv:1912.06359*, 2019.

[23] T.-Y. Chen, K. Vakilinia, D. Divsalar, and R. D. Wesel, "Protograph-based raptor-like LDPC codes," *IEEE Transactions on Communications*, vol. 63, no. 5, pp. 1522–1532, 2015.

[24] M. C. Coşkun, G. Durisi, T. Jerkovits, G. Liva, W. Ryan, B. Stein, and F. Steiner, "Efficient error-correcting codes in the short blocklength regime," *Physical Communication*, vol. 34, pp. 66–79, 2019.

[25] D. J. Costello, L. Dolecek, T. E. Fuja, J. Kliewer, D. G. Mitchell, and R. Smarandache, "Spatially coupled sparse codes on graphs: Theory and practice," *IEEE Communications Magazine*, vol. 52, no. 7, pp. 168–176, 2014.

[26] D. G. Mitchell, M. Lentmaier, and D. J. Costello, "Spatially coupled LDPC codes constructed from protographs," *IEEE Transactions on Information Theory*, vol. 61, no. 9, pp. 4866–4889, 2015.

[27] Z. Si, M. Andersson, R. Thobaben, and M. Skoglund, "Rate-compatible LDPC convolutional codes for capacity-approaching hybrid ARQ," in *Proc. 2011 IEEE Information Theory Workshop*. IEEE, 2011, pp. 513–517.

[28] Wikipedia, "B5G wireless Tb/s FEC KPI requirement and technology gap analysis." [Online]. Available: https://epic-h2020.eu/downloads/EPIC-D1.2-B5G-Wireless-Tbs-FEC-KPI-Requirement-and-Technology-Gap-Analysis-PU-M07.pdf

[29] C. Kestel, M. Herrmann, and N. Wehn, "When channel coding hits the implementation wall," in *Proc. 2018 IEEE 10th International Symposium on Turbo Codes & Iterative Information Processing (ISTC)*. IEEE, 2018, pp. 1–6.

[30] A. Süral, E. G. Sezer, Y. Ertuğrul, O. Arikan, and E. Arikan, "Terabits-per-second throughput for polar codes," in *Proc. 2019 IEEE 30th International Symposium on Personal, Indoor and Mobile Radio Communications (PIMRC Workshops)*. IEEE, 2019, pp. 1–7.

[31] X. Wang, H. Zhang, R. Li, J. Tong, Y. Ge, and J. Wang, "On the construction of G N-coset codes for parallel decoding," in *Proc. 2020 IEEE Wireless Communications and Networking Conference (WCNC)*. IEEE, 2020, pp. 1–6.

[32] L. Huang, H. Zhang, R. Li, Y. Ge, and J. Wang, "AI coding: Learning to construct error correction codes," *IEEE Transactions on Communications*, vol. 68, no. 1, pp. 26–39, 2019.

[33] L. Huang, H. Zhang, R. Li, Y. Ge, and J. Wang, "Reinforcement learning for nested polar code construction," in *Proc. 2019 IEEE Global Communications Conference (GLOBECOM)*. IEEE, 2019, pp. 1–6.

[34] S. A. Hashemi, C. Condo, and W. J. Gross, "Fast simplified successive-cancellation list decoding of polar codes," in *Proc. 2017 IEEE Wireless Communications and Networking Conference Workshops (WCNCW)*. IEEE, 2017, pp. 1–6.

[35] S. B. Korada, E. Şaşoğlu, and R. Urbanke, "Polar codes: Characterization of exponent, bounds, and constructions," *IEEE Transactions on Information Theory*, vol. 56, no. 12, pp. 6253–6264, 2010.

[36] A. J. Ferris, C. Hirche, and D. Poulin, "Convolutional polar codes," *arXiv preprint arXiv:1704.00715*, 2017.

[37] H. Saber, Y. Ge, R. Zhang, W. Shi, and W. Tong, "Convolutional polar codes: LLR-based successive cancellation decoder and list decoding performance," in *Proc. 2018 IEEE International Symposium on Information Theory (ISIT)*. IEEE, 2018, pp. 1480–1484.

[38] P. Trifonov, "On construction of polar subcodes with large kernels," in *Proc. 2019 IEEE International Symposium on Information Theory (ISIT)*. IEEE, 2019, pp. 1932–1936.

[39] N. Presman, O. Shapira, and S. Litsyn, "Polar codes with mixed kernels," in *Proc. 2011 IEEE International Symposium on Information Theory*. IEEE, 2011, pp. 6–10.

[40] R. Mori and T. Tanaka, "Non-binary polar codes using Reed-Solomon codes and algebraic geometry codes," in *Proc. 2010 IEEE Information Theory Workshop*. IEEE, 2010, pp. 1–5.

[41] E. Arıkan, "From sequential decoding to channel polarization and back again," *arXiv preprint arXiv:1908.09594*, 2019.

[42] T. Richardson, "Error-floors of ldpc codes," in *Proc. 41st Annual Conference on Communication, Control and Computing*, 2003, pp. 1426–1435.

[43] Huawei, "Details of the polar code design," 3rd Generation Partnership Project (3GPP), Technical Report R1-1611254, Nov. 2016, 3GPP TSG RAN WG1 #87 Meeting.

[44] B. Li, H. Shen, and D. Tse, "Parallel decoders of polar codes," *arXiv preprint arXiv:1309.1026*, 2013.

[45] N. Van Huynh, D. T. Hoang, X. Lu, D. Niyato, P. Wang, and D. I. Kim, "Ambient backscatter communications: A contemporary survey," *IEEE Communications Surveys & Tutorials*, vol. 20, no. 4, pp. 2889–2922, 2018.

[46] M. Fresia, F. Perez-Cruz, H. V. Poor, and S. Verdu, "Joint source and channel coding," *IEEE Signal Processing Magazine*, vol. 27, no. 6, pp. 104–113, 2010.

[47] V. Kostina and S. Verdú, "Fixed-length lossy compression in the finite blocklength regime," *IEEE Transactions on Information Theory*, vol. 58, no. 6, pp. 3309–3338, 2012.

[48] V. Kostina and S. Verdú, "Lossy joint source-channel coding in the finite blocklength regime," *IEEE Transactions on Information Theory*, vol. 59, no. 5, pp. 2545–2575, 2013.

[49] A. Guyader, E. Fabre, C. Guillemot, and M. Robert, "Joint source-channel turbo decoding of entropy-coded sources," *IEEE Journal on Selected Areas in Communications*, vol. 19, no. 9, pp. 1680–1696, 2001.

[50] L. Pu, Z. Wu, A. Bilgin, M. W. Marcellin, and B. Vasic, "LDPC-based iterative joint source-channel decoding for JPEG2000," *IEEE Transactions on Image Processing*, vol. 16, no. 2, pp. 577–581, 2007.

[51] M. Grangetto, P. Cosman, and G. Olmo, "Joint source/channel coding and MAP decoding of arithmetic codes," *IEEE Transactions on Communications*, vol. 53, no. 6, pp. 1007–1016, 2005.

[52] S. Lonardi, W. Szpankowski, and M. D. Ward, "Error resilient LZ'77 data compression: Algorithms, analysis, and experiments," *IEEE Transactions on Information Theory*, vol. 53, no. 5, pp. 1799–1813, 2007.

[53] Y. Wang, M. Qin, K. R. Narayanan, A. Jiang, and Z. Bandic, "Joint source-channel decoding of polar codes for language-based sources," in *Proc. 2016 IEEE Global Communications Conference (GLOBECOM)*. IEEE, 2016, pp. 1–6.

[54] J. Garcia-Frias and J. D. Villasenor, "Joint turbo decoding and estimation of hidden Markov sources," *IEEE Journal on Selected Areas in Communications*, vol. 19, no. 9, pp. 1671–1679, 2001.

[55] G.-C. Zhu and F. Alajaji, "Joint source-channel turbo coding for binary Markov sources," *IEEE Transactions on Wireless Communications*, vol. 5, no. 5, pp. 1065–1075, 2006.

[56] J. Garcia-Frias and W. Zhong, "LDPC codes for compression of multi-terminal sources with hidden Markov correlation," *IEEE Communications Letters*, vol. 7, no. 3, pp. 115–117, 2003.

[57] K. Bhattad and K. R. Narayanan, "A decision feedback based scheme for Slepian–Wolf coding of sources with hidden Markov correlation," *IEEE Communications Letters*, vol. 10, no. 5, pp. 378–380, 2006.

[58] S. Jakubczak and D. Katabi, "Softcast: One-size-fits-all wireless video," in *Proc. ACM SIGCOMM 2010 Conference*, 2010, pp. 449–450.

[59] B. Tan, J. Wu, H. Cui, R. Wang, J. Wu, and D. Liu, "A hybrid digital analog scheme for MIMO multimedia broadcasting," *IEEE Wireless Communications Letters*, vol. 6, no. 3, pp. 322–325, 2017.

[60] F. Liang, C. Luo, R. Xiong, W. Zeng, and F. Wu, "Hybrid digital–analog video delivery with Shannon–Kotelnikov mapping," *IEEE Transactions on Multimedia*, vol. 20, no. 8, pp. 2138–2152, 2017.

[61] O. Y. Bursalioglu, G. Caire, and D. Divsalar, "Joint source-channel coding for deep-space image transmission using rateless codes," *IEEE Transactions on Communications*, vol. 61, no. 8, pp. 3448–3461, 2013.

[62] E. Bourtsoulatze, D. B. Kurka, and D. Gündüz, "Deep joint source-channel coding for wireless image transmission," *IEEE Transactions on Cognitive Communications and Networking*, vol. 5, no. 3, pp. 567–579, 2019.

[63] C.-H. Lee, J.-W. Lin, P.-H. Chen, and Y.-C. Chang, "Deep learning-constructed joint transmission-recognition for internet of things," *IEEE Access*, vol. 7, pp. 76 547–76 561, 2019.

[64] M. Jankowski, D. Gündüz, and K. Mikolajczyk, "Deep joint source-channel coding for wireless image retrieval," in *Proc. 2020 IEEE International Conference on Acoustics, Speech and Signal Processing (ICASSP)*. IEEE, 2020, pp. 5070–5074.

[65] N. Farsad, M. Rao, and A. Goldsmith, "Deep learning for joint source-channel coding of text," in *Proc. 2018 IEEE International Conference on Acoustics, Speech and Signal Processing (ICASSP)*. IEEE, 2018, pp. 2326–2330.

[66] M. Rao, N. Farsad, and A. Goldsmith, "Variable length joint source-channel coding of text using deep neural networks," in *Proc. 2018 IEEE 19th International Workshop on Signal Processing Advances in Wireless Communications (SPAWC)*. IEEE, 2018, pp. 1–5.

[67] R. Ahlswede, N. Cai, S.-Y. Li, and R. W. Yeung, "Network information flow," *IEEE Transactions on Information Theory*, vol. 46, no. 4, pp. 1204–1216, 2000.

[68] S. T. Başaran, G. K. Kurt, M. Uysal, and İ. Altunbaş, "A tutorial on network coded cooperation," *IEEE Communications Surveys & Tutorials*, vol. 18, no. 4, pp. 2970–2990, 2016.

[69] S. Zhang, S. C. Liew, and P. P. Lam, "Hot topic: Physical-layer network coding," in *Proc. 12th Annual International Conference on Mobile Computing and Networking*, 2006, pp. 358–365.

[70] S. Katti, S. Gollakota, and D. Katabi, "Embracing wireless interference: Analog network coding," *ACM SIGCOMM Computer Communication Review*, vol. 37, no. 4, pp. 397–408, 2007.

[71] D. Nguyen, T. Tran, T. Nguyen, and B. Bose, "Wireless broadcast using network coding," *IEEE Transactions on Vehicular Technology*, vol. 58, no. 2, pp. 914–925, 2008.

[72] Z. Zhang, T. Lv, X. Su, and H. Gao, "Dual XOR in the air: A network coding based retransmission scheme for wireless broadcasting," in *Proc. 2011 IEEE International Conference on Communications (ICC)*. IEEE, 2011, pp. 1–6.

[73] H. Zhu, B. Smida, and D. J. Love, "Optimization of two-way network coded HARQ with overhead," *IEEE Transactions on Communications*, vol. 68, no. 6, pp. 3602–3613, 2020.

[74] Y. Chen, S. Kishore, and J. Li, "Wireless diversity through network coding," in *Proc. 2006 IEEE Wireless Communications and Networking Conference*, vol. 3. IEEE, 2006, pp. 1681–1686.

[75] S. Zhang and S.-C. Liew, "Channel coding and decoding in a relay system operated with physical-layer network coding," *IEEE Journal on Selected Areas in Communications*, vol. 27, no. 5, pp. 788–796, 2009.

[76] A. Cohen, G. Thiran, V. B. Bracha, and M. Médard, "Adaptive causal network coding with feedback for multipath multi-hop communications," in *Proc. 2020 IEEE International Conference on Communications (ICC)*. IEEE, 2020, pp. 1–7.

[77] A. Cohen, D. Malak, V. B. Brachay, and M. Medard, "Adaptive causal network coding with feedback," *IEEE Transactions on Communications*, vol. 68, no. 7, pp. 4325–4341, 2020.

[78] A. Douik, S. Sorour, T. Y. Al-Naffouri, and M.-S. Alouini, "Instantly decodable network coding: From centralized to device-to-device communications," *IEEE Communications Surveys & Tutorials*, vol. 19, no. 2, pp. 1201–1224, 2017.

[79] M. S. Karim *et al.*, "Instantly decodable network coding: From point to multi-point to device-to-device communications," Ph.D. dissertation, Australian National University, March 2017.

[80] C. Hausl and P. Dupraz, "Joint network-channel coding for the multiple-access relay channel," in *Proc. 2006 3rd Annual IEEE Communications Society on Sensor and Ad Hoc Communications and Networks*, vol. 3. IEEE, 2006, pp. 817–822.

[81] X. Bao and J. Li, "A unified channel-network coding treatment for user cooperation in wireless ad-hoc networks," in *Proc. 2006 IEEE International Symposium on Information Theory*. IEEE, 2006, pp. 202–206.

[82] S. C. Liew, S. Zhang, and L. Lu, "Physical-layer network coding: Tutorial, survey, and beyond," *Physical Communication*, vol. 6, pp. 4–42, 2013.

[83] Y. Tan, S. C. Liew, and T. Huang, "Mobile lattice-coded physical-layer network coding with practical channel alignment," *IEEE Transactions on Mobile Computing*, vol. 17, no. 8, pp. 1908–1923, 2018.

24 New Multiple Access

24.1 Background and Motivations

In wireless communications, data are transmitted from one device to another via radio resources. The fundamental physical radio resources are time and frequency. The problem of multiple access (MA) arises when multiple users are going to be served with limited DoF in radio resources. MA scheme design may not be a single module design. Instead, it may involve multiple signal processing modules on physical links, such as coding, modulation, precoding, resource mapping, power control, and even waveforms. In P2P communications, the capacity of a single link is maximized. MA differs from this, in that some of the modules may be jointly designed or may be designed to maximize the system capacity as a whole. This is especially the case when the number of users is large and the total DoF might be shared by all users.

In 6G, as introduced in Part II of the book, diversified applications with distinct and extreme requirements will be served. This entails the need for a scalable MA framework, which will need to pay attention to the following considerations.

- **All packet sizes:** With the diversified service use cases elaborated in Part II of this book, in 6G the MA mechanism must be adequately scalable to deal with all types of packet sizes. These may range from the huge holograms in immersive XR services to simple status updates from industrial or health monitoring sensors. Different packet sizes correspond to different transmission durations and different levels of diversity.
- **All device categories:** As discussed in Chapter 18, in 6G smartphones will no longer be the only dominant devices. Different categories of device will access the same network, ranging from highly intelligent automobiles, to robots, to extremely low-cost sensors. The huge differences in these devices' capabilities in signal processing, time and frequency synchronization, hardware efficiency, and means of power supply may make very different demands of MA. Moreover, once all devices for domestic, public, and industrial use are smartly connected, the number of connections will increase tenfold in 10 years. This will demand super-massive connectivity for 6G MA design.
- **All traffic types:** Different services also bring different traffic types. This is reflected in different traffic arrival patterns, as well as diversified requirements in data rate, latency, and reliability. The design of 6G MA should be scalable enough

to efficiently accommodate users with long-burst, periodic, and sporadic traffic patterns. Efficiently multiplexing users with different extremes of performance requirements is another challenge that has not yet been fully solved in 5G with eMBB, URLLC, and mMTC traffic. Leaving the full burden on the MAC scheduler would not be adequately efficient, hence some cross-layer designs should be considered.

- **All deployment scenarios:** As has been discussed in Chapters 6 and 20, the 6G network will provide 3D coverage with HAPSs, UAVs, and VLEOs as part of the RAN. The extension in potential deployment scenarios will bring new challenges for 6G MA design. One typical challenge for integrated non-terrestrial access is the potentially large timing misalignment and limited link budget due to the long propagation distances. An additional new area to investigate will be MA mechanisms involving joint transmission among multi-layers of the 3D RAN.

24.2   Overview of Existing Solutions

MA techniques fall into two categories from the perspective of resource multiplexing: orthogonal multiple access (OMA) and non-orthogonal multiple access (NOMA). Furthermore, there are two transmission procedure-related schemes from the resource scheduling perspective: grant-based (GB) and grant-free (GF) transmissions. These MA techniques and transmission schemes have received attention in 5G research and standardization [1]. Moreover, the two MA techniques are applicable to both GB and GF transmissions. This section describes progress in these areas.

24.2.1   Orthogonal Multiple Access

In previous generations, orthogonal multiple access (OMA) was the dominant design. "Orthogonal" indicates that there is an independent DoF owned by each user in at least one domain, including frequency, time, code, and space. This entailed that a user's transmission would not be impacted by the other users that are being served at the same time. Based on the resource domains which are divided to distinguish the access of multiple users, OMA schemes include frequency-division multiple access (FDMA), time-division multiple access (TDMA), code-division multiple access (CDMA), space-division multiple access (SDMA), and orthogonal frequency-division multiple access (OFDMA) [2].

- **FDMA:** To facilitate multiple users connecting to the network, an FDMA system allocates non-overlapping frequency bands to each user or data stream as dedicated channels. To fully eliminate the interference to adjacent users, guard band should be used. This reduces the spectrum efficiency of the system as a whole. FDMA is thus not efficient enough to allow massive numbers of users to access the network.
- **TDMA:** A TDMA system divides the time domain into many time slots. The transmitter uses these time slots to transmit different signals for different users.

The payload bits of each user are split up in time and sent as bursts when time slots are available. In this way, only a limited number of users or data streams can be accommodated.

- **CDMA:** In a CDMA system, both time and frequency resources can be simultaneously shared by multiple users or data streams. Different users or data streams are distinguished by orthogonal or semi-orthogonal (also known as near-orthogonal) spreading codes. Very long sequences are needed to achieve good processing gains, and the near-orthogonal property is needed to accommodate a large number of users. These requirements make it inadequately flexible to work with large-bandwidth and massive MIMO.

- **OFDMA:** OFDMA has been developed on the basis of OFDM waveforms. This enables tight and orthogonal frequency-domain packing of subcarriers, with a sub-carrier spacing inverse to the symbol duration. In an OFDMA system, the time and frequency plane is divided into a two-dimensional raster. The minimum unit of the raster is called a resource element (RE), consisting of one subcarrier (or tone) in the frequency domain and one symbol in the time domain. Each RE transmits a modulated symbol that belongs to one user or data stream. Resource flexibility is maximized in this way and can easily be combined with features like massive MIMO. However, it has a high requirement of time and frequency synchronization, which may be too costly to maintain for very low-cost and low-power devices.

- **SDMA**: With the advancement in modern MIMO techniques, SDMA is another option for cases with massive antennas. In an SDMA system, analog or digital beamforming can create multiple focused spatial beams to dedicated scheduled users. With perfect channel state information (CSI), the beams for scheduled users can be almost orthogonal to each other as required in order to minimize inter-user interference. However, the orthogonality of SDMA relies heavily on the accuracy of CSI. This becomes hard to track with an expanding number of users and sporadic traffic.

Though many features have been developed based on OMA to improve system capacity, user experience, and numbers of connections, there are still limitations and gaps to overcome in order to meet the diversified and extreme service requirements in 6G. They include the following.

- **Limitation of the number of users simultaneously being served:** The number of users or data streams is strictly constrained by the number of orthogonal channels in an OMA system.

- **High signaling and resource overhead to guarantee orthogonality:** The orthogonality in an OMA system is usually guaranteed by resource grants sent by base stations to users before any transmission in any direction can start, such as uplinks, downlinks, or sidelinks. When the number of users increases, the signaling overhead increases proportionally, reducing the system capacity as a whole. Moreover, for certain IoT services, the dynamic grant processing may either introduce latency that cannot be tolerated, or take up more than 50% of the payload in each transmission. Neither case is acceptable from the point of view of spectrum efficiency and power efficiency. The orthogonality can be guaranteed with preconfigured resources, but the resource overhead needs to be large to ensure reliable performance, especially when the arrival of traffic is unpredictable and sporadic.

- **Heavy CSI dependency:** In real-world deployment scenarios, the performance of closed-loop multi-user MIMO (MU-MIMO) and coordinated multipoint (CoMP) transmission/reception is still far from the theoretical bounds, especially for high-speed mobile users. The major reason behind this is their heavy dependency on precise CSI for closed-loop precoding. However, if network impairments happen, obtaining precise CSI is not feasible and the performance of such closed-loop MU-MIMO or CoMP degrades accordingly. This may happen in cases of channel aging, feedback delay, or abrupt inter-cell interference, which are common in practical networks.

24.2.2   Non-Orthogonal Multiple Access

To overcome the limitations of OMA, non-orthogonal multiple access (NOMA) has been identified as an MA technology that is tolerant of resource collision in orthogonal channels. From a network information theory perspective [3, 4], for both the uplink MA channel [5] and the downlink broadcast channel [6], the capacity region of multiple NOMA users can be improved compared with parallel transmission for the same set of users on orthogonally partitioned resources. Techniques such as successive interference cancellation (SIC) [5] and dirty paper coding (DPC) [6] at base stations have been proven capable of achieving the optimal capacity regions for the MA channel and BC channel, respectively. Moreover, NOMA can support further signal superposition in the orthogonal subspace in MU-MIMO transmission. One recent study [7] has shown that, by integrating the code domain NOMA design into the uplink MU-MIMO transmission, extra throughput gain can be obtained, even for a system with 64 Rx antennas.

In many application scenarios, it is not the single-user capacity that matters, but the number of users that could be served with a guaranteed target rate. In this sense, NOMA can be seen as a mechanism that trades the peak data rate per connection for a higher number of connections [8]. It does this by introducing NOMA transceivers to suppress inter-user interference. This is extremely useful in IoE, where there are massive devices but where each device does not have a very demanding data rate requirement. When combined with MU-MIMO, NOMA can help to support a larger number of users than the number of Tx or Rx antennas.

In summary, with good transceiver design, NOMA has the potential to provide the following inherent advantages over OMA.

- **Achieving multi-user system capacity:** As proven by network information theory, NOMA further improves system capacity while not introducing extra spectrum or antenna resources.

- **Supporting overloaded transmission:** NOMA further increases the total number of connections by introducing affordable symbol collision on orthogonal channels. In such cases, the system becomes overloaded. Through suitable design of multi-user detection (MUD) and NOMA codebooks, which may include multidimensional constellations [9, 10], spreading signatures [11], and resource collision patterns [12, 13], an overloading factor above 300% can be achieved. Here, an overloading factor is defined as the ratio between the number of simultaneously accessed users and the number of orthogonal REs.
- **Enabling reliable grant-free (GF) transmission:** As will be discussed below, to avoid heavy signaling overhead and reduce the handshake latency associated with grant-based (GB) transmission, GF transmission has been proposed, especially for services with smaller packet sizes and sporadic traffic patterns. With its robustness towards symbol collision, NOMA-enabled GF transmission can support more aggressive resource sharing among more users, while still achieving the same level of target reliability.
- **Enabling robust open-loop MU-MIMO:** To overcome the bottleneck of real-time CSI acquisition in closed-loop massive MIMO transmissions, NOMA provides an alternative of open-loop MU-MIMO. Since it does not rely on precise CSI, it can be more robust towards network impairments such as channel aging and user mobility. Such an open-loop scheme can be further extended to multi-base station cases, in which, instead of gathering precise CSI from the target users and then sending the CSI to each collaborative base station for joint precoding, the collaborative base stations can each select a non-orthogonal code and jointly transmit data streams without exchanging CSI with users or other base stations. This is especially useful when accurate CSI is not available or too costly to obtain.
- **Enabling flexible service multiplexing:** To efficiently serve diversified traffic types, a traditional OMA approach is dynamic resource scheduling. This approach consumes extra signaling overhead and may not be fast enough to meet the latency requirements of some services. With NOMA, due to its superpositional nature, low-latency small packets could be superposed on top of big packets for joint transmission, thus improving latency and reducing overhead at the same time.

Downlink NOMA

For downlink NOMA, base stations transmit the sum of the signals for multiple users which occupy the same frequency and time resources. Usually, base stations will assign multiple users with significantly different SNRs to be in the same NOMA group. For example, for two users who are scheduled via the same resources and allocated different NOMA signatures, one user near the base station and the other far from the base station, the transmitter will allocate high transmit power to the far user and low power to the near user. At the receiver side, the far user can consider the near user's signals as noise of which the power is much lower than the power of its own signals; the near user can distinguish its own signals from the far user's signals by detecting and recovering the latter.

In 3GPP, the study of downlink NOMA was initiated in R13 under a study item named "multiuser superposition transmission" (MUST) [14]. MUST can be characterized into three categories.

- **MUST Category 1:** Coded bits of two or more co-scheduled users are independently mapped to the symbols of component constellations, but the composite constellation does not perform gray mapping.
- **MUST Category 2:** Coded bits of two or more co-scheduled users are jointly mapped to the symbols of component constellations, and then the composite constellation performs gray mapping.
- **MUST Category 3:** Coded bits of two or more co-scheduled users are directly mapped to the symbols of the composite constellation.

For most of the existing downlink NOMA designs, there are only two users in each NOMA group and the total capacity is still considered as the main metric to steer the designs. With the potential addition of code-domain design, it is expected that the required SNR gap between NOMA users can be relaxed, thus further simplifying user grouping. Some initial attempts can be found in [15]. Moreover, new metrics such as scheduling latency may be new dimensions to consider during the design of downlink NOMA for more diversified traffic types.

Uplink NOMA

For uplink NOMA transmission, multiple users transmit their signals to the base station via the same time and frequency resources [16]. The design of uplink NOMA is more challenging, since a random channel is applied to each user before multiple data signals from different users are multiplexed together. Such a property prevents the design of joint constellation with superposition in advance, as in MUST Categories 2 and 3. Additionally, this demands a design from per-user or per-layer aspects that can accommodate the randomness brought by user-specific channels. Uplink NOMA study in 3GPP was initiated in R14 for 5G NR. Many different NOMA schemes have been proposed for uplink transmissions to support massive connectivity and enable reliable GF transmissions. The study of this then continued in Release 15 with a dedicated study item [17]. The key aspects studied for uplink NOMA are the following.

- **Unified transceiver framework:** The NOMA study included transmission schemes on how to mitigate inter-user interference and receiving schemes on how to deal with inter-user interference in order to support more simultaneous transmissions. A unified transceiver framework [18] for uplink NOMA based on OFDM waveforms is depicted in Figure 24.1. At the transmitter side, user-specific operations at the bit and/or symbol levels are designed to facilitate decoding of superposed multi-user data at the receiver side with reasonable complexity. At the receiver side, advanced multi-user receivers are applied to better suppress inter-user interference while taking implementation costs and latency into consideration.

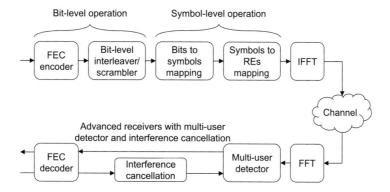

Figure 24.1 Unified transceiver framework for uplink NOMA based on OFDM waveforms.

- **Transmitter-side NOMA signature:** Some typical examples of NOMA signature design are sparse code multiple access (SCMA) [9, 10], multi-user shared access (MUSA) [11], resource spread multiple access (RSMA) [19], pattern division multiple access (PDMA) [12], interleave-grid multiple access (IGMA) [13], and interleave division multiple access (IDMA) [20]. All these schemes can be characterized by different bit-level and symbol-level operations, which are defined as signatures in NOMA research.

  Bit-level operations include UE-specific bit-level coding, scrambling, and interleaving. IDMA utilizes the bit-level interleaving function as an MA signature to distinguish other users. Symbol-level operations include UE-specific symbol-level spreading, modulation, scrambling, and interleaving.

  Symbol-level spreading sequences with the properties of low cross-correlation or low density are an important type of MA signature. Various designs of symbol-level spreading sequences are proposed for NOMA, such as welch-bound equality sequences, complex-valued sequences with quantized elements, and Grassmannian sequences. In addition, the symbol-level spreading and modulation can be jointly designed to improve the performance, such as in SCMA.

- **Receiver-side NOMA receiver:** Due to the non-orthogonal properties of NOMA systems, inter-user interference is inevitable at the receiver side of such systems. NOMA receiver design is the key to suppressing inter-user interference. The unified iterative design of a NOMA receiver [21] consists of two parts: (1) a multi-user detection (MUD) algorithm; and (2) an iterative interference cancellation (IC) structure. Low-complexity and low-latency MUD algorithms and interference cancellation structures are very important for NOMA to be widely applied, balancing performance with implementation costs.

When it comes to interference cancellation structures, recent studies [22] have shown that a hybrid parallel IC (PIC) structure achieves the best trade-off between performance and complexity. In a hybrid PIC structure, there are outer-loop iterations between MUD and channel decoders, and soft and hard IC are performed in parallel. This means that for users whose information streams are successfully decoded, their transmitted signals are reconstructed using the decoded information bits and cancelled from the overall received signals. For users whose information streams are not fully decoded, extrinsic LLRs are fed back as inputs to MUD as the starting point for next-round detection.

In academic circles, there are many MUD algorithms, such as the message passing algorithm (MPA) [23], expectation propagation algorithm (EPA) [24], linear minimum mean square error (LMMSE) [25], and elementary signal estimator (ESE) [20]. These MUD algorithms consist of the following elements.

- **MPA detector:** Conditional probabilities are updated and passed back and forth between edges. On a factor graph corresponding to a given NOMA scheme, they connect a function node (FN), representing an RE, as a pair with a variable node (VN), representing the data layer or user. After a number of iterations (inner loops), the LLRs for the coded bits are calculated and then input to the channel decoder. The MPA detector can achieve performance close to the optimal maximum-likelihood detection but with much less complexity, especially when the factor graph is sparse. Its intrinsic divide-and-conquer feature enables high-parallelism implementation.

- **EPA detector:** This employs the classic approximate Bayesian inference technique [26]. It projects the true posterior distribution of the transmitted symbols into a family of Gaussian distributions by iteratively matching the means and variances with the true posterior distribution. In one sense, EPA is a Gaussian approximation to MPA but with consideration of the non-Gaussian probability distribution of the transmitted symbols as well. The direct benefit of this approximation is linear complexity with respect to modulation size and the number of UEs, while keeping nearly the same performance as MPA in most scenarios of interest [24].

- **LMMSE detector:** This approximates the prior distribution of the signal as being Gaussian, either with mean and variance computed from soft LLRs fed back by the channel decoder, or with zero mean and variance scaled by the signal power if the soft feedback is unavailable. LMMSE can be regarded as a special case of EPA when multiple antennas are jointly processed without inner iterations between FNs and VNs.

- **ESE detector:** This simply approximates inter-symbol interference as Gaussian noise. An ESE detector has to rely on outer-loop iterations (soft feedback from the channel decoder) to achieve an acceptable detection performance. If there is a large number of NOMA users, the number of outer-loop iterations can be very large, and also the decoding latency.

Very recently, a paper [21] revealed the relations between different MUD algorithms from a unified variational inference (VI) perspective, and proposed unified NOMA receivers based on it.

Though it did not reach a conclusion on how to specify a NOMA scheme in NR, comprehensive link- and system-level simulations have been performed by different companies to justify the gain of NOMA over OFDMA, the OMA baseline. It has been agreed that for the evaluated scenarios, significant benefits of NOMA can be observed in terms of uplink-level total throughput and overloading capability. In addition, benefits can be observed in system capacity enhancement in terms of supported packet arrival rate (PAR) at a given system outage level, such as a 1% packet drop rate (PDR). The PAR can then be used to estimate the supportable connections under given per-user traffic patterns.

In summary, NOMA is a promising solution to enhance system capacity by accommodating more users with the same radio resources. However, there are still issues to be solved for existing NOMA schemes. First, many NOMA receivers involve iterative operations, making them more complex than OMA receivers. Second, the NOMA scheme combined with MIMO still needs further investigation to improve overall performance. Last, the related procedures of NOMA transmissions, such as HARQ, signature allocation, and link adaptation, require further investigation.

24.2.3   Grant-Free MA

Base stations perform GB transmission over dynamically scheduled resources. Different from GB transmission, GF transmission is performed over preconfigured or semi-statically configured resources for uplink or downlink users. Resources configured this way can be shared by more than one GF user, which may result in transmission collisions between GF users, and therefore GF transmission is also referred to as contention-based transmission. Due to the semi-static configuration mechanism, the GF transmission scheme – characterized by "arrive-and-go" – is suitable for services and applications demanding low latency, such as URLLC traffic. Moreover, the GF transmission scheme is of special importance to support aperiodic and latency-sensitive traffic of which the arrival time is not predictable but the traffic needs to be transmitted immediately upon its arrival.

The GF transmission scheme can also significantly save power and reduce signaling overhead for uplink transmission, due to the facts that it does not need to send a scheduling request (SR) to the base station upon traffic arrival and that it can avoid the detection process in receiving control information. The power saving resulting from the avoidance of SR handshake and control signaling detection is of paramount importance for power-sensitive devices, such as smart sensors (which require a long battery life).

However, due to its being contention-based, with multiple users possibly sharing time and frequency resources, GF uplink transmission may incur certain collisions. These may lead to unavoidable retransmissions and concerns about reliability. Fortunately, the collision nature of GF transmission can be overcome by the advantages of NOMA. With a combination of GF (uplink) transmission and NOMA, each GF user can be semi-statically assigned an elaborately designed or randomly selected NOMA signature from a preconfigured signature pool. The inter-user interference caused by transmission collisions can then be resolved by the advanced NOMA receiver, which has the ability to separate overlapped signals with high reliability. As a result, the GF scheme with NOMA is a key enabling technology for reliable, fast, and efficient (uplink) data transmission in a wireless network.

As discussed above, a NOMA-enabled GF scheme is able to reduce access latency and control signaling overhead and energy consumption of the user devices, especially in small-packet and latency-sensitive transmissions for cost-effective and power-efficient devices. However, this scheme still has some challenges to overcome, such as optimal pilot design and collision resolution for NOMA, optimal GF HARQ feedback design, trade-offs between GF feedback and retransmission, and trade-offs among achievable rates, packet lengths, and reliability (from the perspective of information theory). The study of these topics has attracted the attention of researchers lately, and an academic review on these topics can be found in [1]. The review [1] also covers early work on addressing the problem of short-packet delivery from the perspective of information theory, as well as recent work on the performance evaluation of the GF scheme with NOMA.

Standardization Progress of GF

GF transmission, a promising transmission scheme to reduce signaling overhead and shorten transmission latency, has been discussed and was introduced in the first NR standard Release (Release 15) for both uplink and downlink transmissions. In GF uplink transmission, the resources of a user are semi-statically configured within one configurable period by the base station, where such configured resources are assumed to be periodic over the time. Once configured, such resources can be immediately employed by the user, thus guaranteeing low-latency transmission. For downlink transmission, the resources of a user can only be used after they are activated by a dynamic signaling from the base station, which can avoid unnecessary downlink transmission detection at the user side and reduce power consumption.

In addition, a user may have multiple GF configurations simultaneously to serve various traffic types with diversified service requirements, which can further improve both latency and reliability performance. To reduce power consumption and latency, NR Release 16 and Release 17 further extend the basic GF transmission to sidelink (device-to-device) transmission, especially in V2X scenarios, and to inactive state (a state between idle and active) transmission to reduce power consumption and latency.

Compressed-Sensing-Based GF Access

Due to the fact that traffic is inherently sporadic, active user detection is an issue in GF transmission for multiple access. In uplink GF transmission, the base station needs to detect active users before decoding and recovering signals. Usually, the detection process is based on the pilots or signatures of data transmission. Hence the design of the pilots or signatures and the corresponding detection algorithms is critical for GF transmission. It has been proven in many works [27, 28] that the signal design and detection task can be modeled as a compressed sensing (CS) problem, by which the existing findings in CS theory can be used to facilitate the design of GF access schemes.

The approximate message passing (AMP) algorithm was proposed to solve the CS detection problem in 2009 [29]. AMP was used as a standard detection algorithm for the CS problem thereafter, and a variety of AMP algorithms were proposed for different application scenarios. In 2011, a generalized approximate messaging passing (GAMP) algorithm was proposed in [30] to deal with more complex models with non-linear relationships between inputs and outputs. In 2017, an improved AMP algorithm based on an MMSE thresholding denoiser was proposed in [27] for detecting active users in massive connection GF transmission scenarios. In addition, this improved AMP algorithm can be used for GF with massive MIMO configurations [28]. Again in 2017, an orthogonal approximate message passing (OAMP) algorithm was proposed in [31], which claimed to achieve Bayesian-optimal performance results with a faster convergence speed than the conventional AMP algorithm.

Non-Coherent GF Access

Although NOMA-enabled GF can mitigate the collision issue for data signals, a good channel estimation is still needed for most NOMA schemes. This being so, an orthogonal or nearly orthogonal pilot design is required for NOMA-based GF. This will lead to large pilot overhead in massive connection scenarios, especially for small packet size transmission. In academic and industrial institutions, researchers have proposed possible solutions to this problem. For instance, in 2017, a framework for non-coherent transmission with common codebook design was proposed in [32] to address the above issue. Some theoretical analysis for this framework was also given therein. Such a framework was named unsourced random access – a solution without any scheduling or pilots and thus avoiding pilot collision issue – and targeted for massive connectivity with small block length. After that, some improved schemes have been proposed under this framework, such as [33, 34]. By schemes such as sub-block partitioning and inter-block coding, or tensor-based modulation [35], the concept of unsourced random access leads to implementable solutions for GF access. Furthermore, combined with CS-based algorithms at the receiver, unsourced random access has the potential to become one of the candidates for large-scale access in the future.

24.3 New Design Expectations and Potential Research Directions

24.3.1 MA for Large-Capacity URLLC Services

URLLC type services have been introduced in 5G mainly to enable vertical applications such as motion control in Industry 4.0 and autonomous driving in V2X. With the features defined in the most up-to-date NR specifications, ultra-high reliability and low latency can be achieved at the same time. However, due to orthogonal resource utilization and conservative link adaptation, there may be only a very limited number of users at the same time under each serving base station. In 6G, such services will continue to evolve towards more application aspects with even higher requirements and larger numbers of devices, such as collaborative robots and teleoperation. In this case, the 6G MA mechanism and related procedures will be further enhanced, with NOMA-enabled GF or GB transmission expected to play a key role.

When it comes to the reliability of data transmission at the physical layer, this generally can be enhanced by exploiting higher degrees of diversity to combat the uncertainties brought by the fading phenomenon of wireless channels. It can also be enhanced by leveraging higher coding gains to combat the noise in channels. Relating to these ways of enhancing reliability, some potential directions for future NOMA enhancement are as follows.

- **NOMA with joint multi-domain design:** Currently, NOMA signatures are designed mainly with a focus on one or two specific resource domains, such as the time/frequency domain, code domain, spatial domain, or power domain. To provide better diversity and tolerate more inter-user interference, more resource domains should be jointly exploited in NOMA signature design to improve the performance in fading channels and boost the total capacity.
- **NOMA with joint channel coding design:** In the iterative NOMA receiver structure described in Section 24.2, there is already an outer-loop iteration between MUD and channel decoders, which leverages coding gains to suppress inter-user interference. It is expected that the reliability of the multi-user access system would be further improved if the NOMA signatures could be jointly designed with channel coding. This will be especially useful for short-packet MA with high-reliability requirements.

24.3.2   MA for Extremely Low-Cost and Low-Power Devices

As discussed in Section 24.1, low-cost device MA will be an important usage scenario for 6G. Currently, NB-IoT and eMTC are used for accessing low-cost and low-power devices in 5G. In 6G, as the target battery life for sensor type IoT devices is expected to be doubled to around 20 years and the number of IoT devices to be increased tenfold, there is continuous demand further to reduce power consumption and cost for these devices. Now, in both NB-IoT and eMTC, time/frequency synchronization is still required. This comprises a large portion of power consumption for these devices, with an infrequent and small payload.

In 6G, maintaining synchronization may not be desirable for some use cases, especially for extremely low-cost and low-power devices. However, cheap components on the RF chains of these devices may also bring phase noise issues to the 6G system, with larger impacts if the devices work on mid-bands rather than low bands. 6G MA design should consider the following research directions for extremely low-cost and low-power devices.

- **MA with robustness towards time/frequency offset and phase noise:** Asynchronous transmission will break the orthogonality between users even if OMA is applied. In this case, NOMA signatures and receivers that are robust towards time/frequency offset and phase noise are expected. In addition, asynchronous waveforms can be jointly designed with NOMA transceivers.
- **MA with low-PAPR waveforms:** In 5G, MA schemes have been studied with low-PAPR waveforms, and such schemes include DFT-based SCMA and single-tone

SCMA [36]. For an extremely low-cost target, MA schemes should be studied jointly with new low-PAPR waveforms in 6G. More specifically, OMA and NOMA schemes should both be optimized to support low-PAPR transmission.

24.3.3 MA for Super-Massive Connectivity

Super-massive connectivity is an indispensable ability required for a 6G system in many applications, especially for short-packet transmissions with sporadic traffic patterns. To enable super-massive connectivity within limited radio resources, NOMA-enabled GF is essential. However, the current applications of NOMA and GF transmission need to address the following challenges in order to achieve super-massive connectivity.

- **Data collision:** Due to limited total radio resources and lack of coordination among transmissions, data collision is inevitable for massive connectivity. To accommodate more simultaneously active devices, research needs to be conducted into improving the NOMA transmission schemes and MUD algorithms.
- **Pilot collision:** Although the data collision issue can be mitigated by advanced NOMA techniques, pilot collision remains an unsolved issue for GF MA. Pilot extension schemes, or even pilot-free non-coherent schemes, should therefore be studied further. In some cases, the schemes need to be considered within the NOMA framework in order to address pilot and data collision issues together, which are two critical issues for super-massive connectivity.
- **Ultra-massive antennas:** The super-massive connectivity problem in configurations with massive antennas at the base station side may not simply be a scaled-up version of the original problem. This may especially be the case when the number of antennas is very large, the antennas are no longer collocated, or new components such as RISs are deployed. Some special channel properties in the case of ultra-massive antennas may be exploited to address collision issues for both data and pilot.

24.3.4 MA for Robust Beamforming

In 6G, mmWave bands and (sub-)THz bands are to be deployed with massive MIMO, especially for large-throughput applications. The signal beams of MIMO will become extremely narrow with the increasing number of antenna elements and higher-frequency range. It will thus be difficult to generate multiple very narrow beams pointing precisely towards multiple users at the same time because accurate CSI will be hard to obtain in many scenarios. In cases such as these, enhanced NOMA schemes can be jointly designed with multi-user precoding for robust beamforming. With NOMA capability in hand, MU-MIMO can employ a modified precoding scheme. Instead of generating very narrow and precise beams targeting each user separately, the MU-MIMO precoder can now generate wider beams to target a group of users which are further multiplexed by NOMA. The increase in the beamwidth makes beamforming more robust to CSI changes, such as those caused by user mobility or delayed measurement and feedback. The inter-user interference within a group will be further suppressed by the NOMA receiver, with the help of the NOMA signatures transmitted. Similar ideas can be further applied in cooperative transmissions with multiple base stations.

24.3.5   MA with AI Assistance

As discussed in Chapter 19, the air interface design in 6G will exploit the advantages and progress in AI and ML to make the PHY design intelligent. The design of many traditional modules in communication links can be facilitated by a data-based AI methodology. For 6G MA, both the transmitter and the receiver may utilize AI techniques, as described below.

- **AI-assisted transmitter:** To design a MA transmission scheme with low-cost, low-PAPR, low-latency, high-reliability, and massive connectivity properties, the traditional model-driven design methodology may be insufficient for this number of MA design targets. Hence a data-driven neural network may be a possible option for addressing these new requirements simultaneously. For instance, the neural network can be used to design the NOMA signatures together with many other modules, such as waveform and MIMO precoding.
- **AI-assisted receiver:** To improve detection performance and reduce complexity, AI/ML techniques can play a role in facilitating the MUD design for NOMA. EPA MUD is one example derived from the VI framework that has been widely used in many Bayesian estimation-related AI tasks. The VI technique increases performance gains by learning prior distributions as the inputs of MUD, other than simply assuming them to be Gaussian. In 6G MA, as discussed earlier, many conventional models may no longer work due to factors such as asynchronous transmission or phase noise. Data-driven AI techniques may be an alternative solution to matching the scenarios automatically. Moreover, with the help of neural networks, different modules in the receiver may be jointly designed, such as those for user activity detection, channel estimation, and MUD.

References

[1] M. Vaezi, Z. Ding, and H. V. Poor, *Multiple access techniques for 5G wireless networks and beyond.*   Springer, 2019.

[2] J. G. Proakis and M. Salehi, *Digital communications.* McGraw-Hill, 2007.

[3] T. M. Cover, *Elements of information theory.*   John Wiley & Sons, 1999.

[4] A. El Gamal and Y.-H. Kim, *Network information theory.*   Cambridge University Press, 2011.

[5] D. N. C. Tse, P. Viswanath, and L. Zheng, "Diversity-multiplexing tradeoff in multiple-access channels," *IEEE Transactions on Information Theory*, vol. 50, no. 9, pp. 1859–1874, 2004.

[6] S. Vishwanath, N. Jindal, and A. Goldsmith, "Duality, achievable rates, and sum-rate capacity of Gaussian MIMO broadcast channels," *IEEE Transactions on Information Theory*, vol. 49, no. 10, pp. 2658–2668, 2003.

[7] K. Senel, H. V. Cheng, E. Björnson, and E. G. Larsson, "What role can NOMA play in massive MIMO?" *IEEE Journal of Selected Topics in Signal Processing*, vol. 13, no. 3, pp. 597–611, 2019.

[8] Y. Liu, Z. Qin, and Z. Ding, *Non-orthogonal multiple access for massive connectivity*. Springer, 2020.

[9] H. Nikopour and H. Baligh, "Sparse code multiple access," in *Proc. 2013 IEEE 24th Annual International Symposium on Personal, Indoor, and Mobile Radio Communications (PIMRC)*. IEEE, 2013, pp. 332–336.

[10] M. Taherzadeh, H. Nikopour, A. Bayesteh, and H. Baligh, "Scma codebook design," in *Proc. 2014 IEEE 80th Vehicular Technology Conference (VTC2014-Fall)*. IEEE, 2014, pp. 1–5.

[11] Z. Yuan, G. Yu, W. Li, Y. Yuan, X. Wang, and J. Xu, "Multi-user shared access for internet of things," in *Proc. 2016 IEEE 83rd Vehicular Technology Conference (VTC-Spring)*. IEEE, 2016, pp. 1–5.

[12] S. Chen, B. Ren, Q. Gao, S. Kang, S. Sun, and K. Niu, "Pattern division multiple access – a novel nonorthogonal multiple access for fifth-generation radio networks," *IEEE Transactions on Vehicular Technology*, vol. 66, no. 4, pp. 3185–3196, 2016.

[13] Q. Xiong, C. Qian, B. Yu, and C. Sun, "Advanced NoMA scheme for 5G cellular network: Interleave-grid multiple access," in *Proc. 2017 IEEE Globecom Workshops*. IEEE, 2017, pp. 1–5.

[14] J. M. Meredith, "Study on downlink multiuser superposition transmission for LTE," in *Proc. TSG RAN Meeting*, vol. 67, 2015.

[15] H. Nikopour, E. Yi, A. Bayesteh, K. Au, M. Hawryluck, H. Baligh, and J. Ma, "SCMA for downlink multiple access of 5G wireless networks," in *Proc. 2014 IEEE Global Communications Conference*. IEEE, 2014, pp. 3940–3945.

[16] L. Dai, B. Wang, Y. Yuan, S. Han, I. Chih-Lin, and Z. Wang, "Non-orthogonal multiple access for 5G: Solutions, challenges, opportunities, and future research trends," *IEEE Communications Magazine*, vol. 53, no. 9, pp. 74–81, 2015.

[17] 3GPP, "Study on non-orthogonal multiple access (NOMA)," 3rd Generation Partnership Project (3GPP), Technical Report (TR) 38.812, Dec. 2018, version 16.0.0. [Online]. Available: https://portal.3gpp.org/desktopmodules/Specifications/SpecificationDetails.aspx?specificationId=3236

[18] Y. Chen, A. Bayesteh, Y. Wu, B. Ren, S. Kang, S. Sun, Q. Xiong, C. Qian, B. Yu, Z. Ding *et al.*, "Toward the standardization of non-orthogonal multiple access for next generation wireless networks," *IEEE Communications Magazine*, vol. 56, no. 3, pp. 19–27, 2018.

[19] Qualcomm Incorporated, "RSMA," 3rd Generation Partnership Project (3GPP), RAN1 (R1) 164688, May 2016. [Online]. Available: https://www.3gpp.org/DynaReport/TDocExMtg--R1-85--31662.htm

[20] L. Ping, L. Liu, K. Wu, and W. K. Leung, "Interleave division multiple-access," *IEEE Transactions on Wireless Communications*, vol. 5, no. 4, pp. 938–947, 2006.

[21] X. Meng, L. Zhang, C. Wang, L. Wang, Y. Wu, Y. Chen, and W. Wang, "Advanced NOMA receivers from a unified variational inference perspective," *IEEE Journal on Selected Areas in Communications*, 2020. (Early Access) [Online]. Available: https://ieeexplore.ieee.org/abstract/document/9181630

[22] X. Meng, Y. Wu, C. Wang, and Y. Chen, "Turbo-like iterative multi-user receiver design for 5G non-orthogonal multiple access," in *Proc. 2018 IEEE 88th Vehicular Technology Conference (VTC-Fall)*. IEEE, 2018, pp. 1–5.

[23] F. R. Kschischang, B. J. Frey, and H.-A. Loeliger, "Factor graphs and the sum-product algorithm," *IEEE Transactions on Information Theory*, vol. 47, no. 2, pp. 498–519, 2001.

[24] X. Meng, Y. Wu, Y. Chen, and M. Cheng, "Low complexity receiver for uplink SCMA system via expectation propagation," in *Proc. 2017 IEEE Wireless Communications and Networking Conference (WCNC)*. IEEE, 2017, pp. 1–5.

[25] X. Wang and H. V. Poor, "Iterative (turbo) soft interference cancellation and decoding for coded CDMA," *IEEE Transactions on Communications*, vol. 47, no. 7, pp. 1046–1061, 1999.

[26] C. M. Bishop, *Pattern recognition and machine learning*. Springer, 2006.

[27] Z. Chen and W. Yu, "Massive device activity detection by approximate message passing," in *Proc. 2017 IEEE International Conference on Acoustics, Speech and Signal Processing (ICASSP)*. IEEE, 2017, pp. 3514–3518.

[28] Z. Chen, F. Sohrabi, and W. Yu, "Sparse activity detection for massive connectivity," *IEEE Transactions on Signal Processing*, vol. 66, no. 7, pp. 1890–1904, 2018.

[29] D. L. Donoho, A. Maleki, and A. Montanari, "Message-passing algorithms for compressed sensing," *Proceedings of the National Academy of Sciences*, vol. 106, no. 45, pp. 18 914–18 919, 2009.

[30] S. Rangan, "Generalized approximate message passing for estimation with random linear mixing," in *Proc. 2011 IEEE International Symposium on Information Theory Proceedings*. IEEE, 2011, pp. 2168–2172.

[31] J. Ma and L. Ping, "Orthogonal amp," *IEEE Access*, vol. 5, pp. 2020–2033, 2017.

[32] Y. Polyanskiy, "A perspective on massive random-access," in *Proc. 2017 IEEE International Symposium on Information Theory (ISIT)*. IEEE, 2017, pp. 2523–2527.

[33] R. Calderbank and A. Thompson, "Chirrup: A practical algorithm for unsourced multiple access," *Information and Inference: A Journal of the IMA*, vol. 9, no. 4, pp. 875–897, Dec. 2020.

[34] A. Fengler, P. Jung, and G. Caire, "Sparcs for unsourced random access," *arXiv preprint arXiv:1901.06234*, 2019.

[35] A. Decurninge, I. Land, and M. Guillaud, "Tensor-based modulation for unsourced massive random access," *IEEE Wireless Communications Letters*, 2021.

[36] 3GPP, "Discussion on the design of NOMA transmitter," 3rd Generation Partnership Project (3GPP), RAN1 (R1) 1812187, Nov. 2018. [Online]. Available: https://www.3gpp.org/DynaReport/TDocExMtg--R1-95--18807.htm

25 Ultra-Massive MIMO

25.1 Background and Motivations

Wireless communication with multiple transmit antennas that surpassed the capabilities of phased antenna arrays [1] became a popular topic of research at the end of the twentieth century. During this period, numerous new transmission and reception schemes emerged, such as space–time block codes (STBC) [2], Alamouti code [3], Bell Laboratories layered space–time (BLAST) [4], and sphere decoding [5]. These greatly improved the link throughput and reliability of multiple transmit antennas.

Not long after, 3GPP embraced these new technologies. Its first release of LTE supported up to four transmit antenna ports at downlink, which were identifiable to the users through cell reference signals (CRSs). To address the ever increasing demand for user density, throughput, and link reliability, MIMO was gradually supported and improved along with new releases of LTE. Later, when NR stepped onto the scene, massive MIMO was natively supported through the beam-based MIMO architecture, especially in TDD bands that facilitated massive MIMO deployment (due to simpler CSI acquisition).

To achieve 6G stringent KPIs, networks must undergo densification, allowing for wider bandwidth allocation over shorter ranges and utilizing even higher orders of MIMO. This will result in distributed 6G RANs with dense remote radio heads (RRHs), ELAAs, and more antenna panels. In addition, each antenna panel will have more antenna elements at mmWave bands, which will be extended to THz bands to utilize a much vaster spectrum. Besides the vast amount of frequency resources, THz MIMO will also facilitate the completion of other 6G KPIs, such as positioning and sensing. In this chapter, we study the state-of-the-art MIMO deployment in RANs, alongside potential MIMO enhancements in the context of 6G.

25.2 Overview of Existing Solutions

While there has been significant progress in NR by improving the reliability and throughput of MIMO links, further enhancement to bring even faster and more reliable MIMO links is both desirable and achievable.

Currently, NR has extensive support for MIMO at both low bands and mid-bands below 7.125 GHz, known as frequency range 1 (FR1) [6], as well as higher-frequency bands from 24.250 to 52.6 GHz (known as FR2) [7]. However, the growing support for MIMO in the 6G era comes with a number of challenges. For example, as the number of antenna elements deployed at each node and the transmit/receive beam increase in the future, so do RS measurement, beam management, and CSI acquisition overhead.

We introduce the existing MIMO technologies adopted in NR for FR1 and FR2 in Sections 25.2.1 and 25.2.2. Section 25.2.3 discusses cooperative MIMO solutions and reviews the NR support for cooperation among different nodes.

25.2.1   MIMO Technologies for FR1

MIMO technology has been a key technical component for wireless communication systems since the advent of 4G, that is, long-term evolution (LTE) and LTE-advanced (LTE-A). In the earliest stages, LTE standardization was designed assuming base stations had only a few antenna ports, mapped to a few sector antennas, where antenna elements were utilized to form the radiation pattern and electronically tilt it. Multi-antenna transmission in LTE/LTE-A utilized different transmission modes to obtain diversity, multiplexing, and array gains, with each mode optimized for a different set of system configurations such as antenna ports, FDD, TDD, carrier frequency, mobility, etc. The large number of transmission modes complicated the implementation and limited the achievable gain of later introduced modes especially when the backward compatibility of earlier low-capability devices is taken into consideration. The 5G NR system unified all transmission modes into just one spatial multiplexing transmission scheme. As LTE evolved to R13 and R14, massive MIMO technology was already mature enough to be deployed in commercial networks. Therefore, since the initial stages of 5G NR standardization, massive MIMO technology has been a basic assumption in designing the NR system. The NR system incorporates multiple active antenna arrays, comprising tens or hundreds of antenna elements at base stations. These arrays are capable of forming beams that are identified by pilots for hundreds of antenna ports, and the number of antenna elements is far more than the number of utilized RF chains.

Channel state information acquisition at the transmitter side has always been a key MIMO component. The basic CSI framework in LTE/LTE-A and 5G NR systems typically needs specific RSs. For example, there could be a CSI reference signal (CSI-RS) in the downlink, or a sounding RS (SRS) in the uplink, based on which a receiver can measure and estimate the channel characteristics of interest. For example, a 5G-based station typically transmits a CSI-RS to a user, which then performs channel measurement, quantization, and feedback. The RS design in NR is very flexible and can be configured on a per-user basis, including periodicity, time, frequency, and spatial resolution, in order to optimize the support for different usage scenarios such as eMBB, URLLC, and mMTC. As discussed in Chapter 1, 6G networks will have even more use cases and will use more antenna elements than 5G. Simply extending the NR mechanism for CSI acquisition into 6G will bring higher overhead and longer delay for configuration, RSs, and feedback.

Codebook design has been a key technical component for CSI acquisition in NR. The 5G NR supports different codebooks known as type I and type II codebooks [8]. NR codebook design is based on the assumption of regular and uniform planar and cross-polarized antenna arrays at the gNBs (5G NR base stations) where DFT-based vectors are well structured to match the steering vectors of such antenna arrays. In general, type I codebooks are relatively simple. They are designed to select a single beam from a large set of DFT-based beam directions. In contrast to type I codebooks, type II codebooks target MU-MIMO scenarios and are designed to report a beam as comprising a linear combination of a few dominant beams. Therefore, the type II codebooks are more extensive, and can provide higher resolution of channel characteristics.

As cellular networks become increasingly dense, network interference becomes an important factor that affects network performance. As a consequence, interference measurement has become an important aspect of CSI acquisition. In general, a gNB configures a user device to perform interference measurement based on certain assumptions. As the network complexity increases, a better interference management scheme beyond the existing solutions should be studied.

25.2.2 MIMO Support for FR2

Communication over FR2 (from 24.25 to 52.6 GHz) is supported in NR. 3GPP plans to expand the support for up to 71 GHz frequency bands [9], and to enhance signaling techniques for FR2. Unlike communication over FR1 (0.41–7.125 GHz), which can operate with or without beam-based MIMO, FR2 requires beam-based communication over massive antenna elements. This is a result of the need to overcome the path loss and small antenna aperture caused by short wavelengths. Such massive MIMO utilizes narrow beams that include only the most significant path (or paths) of the wireless channel. Therefore, beam management in NR is designed to establish and retain a suitable beam pair; specifically, a transmitter beam and a corresponding receiver beam, which, when combined, provide good connectivity for communication. In general, the beam management functionality incorporates different components: initial beam establishment, beam sweeping/adjustment and tracking, beam indication, and beam failure recovery.

With beam-based MIMO, data communication, channel measurement reference signals, physical control channels, and synchronization channels are all beamformed. Hierarchical beam search, refinement, and failure recovery through increasingly narrower beams are utilized to enhance data coverage and control in both uplink and downlink directions. The shape and direction of these beams are not revealed from one side of the communication to the other. Beams are identified implicitly by the associated RSs. The user device is configured to constantly monitor several beams. It can then select one or more of the available beams, and promptly report a link failure if one should be detected through the random-access channel (RACH).

As 6G moves towards even higher mmWave bands and the number of antenna elements and beams increases, at either the transmitter or the receiver side, the overheads of measurement and feedback are also increased if the above beam management procedure is followed.

Another important beam management function for FR2 is beam indication: the gNB could indicate to the user the Tx/Rx beams for transmitting/receiving channels and signals. Beam indication in NR is implemented via quasi-located (QCL) and spatial relationship information. For example, after beam measurement and reporting, the gNB may indicate to the user that the scheduled physical downlink shared channel (PDSCH) would follow the QCL assumption, with a particular CSI-RS, and with respect to spatial Rx parameters. This indication communicates to the user that for PDSCH transmission, spatial filtering can be the same as that for the corresponding CSI-RS reception. The beam indication mechanism in NR is based on transmission configuration indication (TCI); in other words, each measurement will have to refer to its configured RS. This implicit signaling reduces the overhead but could be difficult to extend to narrower beams due to the fact that the measurement assumes the original beam-pair link has been found reliably.

25.2.3   Cooperative MIMO

Cooperation among network nodes has been studied since early releases of LTE. During NR standardization, the study of cooperation was used to try to unleash the potential for huge improvements in performance. Despite these potential gains in theory, cooperation comes with practical difficulties and limitations. However, NR has adopted some solutions to overcome these challenges. In addition, end user cooperation can also significantly enhance user experience. In this section, we study network and user cooperation schemes, and their roles in NR.

Multi-TRP Transmission

In traditional cellular networks, a user is only connected to a single transmit–receive point (TRP) at a time; each TRP has independent scheduling and transmission. With cooperative MIMO, multiple TRPs coordinate their transmissions so that a user can receive signals from them simultaneously. Several cooperative MIMO schemes are discussed here.

One scheme is dynamic point selection (DPS). In DPS, a user is served by one TRP, which is dynamically selected within a TRP cluster at the transmission time interval (TTI). For DPS, the network can dynamically select a TRP that either has the best channel quality or offloads the most traffic from the user.

Another scheme is coherent joint transmission (CJT), where multiple TRPs transmit the same data streams to a user, with coherent precoding. However, the support of high-capacity backhaul such as Cloud-RAN is needed for CJT.

A third scheme is non-coherent joint transmission (NCJT), where multiple TRPs transmit different data streams to a user, with non-coherent precoding. NCJT does not rely on tight synchronization or highly accurate CSI, and it is applicable to both ideal and non-ideal backhaul multi-TRP coordination. NCJT can also be used for URLLC reliability enhancement. Current NR only supports ideal backhaul multi-TRP URLLC on the basis of centralized scheduling. It will be useful for NR to support URLLC transmission in non-ideal backhaul.

It is worth noting that multiple TRPs can also cooperate to receive and decode data from the served users. This will allow significant network coverage and throughput enhancement in the uplink direction. However, network-side joint reception could be implementation-specific. Since most techniques do not rely on a specific procedure or signaling from the user, it can be realized through network implementation for multi-TRP reception.

User-Centric Network

NR cells can comprise various TRPs and RRHs, each possibly serving multiple transmit and receive beams. NR air interfaces follow the so-called UCNC design principle to ensure that users can move freely between different beams and cells, with zero interruption mobility. These beams may belong to the same or different TRPs in the same NR cell. Therefore, many different aspects of NR air interfaces are required to enable this level of flexibility. For example, UEs are expected to measure multiple synchronization signal blocks (SSBs), perform beam sweeping for RACHs, and maintain different QCL states and types. Moreover, all user-specific control and data physical channels after completing initial access can be configured to utilize user-specific sequences and scramblers that are not associated with a specific TRP. Such flexibility allows a network with many virtual NR cells to share TRPs and RRHs. 6G is expected to enhance this design in air interfaces, to ensure that extreme network densification and beam-based communication can realize truly transparent "handover" at least from the PHY perspective.

In [10], the authors discuss cell-free massive MIMO with phase synchronized TRPs. The authors show that, as the number of participating TRPs grows, fading, noise, and interference vanishes, and only the pilot-contamination issue remains. 6G can further reduce reliance on the cell, thereby improving the user experience and overall network performance.

The concept of cell-free MIMO in [10] hinges on some stringent assumptions. Careful design is required to address practical and theoretical issues, and successfully deploy cell-free networks. The authors of [11] have investigated non-coherent cell-free MIMO, which aims to alleviate the reliance on tight synchronization among TRPs. The impact of hardware impairments is studied in [12], and in [13], the authors argue that the channel hardening effect of cell-free MIMO strongly relies on the number of antennas per TRP and the statistical properties of the propagation environment.

Cooperative User Transmission

While massive MIMO is being developed for emerging networks that utilize both FR1 and FR2, certain end users may face constraints in accessing FR2 due to their physical limitations. However, end users with sidelink capabilities can cooperate, pooling their antennas and resources to form a distributed MIMO system. User cooperation has several benefits, including diversity gains, multiplexing gains, power aggregation gains, coverage extension, and power saving. In a typical user cooperation solution, it is assumed that multiple UEs can hear the messages from or for a certain user. Through information exchange, UEs form a virtual multi-antenna array, helping original transmission in a similar way to MIMO.

User cooperation can utilize multiplexing, where the original data are split into multiple parts across multiple UEs, and each user transmits only a part of the data. Much higher capacity is expected in this case, especially for *eMBB* services. Other cooperation schemes aim to improve reliability and availability by introducing cooperative diversity, as described in [14–16]. One simple method of cooperative diversity is to select the best user to help the transmission. This method can be seen as a distributed version of MIMO antenna selection. There is also considerable research proposing to exploit distributed space–time coding for cooperative diversity.

For most cooperation methods, one key issue is acquiring CSI before cooperative transmission starts. For example, in cooperative multiplexing, global CSI may be required for the optimal data splitting. As for cooperative diversity, either global CSI or transmit-point CSI is required before cooperative transmission starts. Another key issue that requires further research is synchronization among multiple cooperating nodes, which is mandatory for some schemes.

In summary, the existing MIMO mechanism in NR has a unified CSI and beam management framework. The codebook design has been taken into consideration for multi-user MIMO. The cooperative transmission has taken into consideration the backhaul restriction. However, as 6G moves towards even higher-frequency mmWave bands, enhancement related to CSI measurement and beam management is expected.

25.3   Emerging MIMO Technologies

MIMO is one of the key technologies that will allow 6G to achieve its target KPIs, in terms of user and network throughput, reliability, agility, and energy efficiency. The success of MIMO in 6G hinges not only on enhancements to the existing 5G solutions, but also on new solutions. Notably, future MIMO systems are expected to include advancements in RF technology as well as new materials, antenna architectures, and signal processing techniques. In this section, we briefly review the emerging MIMO technologies in 6G, and examine both the opportunities and challenges they bring.

25.3.1   THz MIMO

The ever-increasing demand for wireless data, especially Tbit/s communication over short ranges, can only be satisfied by tapping into the vast (but widely unutilized) sub-THz spectrum. The use of THz impacts many different aspects of network structure and design. Much research has been devoted to studying these impacts [17, 18].

The THz frequency band occupies the space between mmWave and infrared bands, which are associated with RF and optical devices, respectively. As discussed in Chapter 16, recent advancements in RF components, signal processing, and antenna technology have opened up the THz band for wireless communication. THz communication will impact different aspects of MIMO systems, including modulators, waveforms, and transceiver designs. The new transceivers may use electronic, photonic, or combined technologies [17, 19, 22]. New antenna technology may utilize arrays of active and passive elements to improve beamforming and coverage.

There are many similarities between the mmWave and THz bands, making their MIMO deployments similar. Therefore, it is natural to extend mmWave solutions and architectures into the THz band. However, there are significant characteristics and practical considerations associated with THz, which differentiate the two bands. These include: channel characteristics, device design and signal generation, and antenna technology [17, 22].

THz waves suffer from high path loss. The path loss is shown to increase quadratically with the carrier frequency in free space (see the Friis transmission formula) [23]. Besides this huge path loss, there are other issues with THz waves, such as the low efficiency of power amplifiers and issues caused by weather. Weather issues, such as molecular absorption, or scintillation due to heat and humidity, limit the uses of the THz band to short-range scenarios, such as indoor networks, smart offices, and smart factories [17, 21]. Moreover, since THz wireless channels are not rich in scattering, their coverage is limited by physical obstacles, such as walls and ceilings. Furthermore, due to insignificant wave diffraction, the wireless channels consist mainly of LOS paths, and possibly a few reflection paths, generated by walls or furniture.

The very wide bandwidth available at THz comes with its own challenges. First, the thermal noise power is linearly proportional to the bandwidth used, resulting in a very low SNR even at medium Tx–Rx distances [17, 21]. Second, wide-bandwidth RF chains are more expensive and potentially dissipate more power, resulting in even lower power efficiency for THz transceivers [17, 21]. Third, the array-of-subarrays (AoSA) structure of THz transceivers, along with the very wide bandwidth, leads to a beam squint or split phenomenon [24]. This phenomenon appears because of the need to maintain a reasonable antenna aperture. The antenna panel dimensions cannot shrink linearly with the frequency band, while the pulse durations do shrink as the available bandwidth increases. As a result, different antenna elements (belonging either to the same or different antenna subarrays) experience a flight time difference, comparable to or exceeding the pulse duration. This time difference will cause the beams in parts of the utilized bandwidth to point in slightly different directions, as shown (with exaggeration) in Figure 25.1. This beam squint phenomenon causes the extremely narrow transmission and reception beams to point in different (and unwanted) directions. A possible solution to this phenomenon is to deploy delay-phase precoding [24].

Another challenge for THz links is the pilot and feedback overheads associated with channel acquisition, required to form and maintain beam pairs between the transmitter and receiver.

Beam search requires extensive sweeping at both ends of the wireless link. A slight movement or rotation at either side may result in a beam failure [20–22, 24]. Therefore, agile beam management that features low overhead and high robustness is key to the success of THz deployment. While hierarchical beamforming is a possible

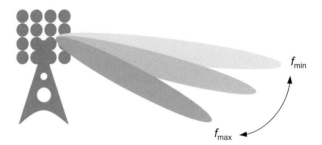

Figure 25.1 Beam squint at THz.

solution to achieve narrow beams at low overhead, it suffers two drawbacks: a very low link budget for wide beam search, and a long delay due to the extended number of measurements at each step. Such delay and overhead, if unresolved, would limit the uses and benefits of THz communication. Given the challenges in beam management and other THz channel characteristics, more efficient designs are required for higher-layer protocols, such as the MAC protocols [18, 21].

25.3.2   Reconfigurable Intelligent Surfaces

Reconfigurable intelligent surfaces (RISs) have recently emerged as a promising paradigm for designing wireless networks and wireless transmission modes. They can also produce smart radio environments (or smart radio channels), which means that the radio propagation properties of the environment can be controlled to create a personalized communication channel [25]. In the general model depicted in Figure 25.2, an RIS network is established among multiple TRPs to produce large-scale smart radio channels that serve multiple users. Without a controllable environment, the wireless system architecture and transmission mode can only be optimized according to the statistical properties of physical channels, and/or by information fed back from the receiver to the transmitter. With a controllable environment, RISs first sense the environment data and feed it back to the system. According to this data, the system optimizes the transmission mode and RIS parameters through smart radio channels, at the transmitter, channel, and receiver sides.

Thanks to the beamforming gains associated with RISs, exploiting smart radio channels can significantly improve link quality, system performance, cell coverage, and cell-edge performance in wireless networks; this has been proven by numerous simulation results in different scenarios [26, 27]. It is worth noting that not all RIS panels use the same structure. This is because they are designed with various phase-adjusting capabilities that range from continuous phase control to discrete control with only a handful of levels. Research has demonstrated that, with a large number of RIS units, limited adjustable phases per unit are enough to significantly improve the overall system performance [26, 28].

Another application of RISs is in transmitters that directly modulate incident radio wave properties, such as phase, amplitude, polarization, and frequency, without the

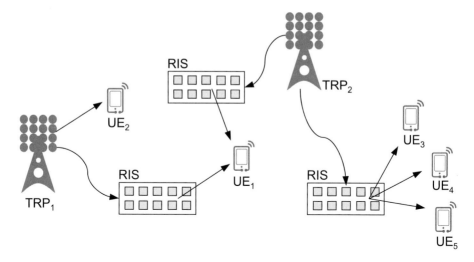

Figure 25.2 Integrated RIS wireless transmission scenario.

need for active components (as is required by RF chains in traditional MIMO transmitters) [29, 30]. For example, RIS-based 256QAM modulation has recently been trialed [31]. RIS-based transmitters have many merits, such as simple hardware architecture, low hardware complexity, low energy consumption, and high spectrum efficiency [32, 33]. Therefore, RISs provide a new direction for transmitter design in radio systems.

RIS-assisted MIMO has other potential applications than those already discussed. For example, it can be used to assist fast beamforming with the use of accurate positioning, or to conquer blockage effects through CSI acquisition in mmWave systems [34]. It can also be used in NOMA in order to improve reliability at a very low SNR, accommodate more users, and enable higher-modulation schemes [35]. RISs are also potentially applicable to native physical security transmission, wireless power transfer (or simultaneous data and wireless power transfer), and flexible holographic radio [36–38].

25.3.3   Extremely Large Aperture Arrays

Over the past 10 years, hardware, algorithms, and system design that will push the limits of MIMO have been developed and tested by academic researchers and industrial engineers in the field. Regarding 6G massive MIMO research, utilizing extremely large aperture arrays (ELAAs) is a promising direction for ultra-massive MIMO techniques.

In [39], an ELAA was defined as an array consisting of hundreds of antenna elements that are jointly serving distributed users. The ultimate goal of ELAAs is to have all the end users using mutually orthogonal channels, with per-user throughput similar to that of an additive white Gaussian noise channel [40, 41]. In addition to enhancing mobile broadband services, which constitute the majority of wireless traffic [42], the great spatial resolution of ELAAs can also be exploited for spatial multiplexing of a large number of communication devices.

A large body of research has focused on new algorithms that could meet the special requirements of ELAAs. In [39, 43, 44], the authors discuss how MMSE processing methods can be used to suppress inter-cell interference, which indicates that ELAAs may work efficiently when deployed in cells.

ELAA research has discovered special features that are significantly different from those seen in traditional massive MIMO, and such features include the following.

- **From far field to near field:** With increased array aperture size in ELAAs, the wavefront becomes significantly spherical and therefore cannot be approximated as a plane wave. As such, the so-called radiative near field begins to dominate electromagnetic propagation. By exploiting the spherical shape of the wavefront, the antenna array can resolve not only the spatial angles of a wave, i.e., the angle of arrival (AoA) and angle of departure (AoD), but also the spatial depth the wave has traveled. This new feature further enables new multiplexing schemes, by using the spatial depth information to separate streams or UEs through precoding.
- **From stationary to non-stationary:** When an array aperture spans hundreds of wavelengths, the wireless channel may become non-stationary. This means that scatterers are only partially visible, or their power contribution may vary considerably over the whole array. The non-stationary property introduces inherent channel sparsity in the system, where the signal from any one user only appears within a small part of the array. This channel sparsity facilitates low-complexity signal processing schemes, such as subarray interference cancellation and message passing methods.

In order to improve system performance, the above unique features need to be considered in ELAA design in the future.

25.3.4   AI-Assisted MIMO

As discussed in Chapter 19, due to its extraordinary ability to extract features from data, machine learning has recently emerged as a promising solution to many real-world problems, including physical layer transmission issues [45, 46]. In general, machine learning may be utilized to improve the performance of existing algorithms, in order to approach optimal algorithm performance with reduced complexity. It can even be used to find modeling solutions to issues that cannot be solved in other ways.

Machine learning has been used in a wide range of applications in MIMO. Some typical examples include MIMO power control [47, 48], CSI acquisition [49, 50], channel estimation [51, 52], MIMO precoding [53, 54], and detection [55, 56].

In machine learning, an AI learns the underlying characteristics of a target feature from the data it is fed, and can then apply this knowledge to a given task. This is seen as an upgrade on conventional approaches to problem-solving, where some parameters are optimized based on a simplified mathematical model of the target feature. These conventional approaches are limited by the precision and complexity of the mathematical models, which often cannot faithfully incorporate the underlying characteristics of the feature in question. Therefore, as proven by current studies, AI-assisted MIMO schemes always outperform conventional ones. The following are some typical instances of AI-assisted MIMO.

- **AI-assisted CSI acquisition:** In [49], the authors utilized pairs of encoder and decoder networks to produce more efficient CSI feedback. Rather than using random projection, the encoder learns the transformation (from original channels to compressed representations) by using training data. Meanwhile, the decoder learns the inverse transformation (from compressed representations to original channels) in the same way. According to evaluations in the test scenario, this data-driven method can obtain more accurate CSI than conventional methods do. Further enhancements to learning performance can be achieved if AI-assisted MIMO additionally uses algorithm modeling in the training process. By integrating model-driven strategies into conventional data-driven training, quicker convergence and more accurate inference can be realized.
- **AI-assisted precoding design:** The article [53] proposed a deep-learning-based mmWave MIMO framework for effective hybrid precoding. This approach aims to exceed the fundamental limits of the existing hybrid precoding schemes. These limits include high complexity, as well as poor exploitation of spatial information. In the given study, each of the precoders required for obtaining the optimized decoder is regarded as a mapping in the DNN. The results show that this approach is capable of minimizing BER and of enhancing the spectrum efficiency of the mmWave massive MIMO, while substantially reducing computational complexity.
- **Machine learning-assisted MIMO detection**: Better performance has also been observed in MIMO detection that is assisted by machine learning. In [55], the authors proposed a deep learning network for high-performance MIMO detection. The conventional maximum-likelihood detector is optimal in terms of minimizing the probability of error when detecting symbols simultaneously, yet it has very high complexity. The proposed deep learning network is derived by employing a projected gradient descent method. Simulations show that this network achieves near optimal detection performance, an accuracy similar to the maximum-likelihood detector, whilst being more efficient and robust; it has a running speed that is at least 30 times faster.
- **AI-assisted MIMO channel estimation:** High-performance channel estimation becomes more and more challenging for massive MIMO cellular networks with an expanding antenna number. The article [52] proposed a deep-learning-based channel estimation method for multi-cell interference-limited massive MIMO systems. The estimator employs a specially designed DNN based on a deep image prior (DIP) network. The proposed method involves only two steps: first, denoise the received signal, and second, estimate the channel with the conventional least squares (LS) algorithm. Simulations prove that this estimator can approach the minimum mean square error (MMSE) for high-dimensional signals, while avoiding

complex channel inversions and not requiring knowledge of the channel covariance matrix. This estimator is also robust to pilot contamination and under certain conditions can completely eliminate it.

Based on the considerable improvements in system performance recorded in ongoing research, it is expected that AI will be a good candidate to enable more reliable and efficient MIMO in 6G. However, various issues need to be addressed before AI can be widely applied to MIMO systems. First, existing AI-assisted MIMO designs can learn from input data, but do not generalize well to modified distribution or non-stationarity. Poor use of learned knowledge is another key deficiency of current AI-assisted MIMO models. In addition, inadequate excavation of data can limit the efficiency and accuracy of AI-assisted MIMO algorithms. All the above factors make it challenging to achieve a balance between training efficiency and inference performance in the widespread application of AI-assisted MIMO.

25.3.5   Other Potential MIMO Technologies

Orbital Angular Momentum (OAM)

The OAM is the angular momentum of an electromagnetic beam, which is dependent on the field spatial distribution, but not on the polarization. In 1992, Allen *et al.* were the first to combine the concept of OAM with the idea of the optical vortex [57]. In 2007, the concept of OAM was creatively extended to radio frequencies by Bo Thidé *et al.*, resulting in radio OAM (referred to as OAM hereafter) [58].

Ways to generate OAM modes have previously been discussed, in Chapter 15. The electromagnetic waveforms carrying OAM modes, called OAM beams, can be classified into several types, such as Gaussian beam, Laguerre–Gaussian beam, Bessel beam, and Bessel–Gaussian beam. In terms of wireless communication in the RF domain, OAM beams increase the non-stationarity of wireless channels, and introduce a helical phase structure into the radial direction of propagation. Previous work has verified the feasibility of OAM, both from theoretical [59] and practical [60] perspectives.

Classical reception methods of OAM-based multiplexing are listed in Table 25.1. Several demonstrations of mmWave OAM communication have been completed, such as [61] at Ka-band, [62] at E-band, and [63] at D-band. Based on 2D beam-steerable circular arrays [64], the pseudo-Doppler interpolation method [65], electromagnetic field fingerprint method [66], and modulo-based robust phase gradient method [67] to adapt OAM for long-distance and low-speed mobility have been proposed.

All current research points to OAM as an effective method of utilizing the spatial domain, especially when the transmitter and receiver are in near field, that is, closer than the Rayleigh distance (defined as the axial distance with a phase difference of $\pi/8$ between axial and edge rays) [68]. The integrated OAM and MIMO framework, described as being multi-mode–multi-spatial (MOMS), is proposed to flexibly obtain spatial diversity gain in dynamic channel conditions using unified hardware [69].

Table 25.1 Classical reception methods of OAM-based multiplexing.

Reception method	Aperture of Rx antenna	Reference
Scalar phase gradient method	Full aperture	[70]
Spectral analysis method	Full aperture	[71]
CNN-based spatial mapping method	Full aperture	[72]
Scalar phase gradient method	Partial aperture	[73]
Vector phase gradient method	Partial aperture	[66]
Pseudo-Doppler method	Partial aperture	[74]

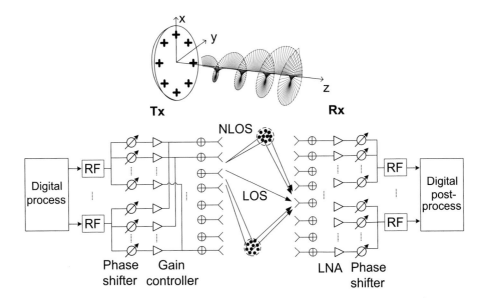

Figure 25.3 An OAM-based wireless communication system.

However, other research [69, 75–79] has uncovered drawbacks of utilizing OAM beams. OAM beams cannot bring new, independent DoF in wireless communication systems, nor can they go beyond traditional MIMO capacity with the same channel environment. Consequently, OAM cannot increase the spectrum efficiency of MIMO systems in classical wireless scenarios, such as cellular radio access networks. Nevertheless, in some favorable scenarios (e.g., in micro cells and indoor small-cell situations), OAM may obtain higher spectrum efficiency than MIMO systems with simple and suboptimal MIMO decoders. This is because multiple OAM modes improve the condition number of the channel.

Despite the drawbacks of OAM beams, the ultimate benefit of introducing OAM transmission is two-fold. First, when the channel satisfies the paraxial limitation and LOS condition, the information carried by orthogonal OAM beams can be efficiently demultiplexed without complex signal post-processing, as shown in Figure 25.3. This low processing complexity has significant value in ultra-high-speed wireless communications (e.g., Tbit/s throughput at the mmWave and THz bands). Second, as previously mentioned, OAM beams can be regarded as a special spatial filtering technique that can be combined with MIMO signal processing to obtain a flexible trade-off between multiplexing gain and diversity gain in more general channel conditions [69, 80]. This combined scheme is especially suitable for short-distance mmWave antenna arrays, where conventional planar wave MIMO is prone to ill-conditioned channel matrices (i.e., rank 1 per polarization).

Practically, OAM can be used for high-speed transmission by introducing mode division multiplexing access. For example, it could be used in backhaul transmission between base stations, device-to-device (D2D) transmission, and satellite-to-satellite communication, where the required axial alignment is relatively easy to obtain.

Holographic MIMO

The theoretical capacity of massive MIMO systems grows with the number of antennas. This means that, theoretically, it is always preferable to have a greater number of antennas, with the optimal amount being infinitely many [39]. In practice, however, there are issues with continually increasing the number of antennas. For example, one issue is how to implement a very large number of antennas in a limited space in the form of a spatially continuous transmitting/receiving aperture. Another issue is the MIMO capacity limit when considering the electromagnetic effect between tightly placed antennas. Research to address these kinds of issues falls under the bracket of holographic MIMO (Holo-MIMO) research. However, such research has been given several other names, including holographic radio or holographic RF system research [81–84], or even holographic massive MIMO or holographic beamforming research [85–88].

Holo-MIMO is defined as the ultimate form of a spatially constrained multiple-antenna system, where the number of antennas could grow infinitely whilst maintaining limited aperture size (i.e., an approximately continuous aperture). This design has opposing principles to the ELAA design, where high DoF comes from an enlarged array aperture. One potential benefit of a Holo-MIMO system is that a continuous aperture may achieve higher spatial resolution in wireless communications than a very large number of conventional discrete antennas. Another benefit is that a continuous aperture enables the creation and detection of electromagnetic waves with arbitrary spatial-frequency components, without undesired side-lobes.

Theoretical study on the DoF of continuous and limited-size apertures can be traced back to 2005 [89]. In this research, DoF was shown to be linearly dependent on the product of aperture size and angular spread according to random non-line-of-sight (NLOS) propagation. Furthermore, the results in [87, 88] indicate that volume apertures could asymptotically yield a two-fold increase in the available DoF compared with planar apertures, regardless of how thin the volume apertures are. In [86], it was shown that mutual coupling of closely spaced antennas can be exploited by precoding in order to achieve super-directivity gain.

Holographic radio is proposed to create an active continuous electromagnetic aperture that will enable holographic imaging-level, ultra-high-density, and pixelated ultra-high-resolution spatial multiplexing. Spatial RF wave field synthesis and modulation can obtain a 3D pixel-level structured electromagnetic field (the high-density multiplexing space of holographic radio) that is different from the sparse beam space associated with traditional massive MIMO. Moreover, holographic interference imaging could be used to obtain the spectral hologram of RF transmitting sources (UEs). This may avoid the need for conventional pilot transmission and CSI estimation procedures.

One challenge to holographic radio systems is the implementation cost of active continuous apertures. A possible solution is to integrate a large number of antenna elements into a compact space in the form of a metasurface. However, this method is limited to passive reflection due to the metasurface property. It is therefore not suitable for active arrays. An alternative (and more promising) solution is to use a tightly coupled array (TCA) of broadband antennas. This technology is based on a current sheet with uni-traveling-carrier photodetectors (UTC-PDs), where the highly dense RF feed network is replaced by fiber links, which have the advantages of being low-cost and consuming little power [90].

Another challenge to holographic radio is the complexity of its signal processing, which is a result of the vast amounts of data generated by the nearly infinite number of antenna elements within the continuous aperture. One solution is to convert radio signals into optical signals and process them directly in the optical domain, where higher processing speed and lower power consumption are possible. Conversion to optical signals can be achieved through optical FFT [90].

The exact benefits of packing more antennas into size-limited apertures remain unclear. Current research shows that the DoF is still limited by aperture size. In [82, 84] it was pointed out that due to the lack of models, holographic radio needs a featured theory and modeling technique to converge the theories of communication and electromagnetism. Moreover, the performance evaluation of Holo-MIMO communication requires dedicated electromagnetic numerical computation, such as algorithms and simulation tools related to computational electromagnetics and computer holography.

25.4 New Design Expectations and Potential Research Directions

It is envisioned that in 6G, there will be a substantial expansion of the spectrum range in which radio access technology can operate. The spectrum range will cover waves from below 1 GHz up to mmWave and THz. There also will be various new types of terrestrial and non-terrestrial network nodes in 6G, including satellites, HAPSs, UAVs, and even RISs. Given these changes, MIMO in 6G will utilize a whole-spectrum aerial/terrestrial architecture (as shown in Figure 25.4) over traditional terrestrial MIMO mounted on fixed towers.

Future TRPs will comprise different antenna types, which can be either active or passive, fixed or moving. All antennas will form a virtual and large antenna array, which will be able to serve mobile users in an intelligent and flexible manner,

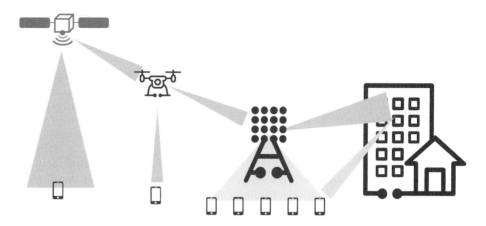

Figure 25.4 Whole spectrum 3D M-MIMO framework.

satisfying 6G KPI requirements. Such a flexible and versatile network is empowered not only by advancements in the existing technologies, but also by numerous emerging technologies, such as ELAA, RIS, AI, sensing, new materials, and new antenna designs and structures.

These new technologies will provide a versatile toolbox to boost network capacity. To employ these technologies in an efficient manner, the MIMO framework in 6G should adopt new design principles and optimization criteria, which will be discussed below.

25.4.1   Sensing-Assisted MIMO

6G is expected to integrate sensing and communication capabilities. Empowered by AI, 6G network nodes and UEs will cooperate to bring powerful sensing capabilities and make the 6G network aware of its surroundings and situation.

Situation awareness (SA) is an emerging communication paradigm, wherein network equipment makes decisions based on knowledge of the propagation environment, user traffic patterns, user mobility behavior, and weather conditions. If the network equipment knows the location, orientation, size, and fabric of the main cluster interacting with the electromagnetic wave in the environment, it can deduce a more accurate picture of the channel condition, such as the beam direction, attenuation and propagation loss, interference level and its source, and shadow fading, in order to enhance network capacity and robustness. For example, the RF map can be used to perform beam management and CSI acquisition with significantly less resources and power than aimless and exhaustive beam sweeping. In the following paragraphs, we explore how sensing can help CSI acquisition and beam management.

- **Real-time CSI acquisition:** A significant challenge for the MIMO framework in 6G is the need for fast and accurate CSI acquisition. Traditional CSI acquisition methods utilized in 4G and 5G pose overheads on time/frequency resources. The overhead increases further as the number of antennas increases. Using traditional methods, increasing the number of antennas also increases measurement delay and CSI aging. This is a significant issue, as it can render the acquired CSI useless due to excessive aging, especially with the presence of narrow beam communication, which is more sensitive to CSI error. Without a smart and real-time CSI acquisition scheme, CSI measurement and feedback will consume all the time/frequency resources. One solution is to use sensing and positioning techniques to assist in determining the channel subspace and identifying candidate beams. This solution can reduce the beam search space while lowering energy consumption for both the user and network. In addition, sensing enables real-time tracking and prediction of wireless channels, which will result in lower beam search and CSI acquisition overheads. Moreover, 6G CSI feedback should be generalized to be agnostic to the antenna structure by quantizing the underlying wireless channels.

Furthermore, CSI acquisition in 6G should utilize the channel characteristics of THz links, as well as the available sensory data, in order to be more efficient and less costly. The THz channel is even more sparse than the mmWave channel in angular and temporal domains, while the available bandwidth and antenna array further enhance the temporal and angular resolutions. As a result, THz transceivers are capable of distinguishing and differentiating different paths with fewer measurements than mmWave transceivers, relative to the number of antenna elements. Additionally, sensory data can be used to compensate for the impact of movement and rotation, or to predict the possible directions of incoming waves. This prediction is enabled by the knowledge of the locations and orientations of access points and end users, as well as the locations of possible reflectors such as walls, ceilings, and furniture.

- **Proactive user-centric beam management:** 6G MIMO is expected to utilize and rely on an increased number of antenna elements for transmission and reception, which makes the 6G air interface predominantly beam-based. To facilitate MIMO technologies in achieving the goals of 6G networks, a reliable, agile, proactive, and low-overhead beam management system is needed. To achieve this state, the beam management system must follow certain design principles.

  A proactive beam management system detects and predicts beam failure, and subsequently mitigates it. The system should facilitate agile beam recovery while also autonomously tracking, refining, and adjusting beams. To achieve this proactivity, intelligent and data-driven beam selection should be assisted with sensory and localization data gathered through air interfaces. Alternatively, other sensors should be supported by 6G to enable handover-free mobility through user-centric beams.

25.4.2 Controllable Radio Channel and Topology

The ability to control the environment and network topology through strategic deployment of RISs, UAVs, and other non-terrestrial and controllable nodes has constituted an important paradigm shift in 6G MIMO. Such controllability is in contrast to the traditional communication paradigm, where transmitters and receivers adapt their communication methods to achieve the capacity predicted by information theory for a given wireless channel. Instead, by controlling the environment and network topology, MIMO aims to be able to change the wireless channel and adapt to the network condition, in order to increase the network capacity.

One way to control the environment is to adapt to the network topology as the user distribution and traffic pattern change over time. This involves utilizing HAPSs and UAVs when necessary. These non-terrestrial access points come with their own challenges and opportunities, as discussed in Chapter 20.

RIS-assisted MIMO utilizes RISs to enhance MIMO performance by creating smart radio channels. To extract the full potential of RIS-assisted MIMO, new system architectures and more efficient schemes or algorithms should be extensively investigated. Compared with traditional beamforming, at both transmit and receiver sides RIS-assisted MIMO has greater flexibility when realizing beamforming gain. It also helps to avoid blockage fading between the transmitter and receiver. Since the link between the TRP and RIS is common for all served users, its condition significantly impacts the overall performance of RIS-assisted MIMO. Therefore, it is necessary to properly optimize the RIS deployment strategy and RIS groups. Moreover, the RIS beamforming gain relies on CSI acquisition between UEs and networks. Typically, the measurement overhead increases with the number of RIS units. Since the distance between two adjacent RIS units is short (from one-eighth to half a wavelength), there are many RIS units in high-frequency bands in any given array area. Using traditional CSI acquisition to optimize RIS parameters will cause a very high measurement overhead for single-user RIS-assisted MIMO, not to mention multi-user RIS-assisted MIMO. To overcome these challenges, we should investigate hybrid CSI acquisition schemes supporting partially active RISs.

25.4.3   MIMO at FR2 and THz

6G will make mmWave mature and exploit even higher-frequency bands at THz; therefore, it is expected to enhance MIMO support by mitigating its limitations. The antenna panels are denser, and beams tend to become narrower, at these higher frequencies. Hence, they are more sensitive to mobility, CSI aging and errors, user and network node orientation, and calibration mismatch. Moreover, RF devices and antennas face unprecedented challenges, such as non-linearity and low efficiency. In this section, we study the challenges associated with mmWave and THz MIMO. We also consider the expected designs of MIMO in 6G.

- **Minimized overhead for beam management:** One of the expected advantages of 6G is to enable large numbers of antennas and beams with limited measurement and feedback overheads. The solution relies mainly on the inherently sparse channels in temporal and angular domains, as well as access to a variety of sensors and other supporting information. Another expected advantage of 6G is enabling a beam measurement and feedback design that utilizes radio resources more efficiently, from a network perspective, while ensuring the network can serve a variety of UEs and user capabilities with minimal adaptation.

- **Natural support for multi-panel communication:** With multiple antenna panels at the user side, the user is well poised to maintain a reliable link with the network, and the link is robust to user rotation and body blockage. Instead of adopting an add-on design after the first release in 5G, native support for multi-panel communication is expected in 6G. The network and user can better coordinate the available user panels and beams. This will be enabled not just by measurements, but with sensory data, gyroscope information, and possibly AI-enabled algorithms. Learning from 5G, the increased power consumption required to maintain multiple active panels at the user side will be addressed in 6G design. As such, the benefits of multi-panel communication will be available to a greater number of UEs.
- **Full utilization of sensory data for beam management/indication:** For FR1, the beam pattern and selection is largely implementation-specific. As such, LTE and NR support for FR1 does not depend on the choice of analog beams at the user side. The same applies for NR support of FR2. The beamforming strategy is concealed at both ends, and is implicitly represented through pilot identification without mandating an exact beam pattern or updating the strategy. This leads to significant efforts being made to ensure interoperability. 6G support for mmWave MIMO can help us better regulate beam patterns, antenna panels, and beamforming strategies over time and space, in order to optimize beams at both sides with lower overhead and latency while employing sensory data more efficiently.
- **Robust design to compensate for hardware constraints in THz:** 6G is expected to expand the current frequency bands to the THz level. Such expansion should be carefully designed to consider the practical and physical opportunities and challenges associated with the THz band, THz RF devices, and signal processing.

As with mmWave, THz communication relies heavily on beamforming that employs very large AoSA. AoSA contains an even greater number of antenna elements at both transmitter and receiver ends to compensate for more severe path loss than for FR2. Thus, antenna elements tend to become very small and exist at close proximity to one another. The resulting antenna panel should accommodate not only the antennas, but also a huge number of tiny-RF-power and low-noise amplifiers, phase shifters, DACs, and ADCs all within a small area, as discussed in Chapter 11. These components should support a very wide frequency band for wireless communication. Moreover, MIMO technologies and solutions should take into account hardware constraints, such as non-linearity, calibration issues, and phase noise, which are often more severe in wider frequency bands. They should also account for beam instability caused by the fluctuating physical and electrical properties of components and the environment (e.g., the adverse effects of heat and humidity).

THz antenna panel design may come with an ADC resolution dilemma. High-order digital beamforming can be enabled by utilizing multiple high-speed ADCs connected to different sub-arrays and RF chains. High-speed and high-resolution ADCs are expensive and dissipate a lot of power. Hence, for many low-cost and low-power devices, an alternative may be the utilization of multiple low-resolution ADCs (with resolutions as low as 1 bit), or a few high-resolution ADCs; perhaps even a combination of the two. Channel estimation and data demodulation for low-resolution ADCs require the careful design of pilots, modulation, and waveform to combat high quantization noise.

25.4.4   Extremely Large Aperture Arrays

ELAAs are expected to significantly improve the spectrum efficiency of wireless air interfaces through enhanced spatial reuse, which is a benefit of an increased number of antennas. In terms of MIMO functionality for 6G, we should consider the following ELAA design expectations.

- **Generalized channel quantization and feedback:** In 6G, new antenna arrays may be deployed. Due to the non-stationary phenomenon (see Section 25.3.3), the codebook design that assumes every channel path is visible across all antenna elements of an antenna array is not suitable for non-stationary channels. Moreover, antenna elements may form a non-regular and non-uniform layout. Therefore, a generalized channel quantization and feedback mechanism should be designed to accommodate for the deployment of diverse antenna structures in the future.
- **Higher-order MU-MIMO and SU-MIMO:** In 4G and 5G systems, MIMO transmission is based on coherent detection. Dedicated demodulation RSs are designed to estimate the effective channels and intra-/inter-user interference, while performing MIMO equalization at the receiver side. Since the overhead of RSs increases proportionally with the order of MU-MIMO and single-user MIMO (SU-MIMO), a trade-off between performance and overhead in 6G will have to be identified.
- **Low-complexity channel estimation:** With an increasing number of antennas, along with the aforementioned near-field non-stationary channel conditions, traditional channel estimation methods such as LMMSE will be unfeasible if the current pilot overhead is not reduced. This is due not only to the increased scale of the problem but also adjustment of the physical channel model. AI-based methods can exploit the partial sparsity pattern to estimate the visible regions of scatterers. Moreover, with spherical array steering, we can localize scatterers and identify visible regions to recover near-field non-stationary channels.

25.4.5   AI-Enabled MIMO

Because of its excellent reasoning ability, accurate feature extraction, and high parallelism, AI has the potential to improve the performance of MIMO systems.

Nevertheless, AI should facilitate the incorporation of MIMO into 6G; this will help deal with large array dimensions and increasingly diverse network scenarios. High-performance channel inference, precoding/beamforming, and detection designs in 6G are the main challenges of extending array dimensions. For example, regardless of exploiting SU-MIMO, MU-MIMO, or beam tracking, MIMO performance relies on highly accurate CSI acquisition and subsequent precoding. However, achieving highly accurate precoding is becoming more and more challenging. This is in part due to the sharp increase in the level of computation required. Similar challenges face MIMO detection. To address these issues, AI-enabled MIMO design should pay attention to data collection and utilization, learning-theory evolution, and innovative scheme designs for different MIMO scenarios. We shall now elaborate these solutions in more detail.

First, mature data collection, verification, and utilization schemes are critical to enabling the accurate assimilation of (potentially) massive amounts of data that are provided by the sensing technologies in 6G. Sensory data may enable AI to directly (or indirectly) infer and predict relevant information for CSI acquisition, beam management, mobility, handover, caching (such as device range, speed, and orientation), Doppler shift, AoA, and AoD. Taking CSI acquisition as an example, with abundant data provided by sensing, AI can extract more useful features from angle and distance information to assist CSI recovery or prediction.

Second, pursuing further research into learning and reasoning theory will help realize more accurate and efficient AI-enabled MIMO designs. For example, studying the interpretability of DNN may enable native AI to design guidelines on how to properly choose AI models and realize efficient training. By improving the ways to reuse and modularize prior information provided by sensing, we can enhance this model- and data-driven AI-enabled MIMO learning and inference, and the model- and data-driven learning will improve MIMO system performance in terms of prediction accuracy and learning overhead (e.g., training data and time).

Third, based on the inherent differences between conventional model-based MIMO solutions and data-driven schemes, AI-enabled MIMO will introduce fundamental changes to the conventional MIMO framework. As an illustrative example, auto-encoder-based systems can effectively replace traditional complex links. The E2E physical layer module design may help us better realize joint optimization between transceivers. On the basis of the potentially massive amounts of data and the potential evolution of learning theory, AI-enabled MIMO is expected to become an inherent feature of 6G. Several design principles and possible research directions for native AI-enabled MIMO are as follows.

- **Extremely high efficiency:** With the assistance of AI, MIMO in 6G should enjoy extremely low overheads for channel measurement, feedback, beam management, and data demodulation. Given its impressive analytical ability, AI can fully explore the inner structure of high-dimensional channels and carry out extremely efficient measurements. Moreover, by employing its reasoning ability to make use of time correlation, AI can further enhance current (or even future) measurements based on historical measurement data. For example, regardless of what form the antenna takes, channel inference will be realized in MIMO systems with extremely low pilot overhead. This is possible due to AI's ability to understand the inner structure of channels. AI may also be incorporated into smart beamforming. By allowing intelligent beam inference and prediction, AI may give rise to more efficient, innovative, and less complicated beam management.

- **Extremely low complexity:** Given AI's complex reasoning and computational capabilities, several conventional MIMO bottleneck problems (e.g., MIMO detection, interference management, and resource allocation) can be efficiently solved to approach theoretical performance limits. For example, maximum-likelihood detection can achieve optimal performance with existing detectors; yet as the number of decision variables increases, its complexity increases exponentially. Though there has been much interest in implementing suboptimal detection algorithms, current algorithms suffer robustness and latency problems. By learning detection rules from data, AI can provide a new approach to attain near-optimal detection with lower complexity.
- **Extremely high accuracy and generality:** AI is well known for accurate feature extraction from various kinds of data. With sufficient training data, AI can obtain more effective CSI representations. As an example of an area that can be improved, CSI acquisition in NR, which is always based on regular basis mapping and codebook design, utilizes a simple mapping and quantization model, leading to a relatively poor mastery of sparsity features. On the other hand, AI enables more accurate CSI feature extraction from data by leveraging data-driven methods to create more effective CSI representations. In addition, thanks to the variety of sensory data, generality can also be guaranteed when the sensory data are used and modularized effectively.

To enable truly intelligent massive MIMO in 6G, AI-enabled MIMO needs to be more reliable, more efficient, and easier to implement. To achieve this goal, explainable AI should be implemented to facilitate training and improve learning efficiency; while data- and model-driven AI should be implemented to ensure more timely and accurate learning. user-centric designs will be required to provide improved and differentiated user experience. Finally, duality frameworks should enable reliable E2E learning.

References

[1] J. Spradley, "A volumetric electrically scanned two-dimensional microwave antenna array," in *Proc. 1958 IRE International Convention Record*, vol. 6. IEEE, 1966, pp. 204–212.

[2] V. Tarokh, N. Seshadri, and A. R. Calderbank, "Space-time codes for high data rate wireless communication: Performance criterion and code construction," *IEEE Transactions on Information Theory*, vol. 44, no. 2, pp. 744–765, 1998.

[3] S. M. Alamouti, "A simple transmit diversity technique for wireless communications," *IEEE Journal on Selected Areas in Communications*, vol. 16, no. 8, pp. 1451–1458, 1998.

[4] G. J. Foschini, "Layered space-time architecture for wireless communication in a fading environment when using multi-element antennas," *Bell Labs Technical Journal*, vol. 1, no. 2, pp. 41–59, 1996.

[5] M. O. Damen, H. El Gamal, and G. Caire, "On maximum-likelihood detection and the search for the closest lattice point," *IEEE Transactions on Information Theory*, vol. 49, no. 10, pp. 2389–2402, 2003.

[6] 3GPP, "User equipment (ue) radio transmission and reception; part 1: Range 1 standalone," 3rd Generation Partnership Project (3GPP), Technical Specification (TS) 38.101, May 2020, version 16.4.0. [Online]. Available: http://www.3gpp.org/ftp//Specs/archive/38_series/38.101-3/38101-3-g40.zip

[7] 3GPP, "User equipment (ue) radio transmission and reception; part 2: Range 2 standalone," 3rd Generation Partnership Project (3GPP), Technical Specification (TS) 38.101, May 2020, version 16.4.0. [Online]. Available: http://www.3gpp.org/ftp//Specs/archive/38_series/38.101-3/38101-3-g40.zip

[8] 3GPP, "Physical layer procedures for data," 3rd Generation Partnership Project (3GPP), Technical Specification (TS) 38.214, July 2020, version 16.2.0. [Online]. Available: http://www.3gpp.org/ftp//Specs/archive/38_series/38.214/38214-g20.zip

[9] 3GPP, "New SID: Study on supporting NR from 52.6 GHz to 71 GHz," 3rd Generation Partnership Project (3GPP), Technical Report RP-193259, Dec. 2019. [Online]. Available: https://www.3gpp.org/ftp/tsg_ran/TSG_RAN/TSGR_86/Docs/RP-193259.zip

[10] H. Q. Ngo, A. Ashikhmin, H. Yang, E. G. Larsson, and T. L. Marzetta, "Cell-free massive MIMO versus small cells," *IEEE Transactions on Wireless Communications*, vol. 16, no. 3, pp. 1834–1850, 2017.

[11] Ö. Özdogan, E. Björnson, and J. Zhang, "Downlink performance of cell-free massive MIMO with Rician fading and phase shifts," in *Proc. 2019 IEEE 20th International Workshop on Signal Processing Advances in Wireless Communications (SPAWC)*. IEEE, 2019, pp. 1–5.

[12] J. Zhang, Y. Wei, E. Björnson, Y. Han, and X. Li, "Spectral and energy efficiency of cell-free massive MIMO systems with hardware impairments," in *Proc. 2017 9th International Conference on Wireless Communications and Signal Processing (WCSP)*. IEEE, 2017, pp. 1–6.

[13] Z. Chen and E. Björnson, "Can we rely on channel hardening in cell-free massive MIMO?" in *Proc. 2017 IEEE Globecom Workshops*. IEEE, 2017, pp. 1–6.

[14] J. N. Laneman and G. W. Wornell, "Distributed space-time coded protocols for exploiting cooperative diversity in wireless networks," in *Proc. 2002 Global Telecommunications Conference*, vol. 1. IEEE, 2002, pp. 77–81.

[15] Y. Jing and B. Hassibi, "Distributed space-time coding in wireless relay networks," *IEEE Transactions on Wireless Communications*, vol. 5, no. 12, pp. 3524–3536, 2006.

[16] A. Bletsas, H. Shin, and M. Z. Win, "Cooperative communications with outage-optimal opportunistic relaying," *IEEE Transactions on Wireless Communications*, vol. 6, no. 9, pp. 3450–3460, 2007.

[17] L. Bariah, L. Mohjazi, S. Muhaidat, P. C. Sofotasios, G. K. Kurt, H. Yanikomeroglu, and O. A. Dobre, "A prospective look: Key enabling technologies, applications and open research topics in 6G networks," *arXiv preprint arXiv:2004.06049*, 2020.

[18] S. Ghafoor, N. Boujnah, M. H. Rehmani, and A. Davy, "Mac protocols for terahertz communication: A comprehensive survey," *IEEE Communications Surveys & Tutorials*, 2020.

[19] Y. Zhao, "A survey of 6G wireless communications: Emerging technologies," *arXiv preprint arXiv:2004.08549*, 2020.

[20] Y. Yifei, Z. Yajun, Z. Baiqing, and P. Sergio, "Potential key technologies for 6G mobile communications," in *SCIENCE CHINA Information Sciences*, Springer, vol. 63, pp. 1–19, 2020.

[21] C. Han, Y. Wu, Z. Chen, and X. Wang, "Terahertz communications (TeraCom): Challenges and impact on 6G wireless systems," *arXiv preprint arXiv:1912.06040*, 2019.

[22] M. H. Alsharif, A. H. Kelechi, M. A. Albreem, S. A. Chaudhry, M. S. Zia, and S. Kim, "Sixth generation (6G) wireless networks: Vision, research activities, challenges and potential solutions," *Symmetry*, vol. 12, no. 4, p. 676, 2020.

[23] H. T. Friis, "A note on a simple transmission formula," *Proceedings of the IRE*, vol. 34, no. 5, pp. 254–256, 1946.

[24] J. Tan and L. Dai, "THz precoding for 6G: Applications, challenges, solutions, and opportunities," *arXiv preprint arXiv:2005.10752*, 2020.

[25] M. Di Renzo, M. Debbah, D.-T. Phan-Huy, A. Zappone, M.-S. Alouini, C. Yuen, V. Sciancalepore, G. C. Alexandropoulos, J. Hoydis, H. Gacanin *et al.*, "Smart radio environments empowered by reconfigurable AI meta-surfaces: An idea whose time has come," *EURASIP Journal on Wireless Communications and Networking*, vol. 2019, no. 1, pp. 1–20, 2019.

[26] Y. Han, W. Tang, S. Jin, C.-K. Wen, and X. Ma, "Large intelligent surface-assisted wireless communication exploiting statistical CSI," *IEEE Transactions on Vehicular Technology*, vol. 68, no. 8, pp. 8238–8242, 2019.

[27] C. You, B. Zheng, and R. Zhang, "Intelligent reflecting surface with discrete phase shifts: Channel estimation and passive beamforming," in *Proc. 2020 IEEE International Conference on Communications (ICC)*. IEEE, 2020, pp. 1–6.

[28] Q. Wu and R. Zhang, "Beamforming optimization for wireless network aided by intelligent reflecting surface with discrete phase shifts," *IEEE Transactions on Communications*, vol. 68, no. 3, pp. 1838–1851, 2019.

[29] B. Xiong, L. Deng, R. Peng, and Y. Liu, "Controlling the degrees of freedom in metasurface designs for multi-functional optical devices," *Nanoscale Advances*, vol. 1, no. 10, pp. 3786–3806, 2019.

[30] J. Yuan, E. De Carvalho, and P. Popovski, "Wireless communication through frequency modulating reflective intelligent surfaces," *arXiv preprint arXiv:2007.01085*, 2020.

[31] T. J. Cui, "Electromagnetic information computing and intelligent control," in *Proc. Huawei Internal Workshop*. Huawei, 2020.

[32] W. Tang, X. Li, J. Y. Dai, S. Jin, Y. Zeng, Q. Cheng, and T. J. Cui, "Wireless communications with programmable metasurface: Transceiver design and experimental results," *China Communications*, vol. 16, no. 5, pp. 46–61, 2019.

[33] W. Tang, J. Y. Dai, M. Z. Chen, K.-K. Wong, X. Li, X. Zhao, S. Jin, Q. Cheng, and T. J. Cui, "Mimo transmission through reconfigurable intelligent surface: System design, analysis, and implementation," *IEEE Journal on Selected Areas in Communications*, 2020.

[34] Y. Cui and H. Yin, "An efficient CSI acquisition method for intelligent reflecting surface-assisted mmwave networks," *arXiv preprint arXiv:1912.12076*, 2019.

[35] V. C. Thirumavalavan and T. S. Jayaraman, "BER analysis of reconfigurable intelligent surface assisted downlink power domain NOMA system," in *Proc. 2020 International Conference on COMmunication Systems & NETworkS (COMSNETS)*. IEEE, 2020, pp. 519–522.

[36] Q. Wu and R. Zhang, "Joint active and passive beamforming optimization for intelligent reflecting surface assisted SWIPT under QoS constraints," *arXiv preprint arXiv:1910.06220*, 2019.

[37] N. M. Tran, M. M. Amri, J. H. Park, S. I. Hwang, D. I. Kim, and K. W. Choi, "A novel coding metasurface for wireless power transfer applications," *Energies*, vol. 12, no. 23, p. 4488, 2019.

[38] J. W. Wu, Z. X. Wang, L. Zhang, Q. Cheng, S. Liu, S. Zhang, J. M. Song, and T. J. Cui, "Anisotropic metasurface holography in 3D space with high resolution and efficiency," *IEEE Transactions on Antennas and Propagation*, 2020.

[39] E. Björnson, J. Hoydis, and L. Sanguinetti, "Massive MIMO has unlimited capacity," *IEEE Transactions on Wireless Communications*, vol. 17, no. 1, pp. 574–590, 2017.

[40] S. Hu, F. Rusek, and O. Edfors, "Beyond massive MIMO: The potential of data transmission with large intelligent surfaces," *IEEE Transactions on Signal Processing*, vol. 66, no. 10, pp. 2746–2758, 2018.

[41] E. Björnson and E. G. Larsson, "How energy-efficient can a wireless communication system become?" in *Proc. 2018 52nd Asilomar Conference on Signals, Systems, and Computers*. IEEE, 2018, pp. 1252–1256.

[42] N. Heuveldop *et al.*, "Ericsson mobility report," *Ericsson AB, Technol. Emerg. Business, Stockholm, Sweden*, Technical Report EAB-17, vol. 5964, 2017.

[43] A. Amiri, M. Angjelichinoski, E. De Carvalho, and R. W. Heath, "Extremely large aperture massive MIMO: Low complexity receiver architectures," in *Proc. 2018 IEEE Globecom Workshops*. IEEE, 2018, pp. 1–6.

[44] E. Björnson and L. Sanguinetti, "Making cell-free massive MIMO competitive with MMSE processing and centralized implementation," *IEEE Transactions on Wireless Communications*, vol. 19, no. 1, pp. 77–90, 2019.

[45] C. Jiang, H. Zhang, Y. Ren, Z. Han, K.-C. Chen, and L. Hanzo, "Machine learning paradigms for next-generation wireless networks," *IEEE Wireless Communications*, vol. 24, no. 2, pp. 98–105, 2016.

[46] Z. Qin, H. Ye, G. Y. Li, and B.-H. F. Juang, "Deep learning in physical layer communications," *IEEE Wireless Communications*, vol. 26, no. 2, pp. 93–99, 2019.

[47] L. Sanguinetti, A. Zappone, and M. Debbah, "Deep learning power allocation in massive MIMO," in *Proc. 2018 52nd Asilomar Conference on Signals, Systems, and Computers*. IEEE, 2018, pp. 1257–1261.

[48] C. D'Andrea, A. Zappone, S. Buzzi, and M. Debbah, "Uplink power control in cell-free massive MIMO via deep learning," in *Proc. 2019 IEEE 8th International Workshop on Computational Advances in Multi-Sensor Adaptive Processing (CAMSAP)*. IEEE, 2019, pp. 554–558.

[49] C.-K. Wen, W.-T. Shih, and S. Jin, "Deep learning for massive MIMO CSI feedback," *IEEE Wireless Communications Letters*, vol. 7, no. 5, pp. 748–751, 2018.

[50] J. Wang, Y. Ding, S. Bian, Y. Peng, M. Liu, and G. Gui, "UL-CSI data driven deep learning for predicting DL-CSI in cellular FDD systems," *IEEE Access*, vol. 7, pp. 96 105–96 112, 2019.

[51] P. Dong, H. Zhang, G. Y. Li, I. S. Gaspar, and N. Naderi Alizadeh, "Deep CNN-based channel estimation for mmWave massive MIMO systems," *IEEE Journal of Selected Topics in Signal Processing*, vol. 13, no. 5, pp. 989–1000, 2019.

[52] E. Balevi, A. Doshi, and J. G. Andrews, "Massive MIMO channel estimation with an untrained deep neural network," *IEEE Transactions on Wireless Communications*, vol. 19, no. 3, pp. 2079–2090, 2020.

[53] H. Huang, Y. Song, J. Yang, G. Gui, and F. Adachi, "Deep-learning-based millimeter-wave massive MIMO for hybrid precoding," *IEEE Transactions on Vehicular Technology*, vol. 68, no. 3, pp. 3027–3032, 2019.

[54] T. Mir, M. Z. Siddiqi, U. Mir, R. Mackenzie, and M. Hao, "Machine learning inspired hybrid precoding for wideband millimeter-wave massive MIMO systems," *IEEE Access*, vol. 7, pp. 62 852–62 864, 2019.

[55] T. D. N. Samuel and A. Wiesel, "Deep MIMO detection," in *IEEE 18th International Workshop on Signal Processing and Advanced Wireless Communications (SPAWC)*. IEEE, 2017, pp. 1–5.

[56] H. He, C.-K. Wen, S. Jin, and G. Y. Li, "Model-driven deep learning for MIMO detection," *IEEE Transactions on Signal Processing*, vol. 68, pp. 1702–1715, 2020.

[57] L. Allen, M. W. Beijersbergen, R. Spreeuw, and J. Woerdman, "Orbital angular momentum of light and the transformation of Laguerre–Gaussian laser modes," *Physical Review A*, vol. 45, no. 11, p. 8185, 1992.

[58] B. Thidé, H. Then, J. Sjöholm, K. Palmer, J. Bergman, T. Carozzi, Y. N. Istomin, N. Ibragimov, and R. Khamitova, "Utilization of photon orbital angular momentum in the low-frequency radio domain," *Physical Review Letters*, vol. 99, no. 8, p. 087701, 2007.

[59] S. M. Mohammadi, L. K. Daldorff, J. E. Bergman, R. L. Karlsson, B. Thidé, K. Forozesh, T. D. Carozzi, and B. Isham, "Orbital angular momentum in radio system study," *IEEE Transactions on Antennas and Propagation*, vol. 58, no. 2, pp. 565–572, 2009.

[60] F. Tamburini, E. Mari, A. Sponselli, B. Thidé, A. Bianchini, and F. Romanato, "Encoding many channels on the same frequency through radio vorticity: First experimental test," *New Journal of Physics*, vol. 14, no. 3, p. 033001, 2012.

[61] D. Lee, H. Sasaki, H. Fukumoto, Y. Yagi, T. Kaho, H. Shiba, and T. Shimizu, "An experimental demonstration of 28 GHz band wireless OAM-MIMO (orbital angular momentum multi-input and multi-output) multiplexing," in *Proc. 2018 IEEE 87th Vehicular Technology Conference (VTC-Spring)*. IEEE, 2018, pp. 1–5.

[62] NEC Corporation, "NEC successfully demonstrates real-time digital OAM mode multiplexing transmission in the 80 GHz-band for the first time," Dec. 2018. [Online]. Available: https://www.nec.com/en/press/201812/global_20181219_02.html

[63] NEC Corporation, "NEC successfully demonstrates real-time digital OAM mode multiplexing transmission over 100 m in the 150 GHz-band for the first time," March 2020. [Online]. Available: https://www.nec.com/en/press/202003/global_20200310_01.html

[64] M. Klemes, H. Boutayeb, and F. Hyjazie, "Orbital angular momentum (OAM) modes for 2-D beam-steering of circular arrays," in *Proc. 2016 IEEE Canadian Conference on Electrical and Computer Engineering (CCECE)*. IEEE, 2016, pp. 1–5.

[65] M. Klemes, "Reception of OAM radio waves using pseudo-doppler interpolation techniques: A frequency-domain approach," *Applied Sciences*, vol. 9, no. 6, p. 1082, 2019.

[66] R. Ni, Y. Lv, Q. Zhu, and M. Debbah, "Electromagnetic field fingerprint method for circularly polarized OAM," in *Proc. 2020 IEEE International Conference on Communications Workshops*. IEEE, 2020, pp. 1–6.

[67] Y. Lv, Q. Zhu, and R. Ni, "Modulo-based phase gradient method for OAM mode detection," in *Proc. 2020 9th IEEE/CIC International Conference on Communications (ICCC)*. IEEE, 2020, pp. 1–6.

[68] J. D. Kraus and R. J. Marhefka, *Antennas for all applications (Third Edition)*. McGraw-Hill, 2001.

[69] R. Ni, Y. Lv, Q. Zhu, G. Wang, and G. He, "Degrees of freedom of multi-mode-multi-spatial in-line-of-sight channels," in *Proc. 2020 IEEE GLOBECOM Conference*. IEEE, 2020, pp. 1–6.

[70] S. M. Mohammadi, L. K. Daldorff, K. Forozesh, B. Thidé, J. E. Bergman, B. Isham, R. Karlsson, and T. Carozzi, "Orbital angular momentum in radio: Measurement methods," *Radio Science*, vol. 45, no. 4, pp. 1–14, 2010.

[71] H. Wu, Y. Yuan, Z. Zhang, and J. Cang, "UCA-based orbital angular momentum radio beam generation and reception under different array configurations," in *Proc. 2014 Sixth International Conference on Wireless Communications and Signal Processing (WCSP)*. IEEE, 2014, pp. 1–6.

[72] S. Rostami, W. Saad, and C. S. Hong, "Deep learning with persistent homology for orbital angular momentum (oam) decoding," *IEEE Communications Letters*, vol. 24, no. 1, pp. 117–121, 2019.

[73] S. Zheng, X. Hui, J. Zhu, H. Chi, X. Jin, S. Yu, and X. Zhang, "Orbital angular momentum mode-demultiplexing scheme with partial angular receiving aperture," *Optics Express*, vol. 23, no. 9, pp. 12 251–12 257, 2015.

[74] C. Zhang and M. Lu, "Detecting the orbital angular momentum of electro-magnetic waves using virtual rotational antenna," *Scientific Reports*, vol. 7, no. 1, pp. 1–8, 2017.

[75] O. Edfors and A. J. Johansson, "Is orbital angular momentum (oam) based radio communication an unexploited area?" *IEEE Transactions on Antennas and Propagation*, vol. 60, no. 2, pp. 1126–1131, 2011.

[76] M. Tamagnone, C. Craeye, and J. Perruisseau-Carrier, "Comment on encoding many channels on the same frequency through radio vorticity: First experimental test," *New Journal of Physics*, vol. 14, no. 11, p. 118001, 2012.

[77] J. Xu, "Degrees of freedom of oam-based line-of-sight radio systems," *IEEE Transactions on Antennas and Propagation*, vol. 65, no. 4, pp. 1996–2008, 2017.

[78] R. Gaffoglio, A. Cagliero, G. Vecchi, and F. P. Andriulli, "Vortex waves and channel capacity: Hopes and reality," *IEEE Access*, vol. 6, pp. 19 814–19 822, 2017.

[79] A. Omar, "Dependence of beamforming on the excitation of orbital angular momentum (OAM) modes," *IEEE Transactions on Antennas and Propagation*, 2020.

[80] S. Saito, H. Suganuma, K. Ogawa, and F. Maehara, "Performance analysis of OAM-MIMO using SIC in the presence of misalignment of beam axis," in *Proc. 2019 IEEE International Conference on Communications Workshops*. IEEE, 2019, pp. 1–6.

[81] D. Prather, "Toward holographic rf systems for wireless communications and networks," *IEEE ComSoc Technology News*, 2016.

[82] N. Rajatheva, I. Atzeni, E. Bjornson, A. Bourdoux, S. Buzzi, J.-B. Dore, S. Erkucuk, M. Fuentes, K. Guan, Y. Hu *et al.*, "White paper on broadband connectivity in 6g," *arXiv preprint arXiv:2004.14247*, 2020.

[83] M. Latva-aho, K. Leppänen, F. Clazzer, and A. Munari, "Key drivers and research challenges for 6g ubiquitous wireless intelligence," 6G Flagship, University of Oulu, Sept. 2019.

[84] Y. Yuan, Y. Zhao, B. Zong, and S. Parolari, "Potential key technologies for 6g mobile communications," *Science China Information Sciences*, vol. 63, pp. 1–19, 2020.

[85] E. Björnson, L. Sanguinetti, H. Wymeersch, J. Hoydis, and T. L. Marzetta, "Massive mimo is a reality – what is next?: Five promising research directions for antenna arrays," *Digital Signal Processing*, vol. 94, pp. 3–20, 2019.

[86] E. J. Black, "Holographic beam forming and mimo," *Pivotal Commware*, 2017. [Online]. Available: https://pivotalcommware.com/wp-content/uploads/2017/12/Holographic-Beamforming-WP-v.6C-FINAL.pdf

[87] A. Pizzo, T. L. Marzetta, and L. Sanguinetti, "Degrees of freedom of holographic MIMO channels," in *Proc. 2020 IEEE 21st International Workshop on Signal Processing Advances in Wireless Communications (SPAWC).* IEEE, 2020, pp. 1–5.

[88] A. Pizzo, T. L. Marzetta, and L. Sanguinetti, "Spatial characterization of holographic mimo channels," *arXiv preprint arXiv:1911.04853*, 2019.

[89] A. S. Poon, R. W. Brodersen, and D. N. Tse, "Degrees of freedom in multiple-antenna channels: A signal space approach," *IEEE Transactions on Information Theory*, vol. 51, no. 2, pp. 523–536, 2005.

[90] D. W. Prather, S. Shi, G. J. Schneider, P. Yao, C. Schuetz, J. Murakowski, J. C. Deroba, F. Wang, M. R. Konkol, and D. D. Ross, "Optically upconverted, spatially coherent phased-array-antenna feed networks for beam-space mimo in 5g cellular communications," *IEEE Transactions on Antennas and Propagation*, vol. 65, no. 12, pp. 6432–6443, 2017.

26 Integrated Super-Sidelink and Access Link Communication

26.1 Background and Motivations

Radio access in wireless systems is fundamentally designed on a P2MP architecture, in which mobile devices connect to one or more base stations. Since the introduction of 4G, wireless connectivity has supported the use of auxiliary peer-to-peer communication links (i.e., sidelinks) for short-range D2D wireless communication. Device-to device (V2V) communication – an example of D2D – has been defined in 4G LTE systems, and enhanced V2V is supported in 5G NR systems.

In the air interface design, NR sidelinks reuse the key features of NR access links (the links between user devices and base stations). As such, NR sidelinks have become an integral part of the NR design. In particular, NR sidelinks inherit key design concepts and the basic design framework from NR access links, albeit with some simplifications, such as bandwidth, flexible frame structure, coding, modulation, and waveforms. User devices can quickly establish sidelinks with each other, regardless of whether they are connected to the same NR cell or not.

As discussed in Part II, many new 6G applications require high bandwidth and low latency. Such applications, including ultimate XR, holographic display, haptic transmission, highly dynamic motion control, positioning, and imaging, typically involve short-distance communication between two devices. Consequently, in order to meet requirements for Tbit/s throughput and sub-millisecond latency, new short-distance technologies such as super-sidelink are needed. THz communication and optical wireless communication (OWC), which offer extremely wide bandwidth, are ideal candidate technologies. As such, 6G super-sidelinks are expected to feature extremely high throughput and extremely low latency, leveraging the new spectrum of THz or OWC, and supporting local mesh networking.

6G super-sidelinks and the associated mesh networking should be an integral part of the overall mobile system. This is necessary to meet both technical and application requirements. In terms of the technical requirements, the high penetration rate of mobile devices makes shorter-distance communication between mobile devices more feasible. In addition, the D2D or V2V approach introduced in LTE or NR as sidelink communication offloads traffic from base stations and facilitates short-distance communication among devices. Furthermore, in 6G, mmWave will be extensively deployed for access links, and, for short-distance communication at higher frequencies, THz and OWC technologies will be used. In terms of the application requirements, the current IMT spectrum is becoming too crowded. As mentioned in previous chapters, 6G will connect huge numbers of intelligent devices. Furthermore, different types of devices will need flexible connections for human-centric content sharing as well as for other types of machine-to-machine communications, such as swarm intelligence and industry-level motion control. These all require a vast amount of bandwidth. As such, to cope with the explosive demand for wireless traffic, an integrated super-sidelink leveraging new wireless spectrum bands is urgently needed.

Integrated super-sidelinks have several advantages over separated access links and sidelinks. For example, a dynamic locally organized super-sidelink network, such as a multi-hop mesh network, can offload the access link traffic, increase the overall cell capacity, and boost the user data rate. It can even reduce the latency. This integrated design is also essential in V2V applications. Specifically, the mesh network can aggregate all sensor information from all adjacent vehicles for autonomous driving. In this mesh network, devices can discover and connect to each other via super-sidelinks even though they are not within the cell coverage. The integrated design will allow the super-sidelink mesh network to strike a balance among flexibility, capacity, and mobility, thereby achieving superior performance.

Figure 26.1 shows an example of a 6G indoor super-sidelink mesh network. The network includes several access links and super-sidelinks for coverage in different rooms. Being closely integrated with 6G access links and leveraging relay techniques, the super-sidelinks are able to achieve extremely high throughput and low latency.

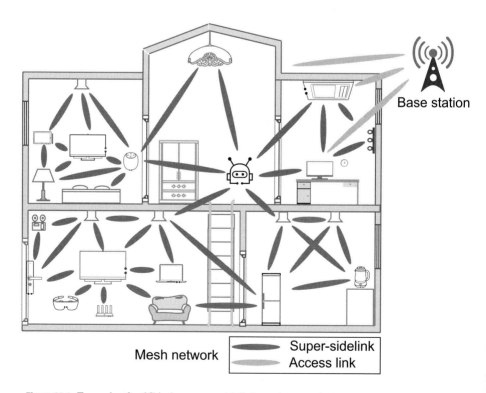

Base station

Mesh network   Super-sidelink
                Access link

Figure 26.1 Example of a 6G indoor super-sidelink mesh network.

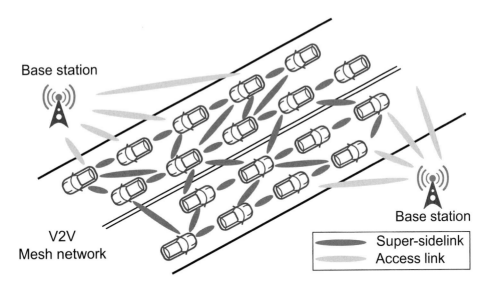

Figure 26.2 Example of a 6G V2V super-sidelink mesh network.

Figure 26.2 provides another example, showing a 6G V2V super-sidelink mesh network. This network includes several access links and super-sidelinks for sharing sensor information and high-resolution images in real time for autonomous driving. In a V2V scenario, employing an extremely-high-throughput and low-latency mesh network is critical. By using integrated access link and super-sidelink design, the mesh network can be created and adapted quickly and dynamically.

Super-sidelinks over THz and optical frequency bands have several advantages compared with conventional RF communication technologies. For example, they leverage a huge spectrum and offer robustness against interference. If the above frequency bands are included in super-sidelinks, 6G prefers a unified PHY design across THz, OWC, and RF communications.

26.2   Overview of Existing Solutions

In 5G NR Release 16, sidelinks reuse the same framework as access links (also known as Uu links in 3GPP). This includes reusing the same waveform, coding, modulation, frame structure, and MIMO. Consequently, NR sidelinks and Uu links can be integrated to simplify UE implementation and even reuse chipsets, achieving significant cost savings for UE vendors.

In addition, 5G sidelinks also employ some new features, including synchronization, resource allocation, HARQ, and interference management. These features, which can be controlled tightly or loosely by the mobile network, are discussed below.

- For sidelink synchronization, different types of synchronization sources, each associated with a synchronization priority, are specified for in-coverage, out-of-coverage, and partial-coverage scenarios in order to obtain the same timing reference. For example, in an in-coverage scenario, the network can function as the synchronization source. Or if one UE is within coverage while another UE is not, the in-coverage UE can act as the synchronization source.

- For sidelink resource allocation, two modes have been defined. Mode 1 applies to in-coverage scenarios where the network allocates sidelink resources. In such scenarios, the Uu link controls the sidelink. Mode 2 applies to both in-coverage and out-of-coverage scenarios, in which UEs autonomously select resources for sidelink communications. These resources can be configured by the network if the Tx UE is in-coverage or preconfigured when UEs are out-of-coverage.

- Sidelink HARQ retransmission enhances the reliability of groupcast and unicast transmission. A case in point is that vehicle platooning expands the V2X service on groupcast transmission in NR. As the name suggests, all the vehicles can obtain information from the lead vehicle in order to manage the platoon. This information allows the vehicles to drive more closely than normal in a coordinated manner, with all vehicles traveling together and heading in the same direction. In this case, the HARQ feedback and retransmission resources can be configured by the network when UEs are in-coverage or preconfigured when UEs are out-of-coverage.

- In interference management for in-band sidelinks, open-loop power control based on the path losses of Uu links and sidelinks is used to control the interference to the network or sidelinks. For example, a minimum path loss from a Uu link and a sidelink is used to determine the Tx power. In sidelink resource allocation mode 2, a Tx UE can rely on SCI decoding and reference signal received power (RSRP) measurements to determine the candidate resources potentially available to be utilized for autonomous resource selection, while avoiding collisions with resources occupied by other UEs. This helps the UE to manage the interference it may cause and receive.

In terms of sidelink spectrum usage, beam operations, and relaying, some of the associated limitations are described below.

- For sidelink spectrum, up to 7 GHz licensed bands and unlicensed intelligent transportation system (ITS) bands are supported up to 5G Release 16. NR sidelinks support both in-band shared carriers with Uu links and out-of-band dedicated sidelink carriers. The former need tighter network control, employing time-division multiplexing (TDM) between Uu links and sidelinks at the cost of increased latency, whereas the latter offer lower latency but need more spectrum resources.

- Because the sidelink design in Release 16 mainly focuses on bands below 7 GHz, beam-based sidelink transmission is not yet fully supported. However, with more focus on mmWave bands, beam-based sidelink transmission for both in-coverage and out-of-coverage scenarios is expected to be further optimized in future releases.

- For sidelink relaying, NR standardization has not yet to reach consensus on Layer 2 and Layer 3 sidelink relaying. Currently, NR sidelinks do not support multi-hop relaying, which is important especially when UE density is high and the sidelink distance between UEs is very short.

26.3   New Design Expectations and Potential Research Directions

In 6G, it is necessary to consider the enabling technologies for integrating super-sidelinks with access links. The design expectations and potential research directions are discussed in the following sections.

26.3.1   Enabling Technologies for Super-Sidelinks

6G super-sidelinks need to involve more high-frequency spectrum, especially for short-distance communications. For example, the use of super-high frequencies, such as THz and optical wireless communication (OWC), has already been proposed [1, 2].

Over the past few decades, THz communication has received a great deal of research interest, particularly in terms of THz modulators, antennas, and THz channel modeling and estimation [3]. The THz band, which ranges from 100 GHz to 10 THz and lies between mmWave and optical frequency, has distinct characteristics. For example, it offers 100 times higher bandwidth than the conventional RF spectrum, meaning it can achieve ultra-high throughput for communications. However, challenges exist in THz communication components (e.g., power amplifiers, mixers, and antennas) and circuits in terms of power efficiency and device limitations. As such, further research is needed to increase device performance. Also of importance is the system-level design (e.g., the joint design of PHY and waveforms to compensate for device limitations).

OWC technologies have made significant advances over the past few years. For example, visible light communication (VLC) and light fidelity (Li-Fi) have already been standardized or demonstrated [4, 5], and free space optical (FSO) has been used for long-range D2D communications [6], where light sources are directly exposed to wireless channels. VLC technology uses an LED and photodiode as the source and receiver, respectively, and modulates the light by varying the intensity in order to map 0s and 1s directly into the optical wave; this modulation must be performed fast enough to avoid flickering effect. The use of VLC technology in 5G preliminarily tested in [7]. However, weather conditions and atmospheric turbulence significantly affect OWC technology. For example, fog attenuates optical signals, leading to OWC link unavailability. In order to mitigate these issues and cope with the varying degrees of reliability, hybrid RF/OWC technology combines the advantages of RF (reliability) and OWC (capacity) and is a suitable complementary option [8].

26.3.2   Integration of Super-Sidelinks with Access Links as One Design

The integration of super-sidelinks and access links offers large amounts of spectrum and extremely high bandwidth, making it possible to achieve ultra-high capacity and data rates. This section focuses on how to achieve efficient integration (e.g., a unified air interface) and implement super-sidelink networking (e.g., mesh networking).

For 6G super-sidelinks, research needs to be conducted into unified physical layer procedures, such as broadcasting, random access mechanisms, and resource sensing and selection. The air interface design, including waveforms, modulation, multiplexing, coding, and MIMO schemes, may also need to be considered at the beginning of integration. Over the past few years, some promising progress has been made. For example, the authors of [7] considered the integration of 5G and VLC, utilizing VLC for communications between traffic lights and vehicles. And in [9, 10] the MA performance of VLC systems analyzed on the basis of the 5G OFDMA or NOMA–OFDMA scheme.

As mentioned earlier, huge numbers of devices will connect to 6G networks. However, due to the limited coverage area of super-sidelinks and the obstruction of signals (e.g., walls, furniture, and even humans can obstruct signals), D2D multi-hop mesh networking is an important consideration in designing super-sidelinks. Determining how to integrate the new air interface design and networking technologies (such as access link and sidelink mesh networking) is also an important area of research.

Spectrum allocation and sharing among different super-sidelinks and access links should also be considered. Due to the narrow beam and short transmission distance of THz and OWC, it is relatively easy to share the same spectrum or channel among different super-sidelinks and access links. This, however, raises the requirement for beam-based interference management in super-sidelinks. The integrated design should maximize network capacity by leveraging the personalized design of access link air interface, including different duplexing capabilities and agnostic link direction. The design of 6G super-sidelinks may also need to achieve license-free spectrum awareness or sensing with the help of licensed spectrum.

The 6G super-sidelinks technology is expected to support beamforming for both single-hop and multi-hop transmission. Because beams may be very narrow in the super-high frequency used by super-sidelinks, research is needed into performing beam management with low overhead and minimal impact on mobility. This should also take into account the fact that beam transmission between a transmitter and receiver may be asymmetric, especially in OWC, due to the power, size, and complexity limitations of devices [2]. Research is also needed into the integration of super-sidelink and access link in terms of RF, THz, and OWC beam management in order to solve issues with asymmetric beam transmission and blockage for certain links. This should also consider the integration of multiple super-sidelinks and access links. Chapter 25 provides greater insight into beam-related topics.

The report [11] discussed Layer 2 and Layer 3 relaying, but neither approach is suitable for latency-sensitive traffic forwarding or multi-hop transmission. Consequently, research is being conducted into the use of flexible forwarding technologies, such as decode and forward (DF), compress and forward (CF), and amplify and forward (AF) relaying, for separate or combined use in mesh networking. Furthermore, PHY network coding can be used in mesh networking in order to enhance efficiency and reliability while also lowering latency. Chapter 23 explores this in more detail.

If air interface and networking techniques are implemented effectively, we believe 6G integrated super-sidelink and access link will lead a paradigm shift. Given the flexible network topology and Tbit/s-level communication capabilities provided by 6G super-sidelink mesh networking, this opens up the potential application space for diversified devices.

References

[1] Y. Corre, G. Gougeon, J.-B. Doré, S. Bicaïs, B. Miscopein, E. Faussurier, M. Saad, J. Palicot, and F. Bader, "Sub-THz spectrum as enabler for 6G wireless communications up to 1 Tbit/s," in *Proc. 6G Wireless Summit, Mar. 2019, Levi Lapland, Finland.*

[2] M. Z. Chowdhury, M. Shahjalal, M. Hasan, Y. M. Jang *et al.*, "The role of optical wireless communication technologies in 5G/6G and IoT solutions: Prospects, directions, and challenges," *Applied Sciences*, vol. 9, no. 20, p. 4367, 2019.

[3] Z. Chen, X. Ma, B. Zhang, Y. Zhang, Z. Niu, N. Kuang, W. Chen, L. Li, and S. Li, "A survey on terahertz communications," *China Communications*, vol. 16, no. 2, pp. 1–35, 2019.

[4] A. Bensky, *Short-range wireless communication.* Newnes, 2019.

[5] D. Tsonev, S. Videv, and H. Haas, "Light fidelity (Li-Fi): Towards all-optical networking," in *Proc. Conference on Broadband Access Communication Technologies VIII*, vol. 9007. International Society for Optics and Photonics, 2014, p. 900702.

[6] A. K. Majumdar, *Advanced free space optics (FSO): A systems approach.* Springer, 2014.

[7] F. Nizzi, T. Pecorella, S. Caputo, L. Mucchi, R. Fantacci, M. Bastianini, C. Cerboni, A. Buzzigoli, A. Fratini, T. Nawaz *et al.*, "Data dissemination to vehicles using 5G and VLC for smart cities," in *Proc. 2019 AEIT International Annual Conference (AEIT).* IEEE, 2019, pp. 1–5.

[8] M. Z. Chowdhury, M. T. Hossan, M. K. Hasan, and Y. M. Jang, "Integrated RF/optical wireless networks for improving QoS in indoor and transportation applications," *Wireless Personal Communications*, vol. 107, no. 3, pp. 1401–1430, 2019.

[9] H. Marshoud, P. C. Sofotasios, S. Muhaidat, G. K. Karagiannidis, and B. S. Sharif, "On the performance of visible light communication systems with non-orthogonal multiple access," *IEEE Transactions on Wireless Communications*, vol. 16, no. 10, pp. 6350–6364, 2017.

[10] B. Lin, W. Ye, X. Tang, and Z. Ghassemlooy, "Experimental demonstration of bidirectional NOMA-OFDMA visible light communications," *Optics Express*, vol. 25, no. 4, pp. 4348–4355, 2017.

[11] 3GPP, "Study on NR sidelink relay," 3rd Generation Partnership Project (3GPP), Technical Report (TR) 38.836, Sept. 2020, version 17.0.0. [Online]. Available: https://portal .3gpp.org/desktopmodules/Specifications/SpecificationDetails.aspx?specificationId=3725

Summary of Part V

In this part, we discussed the potential features of the 6G air interface and the potential enabling technologies that will support versatile usage scenarios, stricter KPIs, and a wider spectrum range in 6G. AI/ML technologies will be used to make the 6G air interface customizable and intelligent. This in turn will enable the air interface to support the customized optimization of transmission parameters so that ultimate user experience can be achieved without sacrificing network capacity.

In terms of the classic physical-layer modules of wireless communication systems, fundamental improvements are needed to maximize the performance of 6G links and meet extreme service requirements. Following this direction, we discussed new expectations and potential design directions for coding, waveforms, modulation, MA, and ultra-massive MIMO. 6G integrates multiple elements: highly integrated sensing and communication, integrated terrestrial and non-terrestrial networks, integrated passive and active nodes, and integrated super-sidelinks. As such, it will require careful design and comprehensive consideration of sensing-assisted communication, communication-assisted sensing, multi-connection joint operations, fast cross-connection switching, and a smart controllable radio environment.

Native power saving and super-flexible spectrum utilization are also vital aspects for 6G design, which is expected to enable a default power-saving mode and a unified mechanism for different duplex modes and to support the co-existence of UEs with different duplex capabilities.

As was the case for previous generations, the technical advances necessary for the next generation must overcome many challenges. We believe the intelligent nature of the 6G air interface will enable a wide range of opportunities to address these challenges.

Part VI

New Features for 6G Network Architecture Design

Introduction to Part VI

Despite the extraordinary success of mobile networks, many underlying design principles and service assumptions that have served us well over the past decades have benefited from fast adoption and adaptation to new emerging technologies and services. From an evolution perspective, mobile networks have evolved towards higher capacity and delivering diversified services to meet continuously increasing demands. Remarkable progress has been made to continuously improve the wireless links supporting ultra-high data rates, low latency, and massive connections. The mobile network architecture, however, continues to evolve at a slower pace. More specifically, incremental changes accompany increased complexity, which is brought about by backward compatibility, service interworking requirements, the disaggregation of network functions, and the introduction of more features that support new application scenarios. The difficulty involved with changing the network architecture arises from issues related to costs and system stability rather than conventional thinking or dogma from the telecom industry. Take the IP network for example. Achieving IPv4 to IPv6 migration is no easy feat. As for the mobile communication system, which is used to serve nation-wide subscribers and services, minor modifications may have a major system-wide impact.

The mobile network was originally designed to deliver voice services. Since then, both the architecture and deployment of mobile networks have followed a centralized and hierarchical paradigm that reflects the nature of voice traffic and Internet applications. New services and capabilities have been introduced for each new generation of mobile networks. It is within this context that proliferation intelligence, joint communication, and sensing are deemed to be the new capabilities that will support innovative applications when 6G arrives on the scene. It is also worth noting that new requirements are emerging for privacy protection, global coverage, and data governance, while extensive research is being conducted on new applications and new devices. New technologies such as AI/ML and blockchain are also being integrated into the new architecture as enablers.

In addition to offering conventional connectivity services, 6G systems could also serve as distributed platforms for executing user workloads in all industry scenarios. This is possible due to the fact that the 6G network will be based on a decentralized and user-centric architecture with native AI capabilities.

6G will break and shift traditional paradigms towards a novel architecture that meets new requirements with integrated new capabilities empowered by new enabling technologies.

Data ownership: from third-party control to self-sovereignty.

Data is perhaps the most valuable asset in the digitalized society, and the key to resolving our privacy concerns lies in how we control data. In today's mobile communication systems, user data is spread across numerous network entities, but is centrally owned and controlled by a single third party. For this reason, end users do not control their own data. Even though the mobile communications systems are equipped with mechanisms for protecting user privacy in accordance with laws and regulations, personal information can be easily shared with third-party service providers, without user awareness. On top of that, it is also difficult for telecom operators to monetize their abundant data sources.

The key privacy and data protection requirements are reflected by the General Data Protection Regulation (GDPR) in Europe, which is arguably the world's first legislation designed to protect consumers from the proliferation of data theft and misuse. As such, it advocates the self-sovereign management of personal data, meaning data ownership should be returned to end users without any intervening authority.

Deployment: from 2D terrestrial to 3D global coverage.

One of our aims when deploying mobile networks is to continuously expand the breadth and depth of coverage. 2D "population coverage" is the deployment model of mobile networks, and it focuses on delivering services to terrestrial-based mobile subscribers. 6G is expected to shift to a 3D global full-earth coverage model that helps mitigate the "digital divide" and efficiently supports new application scenarios, such as vehicular networks, sensor networks via satellites, HAPSs, or drones, as mentioned in Chapter 6.

Network service: from operator-centric to user-centric.

The operator-centric approach results in network designs that need to handle a large scope of services (e.g., regional or national level). This has led to monolithic network entities like mobility management entities (MMEs), serving gateway (SGW), and packet data network gateway (PGW) in 4G and network functions like access management functions (AMFs) and user plane functions (UPF) in 5G, which are expected to serve a large number of end users simultaneously.

From a functional perspective, the network manages the state of each UE or end user. In that sense, the network is essentially a large distributed state machine, which means it maintains consistent states across different network functions. This causes complex signaling message exchange that may limit the extent to which network performance can be improved (e.g., latency), and may also lead to increased attack points. As connected devices/users increase in number, monolithic network functions (both physical and virtual) also become potential sources of serious bottlenecks. Given the fact that the network inherently manages the state of each UE or user, we can understand why a network design based on the per-UE or per-user perspective is needed. More specifically, a user-centric design is capable of delivering a virtual private network (VPN) for each user, and the network service from such per-user level VPN contains features like mobility management, policy control, session control, and personal data management. In addition, signaling overhead and the corresponding network performance can be optimized at the per-user level.

Network: from "the" network to "my" network.

The user-centric paradigm redesigns network service delivery in a per-user manner, meaning network services will be re-centered around users. To provide personalized services, the architecture should avoid the "one-size-fits-all" approach while also allowing users to have and self-manage their own networks, as they currently do with their home networks and Wi-Fi. As such, end users will be able to choose network resources for the creation of their own VPNs. The user-centric approach should go beyond network functions, and will be a prerequisite to making the "my network" vision a reality.

Connectivity: from connection-oriented to task-oriented systems.

Conventional communication systems are connection-oriented, and were originally driven by voice and then data communication. The communication source and destination are clearly defined by end users and the services that they intend to use or the other users with whom they plan to communicate. As such, the entire communication mechanism, e.g., session management and mobility management, should be designed to provide sufficient support for this connectivity model.

The 6G mobile communication system is expected to consist of numerous distributed nodes (e.g., terminals, radio access nodes, and network equipment) with intelligent features (to provide native support for intelligent services or to utilize intelligence for self-improvement). The two key services that 6G will provide are AI and sensing. In order to provide these services, the same task may be executed across numerous distributed nodes with proper coordination; this is referred to as task-oriented communication. Wireless communications technology in the future should support diversified device types and time-varying topologies to achieve optimal performance for task-oriented communication.

Terminal: from separated TPC to integrated TPC.

The three essential components for an end user to access services or applications are the terminal, (network) pipe, and (applications in) cloud, collectively abbreviated as TPC. In earlier generations, each individual component served as a stand-alone technology with its own ecosystem. As networks and applications become increasingly virtualized, both can be realized as virtual entities running in the cloud. With the emergence of data-demanding and computation-intensive applications such AR/VR and holographic communications, data rates are expected to reach the Tbit/s level (as introduced in Chapter 1). However, this will make terminals hit a performance bottleneck because their computational capability is lower than that of cloud edges. Therefore, collaboration between terminals and networks (especially cloud edges) becomes necessary. Terminals could be flexibly designed according to different use cases and capabilities. For example, more user functions could be moved from devices to the cloud, or system tasks and network functions could be brought down to terminals without constraints. Another benefit is that it will be possible to decouple user identities from terminals due to the elimination of the physical boundaries between TPC.

Intelligence: from cloud AI to network AI.

AI services remain at the application level in today's networks, and are located in the central cloud. The 5G core network has a limited usage scope even though it is designed to provide intelligence support through a new type of network function (i.e., a network data analytics function), and its architecture is not AI native. In the 6G era, network architecture and AI will go hand in hand. Put differently, native AI support will be one of the fundamental factors driving network architecture innovation. As such, deeply converged communication and computing resources with fully distributed architecture will lead to a transformation from cloud AI to network AI. The benefits of this will go beyond the superior performance (e.g., ultra-low latency) achieved by bringing AI services closer to end users, as privacy concerns could also be locally resolved. This is one of the primary drivers of network architecture development for 6G, which will be significantly impacted by privacy and data governance requirements such as GDPR. It holds especially true in terms of realizing real-time AI functionalities, training big data for machine learning, and executing AI inference, which are inefficient within the centralized cloud AI architecture and will be incorporated into the network AI architecture.

Security: from security-only to natively trustworthy.

5G security is viewed quite separately from the other network services, and is implemented through a stand-alone framework. In 6G, one of the major paradigm shifts is from a security-only focus to the broader scope of native trustworthiness. This shift involves dealing with a security-by-design framework and a wide range of topics, such as the trust model and security thread from the promising development of quantum computing and application of new technologies (e.g., AI and machine learning) in security design. To guide the design of a trustworthy 6G architecture and define the corresponding key capabilities, new use cases that yield new requirements, as well as new enabling technologies, should be taken into account.

27 Technologies for the Network AI Architecture

27.1 Background

Mobile communication has stepped into a new and exciting era, where personal devices such as smartphones and tablets are becoming the primary computing platform for many applications. Since these devices are used to facilitate our daily lives (from professional activities to leisure and education to entertainment), they have access to an unparalleled amount of personal and private data. As such, businesses that focus on AI can explore this area to deliver more convenient and meaningful services to end users.

It is anticipated that AI applications will not only benefit consumer business, but also touch every economic sector and affect all aspects of society. All types of AI applications and services need to collect and analyze various types of data, and then apply the corresponding results to execute a particular action or set of actions. The majority of current AI/ML implementations perform centralized learning, where data are gathered from all over the system but training is performed in a single location, usually in a power-intensive data center dedicated to these types of computations (e.g., with special hardware). Currently, AI is decoupled from the network architecture design, which is mainly referred to as cloud AI, meaning the underlay network is simply used to transport data to the cloud, where the main intelligence for data processing and inference is located. However, we have yet to witness these two aspects being considered together. For example, when 6G provides AI and sensing as new services, the challenge becomes providing secure and reliable network infrastructure to transmit data from information sources for the purpose of further enforcing the learning method hosted in the cloud or other types of available resources. Therefore, it is imperative that we understand the types of network architecture capable of providing native support for AI. To do this, we also need to grasp the three pillars that form the entire AI empire and their potential requirements on the mobile communication system.

- **Data are the key asset** of the AI industry and are being referred to as "the new oil." As the first wave of AI services focuses more on business-to-consumer (B2C) cases, end users serve as direct data sources. The value of data is further monetized by business-to-business (B2B) cases from vertical industries, which entail different use cases, business models, and technical requirements. Due to performance constraints as well as security and privacy concerns, we expect industry data to be processed at the system edge, which is normally within enterprise premises. As such, AI services geared towards industry will be provided in a more distributed manner with a local focus. This trend will lead to a series of discussions regarding the requirements for data management, processing, ownership handling, etc. A mobile communication system that can meet these requirements while also fully complying with data governance policies will be highly valued.

- **Computing capability** fundamentally powers the AI industry. The more powerful an AI application is, the more computing resources it requires. Given this, the computing resource pooling model based on the central cloud may lack the scalability required to be future proof. When we especially take into account the preceding edge-processing trend from the vertical industry, running AI applications only in a central cloud may not work. As such, a new synergy is pushing AI from the cloud deep into the mobile communication system, which is an ultra-high performance infrastructure that can effectively manage heterogeneous resources with scalability and elasticity. This research domain is both exciting and challenging as it may lead to complete restructuring of conventional system architectures and design philosophies.

- **Algorithms** are the core of the entire AI business, and define the type of intelligence provided by AI applications, as well as the type of data and amount of computing power that AI applications use and consume, respectively. The infrastructure does not know how AI algorithms are defined, but it can provide better support for running these algorithms. For instance, the implementation of deep learning (e.g., federated learning) depends on communications, which may relate to algorithm scalability, bandwidth, and latency requirements. Given this, the design of the network system architecture may influence how AI algorithms are trained and how AI inference is performed.

27.2   Design Considerations and Principles

A new system architecture that aims to proliferate intelligence by providing native support should consider the design demands from the above three pillars. As a result, cloud AI will be transformed to network AI.

27.2.1   Key Requirements

Data Governance

The 6G system will inherently generate a huge amount of data, which will vary from operation to management, control plane to user plane, and environmental sensing to terminal. As these data will be derived from different technical or business domains, the new challenge of the 6G system design will involve efficiently organizing and managing data while also considering privacy protection.

Advanced Infrastructure

In the 6G era, information, communication, and data technologies will be fully intertwined to construct an advanced infrastructure. For AI proliferation to become a reality, real-time learning and inference on ubiquitous connectivity and computing resources should be supported. This will lead to the mobile communication system having high-performance requirements (for example, ultra-low latency and ultra-high data rates) especially for data collection and processing. On top of that, to achieve optimal performance while adhering to privacy and security constraints, distributed and ubiquitous computing resources require the deep convergence of all resource types from all types of stakeholders. AI services could leverage this advanced infrastructure to fully unleash the potential of the three major pillars (i.e., data, computing capability, and algorithms).

Flexible AI Deployment

Easy deployment is essential to attract services from external partners, especially when it requires a certain level of networking and IT expertise. This requirement mainly relates to the design of platform's capabilities. Put differently, AI services should be easily deployed in 6G (at the edge or centrally) and the platform should not limit service migration.

Ecosystem Openness

By providing native AI support, 6G will embrace a much broader ecosystem than previous-generation communication systems. Other than traditional telecommunication players (e.g., vendors, operators, subscribers, etc.), more enterprise customers and vertical industries will be attracted by the new types of AI services provided by the mobile communication system. 6G is envisioned to be built on a multi-player ecosystem serving as its prerequisite, and, as such, system openness (in terms of both business- and technical-level cooperation) should be easy, flexible, and trustworthy. Even though 5G has already begun the journey towards expansive ecosystem openness so as to establish a new ecosystem with vertical industries (e.g., through network capability exposure [1] and application program interfaces (APIs) [2]), it still relies on the conventional telecommunication initiative, which is based on standardized network functions, procedures, and interfaces. In 6G, these considerations should be regarded as an endogenous cause of architecture revolution.

27.2.2 Gaps

5G also aims to provision a fitting network design that supports AI services, especially at the core network side. In that regard, a network data analytics function (NWDAF) [1] has been introduced with the main purpose of facilitating data collection and analytics in 5G. The NWDAF is a 5G network function specifically designed for data acquisition and analytic information provisioning. For instance, it provides analytic information for other network functions to assist network service provisioning. The NWDAF supports data collection from 5G network functions and from OA&M [3]. To do this, it also provides a service for registration and metadata exposure to the corresponding network functions.

However, NWDAF does not make 5G support AI natively, for the following reasons.

- **Limited data source:** NWDAF mainly involves data collection and analytics based on the data received from the 5G network functions. It does not, however, consider the data from the infrastructure, environment, terminals, sensors, etc.
- **No data privacy protection:** The data source considered in 5G is mainly from the same business domain; therefore, data privacy protection is not considered in the fundamental design.
- **No support for external AI services:** NWDAF is a 5G core network function, and external AI services cannot be used directly in the 5G core network or RAN.
- **Infrastructure is not leveraged:** The key features of the 5G architecture (e.g., network slicing, URLLC, and mMTC) are designed to meet vertical requirements from performance, functional, and operational perspectives. No special consideration has been given to providing native AI support for 5G (e.g., in terms of data management and distributed architecture).
- **No data governance:** AI involves more than just data collection and analysis. In order to provide native support for AI, a distinct data governance design should be considered, and this is not within the scope of 5G.

Therefore, native AI support is one of the fundamental architectural differences between 5G and 6G.

27.3   Architectural Features

27.3.1   Overall Design Scope

The essential role of 6G involves building an ultra-high performance infrastructure, which features computing resources that are very close to end users. This infrastructure will meet the high industry standards for resilience. The 6G system will seamlessly orchestrate all types of resources with connectivity to fully converge computing and communication, and the scope of the overall architecture design is shown in Figure 27.1.

6G could provide native AI support features that may lead to three different business models, as shown in Figure 27.1: IaaS (infrastructure as a service), PaaS (platform as a service), and AIaaS (AI as a service), which are concepts borrowed from cloud services.

- **IaaS:** In this model, the mobile communication system orchestrates the computing capabilities and connectivity resources (e.g., from clouds, edges, base stations, and devices) required to run AI services. The fundamental goal is to deliver high-performance infrastructure capabilities that meet the requirements of AI services. That said, in addition to transmitting more data faster, the design of such a

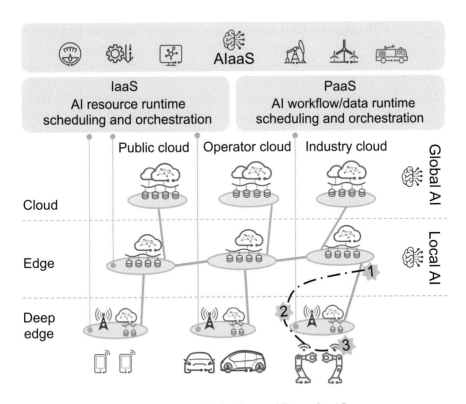

Figure 27.1 Scope of system architecture design for providing native AI support.

high-performance infrastructure involves deeply integrating AI services into the system and ensuring they run seamlessly. It is important to determine whether this will involve a re-engineered edge computing concept or fundamental breakthroughs.

- **PaaS:** In this model, the 6G infrastructure serves as an AI platform in and of itself, and it can orchestrate AI workflows, manage data, and perform other tasks. For instance, to perform an AI service, data from various sources may be required, and aggregating data from different technical domains could be the main difficulty for AI service providers. In this case, the PaaS model could provide added-value for AI by utilizing ICT players' expertise on infrastructure operation and management (specifically gained from traditional multi-vendor and multi-operator scenarios).
- **AIaaS:** In this model, the 6G infrastructure provides AI services for external customers. Example of such services could include AI-enabled high-precision localization and user mobility trends, etc.

These three models could very well co-exist as they cover diversified AI service requirements from very different customers. AI services that run on this advanced infrastructure will bring many advantages, as follows.

- **From global AI to local AI:** From a nation-wide networking perspective, centralized learning is simply too costly as it involves collecting and sending data over the entire network to the central entity. That said, local AI or combined local and global AI may help reduce power consumption, thereby lowering costs. Be that as it may, this data collection method and central training will quickly lead to bottlenecks and single points of failure in or around the central location (e.g., the availability of the data center used for training, the usage of the critical paths towards the data center, and the availability of the nodes on those critical paths). Moreover, it may not be feasible to train or use private data via a centralized learning method, and authorization may not be granted.

- **From offline AI to real-time AI:** Furthermore, traditional ML solutions rely largely on offline training, which involves accessing the training data or modeling and simulating an abstract version of the environment. However, the drawback of this process is that it simplifies many details while also ignoring numerous important metrics, which in turn compromises the performance gains. Fortunately, a high-performance distributed infrastructure may open up a new set of possibilities regarding real-time AI (including training and inference), especially for the use cases with strict latency requirements (e.g., closed-loop control for industry scenarios).

27.3.2   Task-Oriented Communication

One of the key 6G paradigm shifts involves the task-oriented communication model, which is designed to support new services like AI and sensing. Traditional mobile communication services (e.g., voice and data) are all connection-oriented, meaning communication resources are allocated on the basis of proactive requests from end users and the expected services, regardless of whether the users intend to talk to other users or whether they intend to connect to a server in the cloud. As a result, connections are established based on the users' intentions.

In comparison, task-oriented communication involves a vastly different design philosophy where connections are established based on what the network provides users with. For example, an AI task would be to collect and process data over a certain geographical area according to user mobility, real-time population distribution, or terminal usage intensity. Public transport companies could then use this information to understand urban mobility, especially during rush hours. The companies would therefore use the AI service in 6G, without defining how data are collected from specific users to obtain certain information.

In another example, a sensing task could perform 3D mapping of the radio environment across different frequencies. This capability could enable new use cases, such as real-time sensing with high-accuracy localization and tracking, and gesture and activity recognition. Such tasks are done by coordinating multiple base stations, terminals that include connectivity, and computing resources, as shown in Figure 27.2.

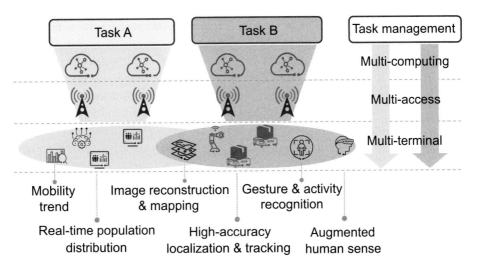

Figure 27.2 Task-oriented connectivity concept.

The creation of a task-oriented communication solution may include four major aspects: task management, resource orchestration or runtime scheduling, data management, and communication mechanism.

Task Management

From an architectural perspective, new network services and APIs may be introduced for task management, which involves defining, operating, and managing tasks throughout the entire lifecycle. Task management could break a task into functional sub-tasks, which need to be implemented through various methods due to the heterogeneous type of access and terminal formats. Task execution on a deeply converged computing and communication platform could mean various sub-tasks are locally executed.

Resource Orchestration/Runtime Scheduling

In the conventional mobile communication system, resources are managed (especially at the radio access side) on the basis of proactive connection setup requests, and are scheduled on the basis of device subscription information and service types.

To perform a task, this system needs to coordinate and access computing resources across different communication equipment in various geographical locations. From a resource perspective, one task could be broken down into a computation task and a communication task. The former could be established across different clouds, cloud edges, base stations, and devices, whereas the latter could be established among different types of devices across multiple cells. With a concept that is considerably different from that of the conventional communication system, resource orchestration or runtime scheduling would be based on tasks provided by the system, as opposed to requests from end users. Instead of reserving radio resources to establish a session for two parties to communicate with each other, computing and communication resources are scheduled to perform a task in region X.

Resource orchestration or runtime scheduling for a task should provide a holistic view of resources depending on the geographical and working scope of the task. If a task is performed within a single technical or business domain, all the existing network management experience could be leveraged. As for multi-player scenarios or systems that are distributed at a large scale, E2E resource orchestration and runtime scheduling will become challenging (e.g., in terms of trust establishment, charging mechanisms, and auditing) and new methods are likely to emerge. Chapter 31 delves into more details on these new methods.

Data Management

With regard to performing tasks, some new areas to explore are data collection, privacy protection, information storage, network capability, and knowledge exposure. Data governance is a dedicated architecture technology for managing data within the broader scope, presented in Chapter 30, and it can be leveraged to provide native AI support.

Communication Mechanism

To support a task, numerous connections may be established, for example, between devices and environmental sensors as well as communication and computing equipment. As for task performance, it involves flexible and efficient multi-user/terminal, multi-access, and multi-computing collaboration. As such, conventional mechanisms (such as session and mobility management) may need to be fundamentally changed.

RAN Architecture

To perform tasks efficiently, it may make sense to separate task control and user plan functions at the network service layer, so that task control can be executed in a large area. Hence, a 6G base station could be logically separated into two parts: a control node (denoted as cNB), and multiple service nodes (denoted as sNBs). The cNB provides task control function over several sNBs spread in a large area. Such design could reduce the overhead of common control information. The sNB provides connectivities between variant terminals and network equipment. cNB and sNB are logical entities, which go beyond the format of physical equipment. Using such a design, the resources in the RAN could be easily grouped as virtual clusters, each of which could be controlled, for instance to support varying tasks. The advantage of such a cluster is that it controls all relevant resources over a manageable set of sites for better resource management. The cluster area size should be sufficient to support the performed tasks.

27.3.3 Deeply Converged Computing and Communication at Edges

The 6G system is expected to serve industry scenarios that require optimal performance and data processing at the local level; therefore, edge nodes will become the key

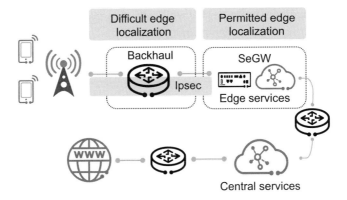

Figure 27.3 Exploration of the edge limit towards the RAN.

innovation platform for future mobile communication networks. Mobile edge computing or so-called multi-access edge computing (MEC) is a well-established topic that is
gaining more importance in the mobile communications domain. It extends the computing capabilities from centralized locations to the edge of the mobile communication
system, thereby delivering flexibility in terms of network function deployment and
operation, as well as better network performance (e.g., latency). However, numerous
practicality issues limit the deployment of edge nodes. For instance, when a security
gateway (SeGW) is installed, deploying an edge node between the RAN and SeGW
involves a large amount of difficulty. This is because IP security (IPsec) sessions cannot be interrupted, as illustrated in Figure 27.3. On top of that, we cannot deploy edges
deep in the RAN due to limitations in the radio access architecture and corresponding
communication logic.

In 6G, deeply converged computing and communication at the RAN level is a fundamental factor to consider, and we refer to this as "deep edge" within the following
context. Deep edges are located close to the data sources. By using deep edges, AI
services could be natively integrated into the RAN architecture design.

Deep edges involve more than just running a cloud IT server at the edge of the
mobile network. We also need to put thought into reconstructing the RAN architecture
in order to break the edge limit, minimize impact on protocol stacks, and improve the
overall system scalability. Research on deep edges could include the following aspects.

New Computing Scheme
Currently, it is already possible to leverage the AI capability in the RAN. For instance,
we can optimize resource scheduling and mitigate interference. That said, this is the
main reason mobile networks utilize AI instead of natively supporting it, which are
two different concepts (i.e., AI for networks versus networks for AI). We will discuss
the latter in detail.

To deeply integrate the edge computing capability into the RAN, a new type of
radio equipment may emerge, which could be referred as a radio computing node
(RCN) to distinguish it from the radio access node, as shown in Figure 27.4. Within

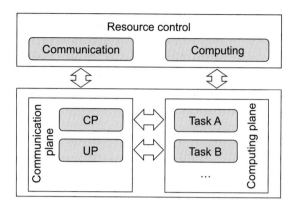

Figure 27.4 Deep convergence of the communication and computing capabilities in an RCN.

an RCN, an independent computing plane could be introduced to host variant tasks (i.e., AI and sensing) or other extended control planes (CPs), user planes (UPs), etc. This type of computing resource not only involves placing a blade into the radio access node, but also requires the overall design of the RAN architecture to be considered, which in turn will allow the computing resources to be seamlessly converged with the communication capability in the RAN. For instance, a local controller could be needed to control the wireless communication resources, functions, and AI pipelines.

Moreover, the corresponding interfaces, applied procedures, and protocols should all be specified in order to support dynamic deployment of and efficient communication between the services deployed at deep edges. Such services could exploit the advantages of reduced latency, transportation costs, potential security risks, and privacy concerns.

New Protocol Stack

Deep edges may also lead to a re-engineered protocol stack, which for example might be simplified to explore high-performance solutions that support task-oriented services (e.g., AI workflows). In the 5G system, data connections are established between UEs and servers in the cloud through the base stations and UPFs at the core network. In order to deeply converge the computing and communication resources at the RAN level, connections are directly established between UEs and Pods or containers hosted in the computing plane of deep edges, enabling the communication mechanism to be significantly simplified. As such, it is possible to provision highly efficient communication that supports *critical* real-time services.

Higher System Scalability

Mobile communication systems are used for both national-wide coverage (e.g., B2C type of services) as well as local-scope coverage (e.g., the B2B type of services). Hence, the amount of deep-edge nodes could be in the order of millions, and they all should not be isolated resources. It will likely be possible to establish a large-scale network based on a huge amount of independent but controllable deep-edge nodes, which potentially cross different technical or business domains for hosting proliferated intelligence. From one perspective, this will be a high-performance network, and from another it will be scalable, open, and transparent. The efficient synchronization of data and services, runtime scheduling, selection of suitable resources, and identification of local (e.g., industrial) applications are essential aspects of the system architecture design.

By forming a large-scale network via deep-edge nodes, it will be possible to perform combined local and global AI, as opposed to purely global AI. As a direct consequence, industry digital transformation could be leveraged at a much larger scale and scope.

27.3.4   AI Service Operations and Management

The AI service operation and management framework is another essential aspect to consider for the design of the native AI support architecture. This framework is used to facilitate the seamless integration and deployment of AI services (especially from external providers).

- **AI service operations:** 6G will be built on a multi-player ecosystem that inherently involves multiple technical and business domains. With regard to AI service operations, this ecosystem should work smoothly and efficiently, especially in multi-operator and multi-vendor scenarios. In order to establish business relationships even in zero-trust situations, technologies such as blockchain may be needed for AI operations.
- **AI service management:** The management of AI services includes AI workflow orchestration and data management, as well as computing and communication resource orchestration. To operate an AI service across different business and technical domains, as shown in Figure 27.5, it is essential that we define the key interfaces used within a single management domain or across different management domains. Chapter 31 expands on the establishment of a platform that supports multi-player collaboration.

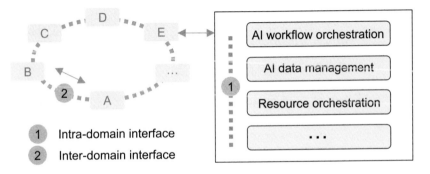

Figure 27.5 AI management framework across different business domains.

The conventional method of defining interfaces within the mobile communication system is likely to be standardized. Nevertheless, building a new framework that fully relies on this standardization is not recommended. Instead, the best practice in the field, for instance, the AI framework itself, could be leveraged on advanced open-source components [4, 5]. With this cross-domain framework, E2E AI pipelines can be programmed and deployed within the 6G network.

The AI service operation and management framework is a new component on top of the conventional mobile communication system architecture. As such, new types of network functionalities or design principle will be introduced, for instance, the following.

- **AI service operations:** For the sake of privacy protection, the architecture's de-privacy functionalities may keep some information from being visible to AI service operations, which are inevitably impacted. That said, the architecture will provide support for AI service operations to mitigate this impact.
- **Resource management:** AI services are compute- and communication-intensive. Therefore, resource management involves properly managing both computing and network resources for devices, servers (or service locations, e.g., data centers), and the network. Improper management can lead to high power consumption and also a deterioration in network and service performance (e.g., latency).
- **AI service deployment:** The architecture will facilitate the deployment of AI services. This includes determining the deployment locations and corresponding interconnections, according to the service-level description.

References

[1] 3GPP, "System architecture for the 5G system," 3rd Generation Partnership Project (3GPP), Technical Specification (TS) 23.501, Aug. 2020, version 16.5.1. [Online]. Available: https://portal.3gpp.org/desktopmodules/Specifications/SpecificationDetails.aspx?specificationId=3144

[2] 3GPP, "Common API Framework for 3GPP northbound APIs," 3rd Generation Partnership Project (3GPP), Technical Specification (TS) 23.222, July 2020, version 17.1.0. [Online]. Available: https://portal.3gpp.org/desktopmodules/Specifications/SpecificationDetails.aspx?specificationId=3337

[3] 3GPP, "Architecture enhancements for 5G System (5GS) to support network data analytics services," 3rd Generation Partnership Project (3GPP), Technical Specification (TS) 23.288, July 2020, version 16.4.0. [Online]. Available: https://portal.3gpp.org/desktopmodules/Specifications/SpecificationDetails.aspx?specificationId=3579

[4] "Kubeflow." [Online]. Available: https://www.kubeflow.org/

[5] "Kubernetes." [Online]. Available: https://kubernetes.io/

28 User-Centric Architecture Technologies

28.1 Background

In previous mobile communication systems, user devices that access the telecom operator domain have zero control of the network itself. User-centric networks (UCNs), as the key feature of the 6G architecture, refer to systems with a native design which allows them to be user-definable, user-configurable, and user-controllable. As implied by the name, user-definable means users can define (with the help of AI) the kinds of services they would like to receive, and how to operate and manage them. For up-running services, users can configure policies for resource usage as well as sub-terminals that belong to the same user domain. The users control all the data that are generated by or belong to them, as well as the corresponding process rights (e.g., identification, access authorization, and user status information). The users could be consumer individuals, enterprises, or industry users, and the services will be composed in an intelligent and personalized manner in the user domain based on the users' profile, behavior, and preferences.

The UCN design will change the way users, network services, and applications currently interact, thereby impacting the corresponding network access, mobility management, security, and personal digital asset ownership. In line with the current trend of network function modularization and cloudification, the softwearized telecom service makes it possible to deploy core network functions (such as forwarding, session management, and policy management) without location constraints. The design of the per-user network will be modular, with a shared context, which will eliminate message exchange among traditional network functions.

One major trend, as we previously identified, is the increasing interest in the local/private scope of communication. In light of this, a fully distributed and interconnected edge platform is the prerequisite to constructing UCN. This will also bring advantages. For instance, it will inherently feature resilience against attacks due to its decentralized system, which will be capable of limiting the scope of damage. Moreover, UCN advocates a "system of systems" model, where different systems and users with respective resource pools interact to obtain the final, expected, results, while every user organizes their own resources through various interactions. Based on this principle, both system resilience and traffic handling efficiency can be easily improved.

In UCN, the user domain will not be limited by physical user devices because the network domain will serve as its natural extension, in which services and applications could be deployed based on different criteria, e.g., services requiring more computing power or more *critical* network performance.

Last but not least, UCN will be the optimal method of providing truly personalized services on a per-user level, which in turn will facilitate the paradigm shift from "the network" to "my network." Put differently, it will transform 6G from a pure access network to a service-execution environment.

28.2   Design Considerations and Principles

28.2.1   Lessons Learned from Current Networks

The mobile communication system has evolved through several generations. The distinct features and supported services of each generation decide the fundamental design philosophy of the system architecture. Before transitioning from 5G to 6G, it is important that we comprehend the limitations of the current system design, on the basis of the vision laid out by 6G and potential use cases.

Network-Function (NF)–Centric Design Philosophy

From a functional perspective, a mobile network consists of a control plane and user plane. The control plane mainly instructs the user plane, ensuring that network resources are provisioned to provide services according to user requests, whereas the user plane mainly follows and implements those instructions. The current mobile communication system applies a design principle centered on the NF, which has resulted in monolithic network entities. A control plane consists of a set of NFs responsible for access authentication and authorization, mobility, session management, data management (e.g., NF repository, UE context, and policy) and operation control (e.g., mobility, session, and QoS). On the other hand, a user plane consists of a set of forwarding elements that transport user data packets from the source to destination, and it reports flow metrics back to the control plane.

When the number of subscribers is in the order of millions, and the provided services focus on voice and data, this design has evident advantages in terms of deployment and OA&M costs. That said, the increasing complexity of this design is intuitively reflected in the fast growing number of subscribers, network equipment, protocols, interfaces, and interconnection/interactions among network equipment. The current cellular network architecture employs many protocols that engage several network functions and numerous message exchanges. The accuracy of any implementation is doubtful at best because the messages carried in these signaling procedures are complex. As the protocols are designed to maintain the consistency of the UE's states across network functions, the high number of states that must be maintained across several network functions become a barrier to scaling.

Issues with a Centralized Architecture and Deployment

The cellular network architecture was primarily designed for long-lived-state sessions (e.g., voice) of power consumption and resource-sufficient user equipment. The previous generations of cellular network architecture all heavily draw on the voice network architecture's concept, i.e., a highly centralized design and hierarchically deployed network. One of the design principles of the centralized architecture is NF-centric, which results in monolithic network function entities realized either in physical equipment or virtual entities serving a large number of end users, such as mobility management, device/subscriber data management, session management, authentication, data packet forwarding, and policy enforcement. In terms of functionality, the cellular network manages the state of the UE or end user; it is essentially a large distributed state machine, i.e., the network as a whole has to maintain consistent states across different network functions by exchanging messages.

Although there is an ongoing debate regarding the pros and cons of distributed versus centralized networks, it is obvious that when the number of connected users/devices increases, and the devices become both heterogeneous and smarter, security and scalability concerns arise.

- **Single point of failure and vulnerability to denial-of-service (DoS) attacks:** Due to the IP-based open architecture, cellular wireless systems are vulnerable to common attacks over the Internet. A case in point is the mobile core network, which is vulnerable to DoS attacks due to its centralized nature. DoS or distributed DoS (DDoS) attacks against centralized network functions entail severe consequences, since mobile networks are the critical infrastructure ensuring the reliable operations of a whole society.
- **Scalability issues:** Since 4G/LTE, a hierarchical gateway-based network has become the user plane of the 4G and 5G core networks. It is designed to manage the sessions for each UE, meaning that whenever an UE requires data connectivity, a sequence of message exchanges between different core and RAN components needs to occur before any application-level data packet can be transmitted. As most connected devices are smartphones, which maintain long-lived connectivity sessions without stringent latency and control packet processing limitations, the current architecture works if a reasonable number of these devices are connected through the core network. However, once the number of connected devices increases drastically and devices become more diversified (such as sensors in a smart vehicle, industrial robots, as well as ubiquitous wearable and implantable devices), the current network architecture may cause serious bottlenecks.

Difficulties in Monetizing Data

The mobile communication system is facing a dilemma regarding data. On the one hand, it features abundant data sources, which telecom operators struggle to monetize, especially data that are directly or indirectly related to the end users due to privacy regulations. On the other hand, end users have no control of their own data, which means they must trust that service providers do not abuse it. On top of that, it is not easy to identify and trace violations of individual privacy due to lack of awareness and knowledge.

Issues with the Predominant Operator-Centric Model

While 5G supports verticals better by introducing support for unlicensed access and campus networks, the core network design and architecture are almost exclusively built for operators. This means there are no special considerations given to the functions required by different vertical industries. Rather, in the operator-centric 5G logic, the verticals should ask telecom operators, who are meant to serve them best. This is problematic for all players, including the telecom operators themselves: supporting such a vast variety of unclear functional and extra-functional service requirements is a big challenge. Recently, new industrial forums were established to better capture such requirements, e.g., 5GAA and 5G-ACIA for the automotive and manufacturing industries, respectively. Yet, even with a good understanding of these requirements, telecom operators will have to make tough decisions, especially because under the usual market pressures, not everything that is technically feasible is economically reasonable.

28.2.2   Key Requirements

Requirements for User-Centric Architecture

The network has been designed based on a per-network function principle, as shown in Figure 28.1. This results in several monolithic network functions serving a large number of users/UEs, with each network function, such as AMF and the session

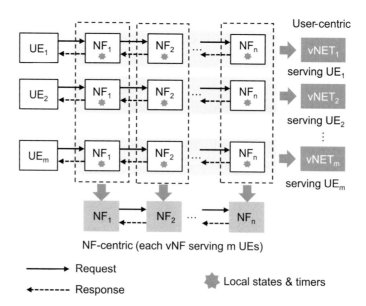

Figure 28.1 User-centric vs. NF-centric systems.

management function (SMF) in 5G [1], now having very specific tasks. It is therefore safe to say that the NF-centric design principle naturally results in a centralized architecture.

The user-centric design means that each user will have its own dedicated network consolidating all necessary functions for service delivery, as shown in Figure 28.1. With this design principle, the UCN is responsible for mobility management, policy management, session management, and personal data management (with the help of decentralized technologies). One of UCN's selling points is the dramatically reduced exchanged signaling, and the reduced latency that follows. On top of that, a reduced number of protocols are needed for communication, which in turn reduces the complexity and therefore results in implementations that are less costly and easier to manage. The user-centric design naturally and perfectly matches with decentralized technology, which is invulnerable to DDoS attacks and single points of failure (SPOFs).

Decentralization Requirements

- **Availability:** Service availability and stability should be ensured by leveraging both the decentralized and distributed cloud technologies, which helps to distribute the system load across network nodes in a balanced manner.
- **Homogeneity:** The number of network node types should be minimized to improve the scalability, complexity posture, and reliability of the network as well as to ultimately reduce the cost of network implementation.
- **Invulnerability:** A decentralized network architecture is inherently invulnerable to attacks. Since the implementation of all mobile network services is distributed, when some of the nodes or network are damaged or attacked, the rest of the network is unaffected.
- **Self-organized:** In a decentralized network, network nodes can collaborate with each other through specific algorithms, e.g., distributed hash tables [2] or a blockchain.

Resource Efficiency Requirements

In every architecture, it is essential that the available resource pool of a communication system is efficiently used in order to control costs and OPEX while offering communication services. With total cost of ownership (TCO) as a major design target, and the strong recent push for ecologically sustainable ICT, the role of efficient system-wide resource management will increase significantly in the future. This will lead to several new problem spaces which are currently unaddressed, underestimated, or completely overlooked by both industry and academia. For instance, the mixed compute/storage/networking environments that are common today require suitable solutions with respect to resource management: the heterogeneity of resources makes it harder to rely on single mechanisms, as different domains apply their own approaches internally, and often do not exhibit this knowledge externally. Also, unique approaches will most likely not meet the requirements of different resource types.

Within this context, runtime service scheduling is paramount as it permits the provision of superior extra-functional properties of the supported allocations (e.g., "slices") and to lower TCO. Dynamic resource assignment, as a quest for more efficient infrastructure sharing (including computing, networking, and energy resources), is difficult to achieve due to heterogeneous, partial, or outdated information, its runtime nature, and the absence of any central governing body or mechanism.

The answer to job scheduling in large networked systems requires extensive fundamental research. For example, we need to discover how to go about leveraging existing solutions derived through data center research and applying them at the network scale with multi-tenancy and concurrency. This will require suitable conflict handling mechanisms, especially if we aim for guaranteed execution. As indicated by the insights gained from research on distributed systems, the major goal should be improved efficiency, as opposed to optimality.

Requirements on New Boundaries Between Different Realms

In all the previous architectures of mobile communication systems, boundaries were always drawn with respect to physical devices (from the UE to the network, from RAN to Core, etc.), resulting in partly unexpected service provisioning constraints. In the future system, boundaries are to be drawn in the service space: separations are to be established between the realms of the executed services. To enforce meaningful boundaries in the service space, the service typically needs to be decoupled from the infrastructure, and this trend has already been confirmed in the IT service space, where infrastructure assumptions have now been relaxed. In this scenario, however, the service runtime implements the required resilience (i.e., security and reliability) measures. We argue that this is the only reasonable way forward for multi-purpose ICT infrastructures, as it leads to intrinsically adaptive services with minimized required trust levels. This means that it is possible to run the same service in different conditions with essentially the same properties (graceful degradation instead of abrupt stop), without needing to translate service trustworthiness requirements (which are non-negotiable) directly to the infrastructure trustworthiness (which would be prohibitive in terms of costs). This redefined physical boundary will also facilitate the paradigm shift from "separated TPC" to "integrated TPC."

As another concrete consequence of the preceding points, the tension between the optimality and size of the resulting multi-tenant resource interconnection should be resolved. In the view suggested above, several stakeholders would be involved as "prosumers" who independently contribute resources and consume services. This is not compatible with a global view with regard to authority, and, as such, the system would have to be strongly decentralized. On top of that, given the system's potentially large scale, it would appear to be intrinsically suboptimal compared to the current, centrally managed, infrastructures of operators. However, centralized approaches do not scale better per se: given their "distance" from the resource layer, they are arguably even less suitable for rapid adaptations in runtime. Network operators typically use decentralized management or rely on the increased intelligence in, or closer to, the resources (e.g., hierarchical, policy-based management). Today, they are mostly used for optimal pre-provisioning; yet, pre-provisioning faces limitations when it comes to mixed services. Besides, multi-tenancy remains an unsolved challenge in this context: it is unclear how to reunite different policies to be applied within the same resource pool. For these reasons, integrating autonomics and AI with distributed systems concepts may need to combine consistency (required for optimality) with partition tolerance and availability (required for scaling) on a per-task basis. For instance, this would be applicable in scenarios where runtime adaptation needs to be enabled for optimality while also maintaining availability at the expense of possibly higher overprovisioning.

Requirements on Digital Asset Ownership

As a new asset within the digital realm, we are gaining an increasing amount of awareness on privacy and data ownership. Phrased differently, users should be able to fully control their personal digital identity as well as monetize their own data. As such, they should be able to choose which data to share with other parties, and trust that these data are not used without consent. Moreover, users should have the ability to isolate themselves from data breaches. Just as importantly, they should be able to revoke third-party access. These are the ways in which the 6G system architecture could guarantee that users fully control their data.

28.3   Architecture Features

28.3.1   Decentralized Architecture for User-Centric Design

Overall Architecture

By leveraging the flexibility of SDN and NFV, the user-centric design of the core network could allow each UE to obtain its own "private" virtual instance of the core network via the decentralized design of virtual instances implemented as *VMs* or containers, as illustrated in Figure 28.2. These virtual instances of functions could be classified into two categories: network-level service nodes (NSNs) and user-level service nodes (USNs). They can realize the typical functions of the core network, such as mobility management, policy management, and subscriber/device management. An NSN, which could be distributed or centralized lightweight core network functions, serves as the UE's first point of attachment to the network, and is responsible for UE authentication during access registration (i.e., attach and default bearer establishment) procedures. A USN, which could be fully distributed and self-organized according to the virtual network, is dedicated to a single UE and handles all the functionalities of the core network, including user plane and control plane functions.

UCNs outshine the current architecture in numerous significant ways. More specifically, thanks to their highly distributed and personalized nature, they stop botnets from performing DDoS attacks and simultaneously allow per-UE custom policy enforcement (e.g., security and QoS). The decentralized user and data management enables end users to own and control their personal digital assets.

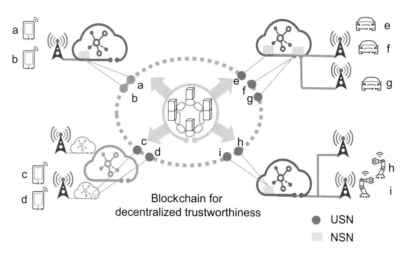

Figure 28.2 Decentralized and user-centric architecture.

Another essential point worth considering is the fact that UCNs are designed to break the physical boundary of conventional TPC, where user space is constrained by physical terminals. This means that by adopting the UCN concept, the user realm could be extended to any resource scope of the user-level service nodes (USNs) in the mobile communication system.

Simply put, the UCN architecture is mainly aimed to be a fundamental network architecture that is resilient, secure, and reliable, while also incorporating a simple design with complete functionalities. All these merits will make 6G trustworthy.

Evolution or Revolution

One of the major technology features of the 6G system, especially in terms of core network design, is the paradigm shift from "NF-focus" to "user-focus." Needless to say, based on current 5G core network, one can argue that slicing mechanisms could be used to realize this vision, which is referred to as "per-user" slicing. However, this patching method is based on the fundamental underlay architecture, which is "NF-focused." Given this, each user may feel as though they have their own network services, which are composed by network functions that possibly do not need to be drastically changed from the service-based architecture, but there are still unresolved issues like digital asset ownership, etc.

The fundamental architecture design could benefit from a completely new approach and implement user centricity. The comparison between the 5G system architecture concept and UCN is shown in Figure 28.3. For 5G, NF is completely defined from the network-level perspective, which UCN will shift from by using the user-level approach to design network functions. The interaction between NSN and USN or within the NSN domain will still need standardization, but it will be purely composed of services within the USN domain.

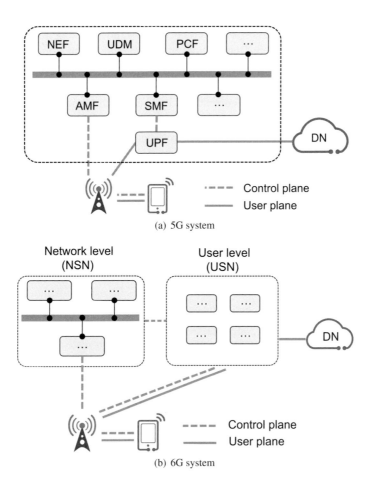

(a) 5G system

(b) 6G system

Figure 28.3 Possible architecture evolution.

Fundamentally, the UCN architecture does not dictate how network services should be designed (e.g., their mobility management and session management). This is because, in future communications systems, these services will be just like IT applications. If a robust infrastructure with an efficient architecture design has been laid out, the services on top will naturally leverage the corresponding advantages in terms of deployment, operation, and scalability. As such, UCN will not diverge from traditional network services, but will instead implement them in a decentralized manner that is more reliable, secure, and resilient. At the same time, it will ensure that the management of subscriber data privacy complies with regulations.

28.3.2   Fusion of the Physical and Cyber Worlds

Reflection of Physical Entities in the Digital Realm

Digital twin has become a broad concept applied by many industry sectors. A digital twin represents a digital or virtual replica of a physical entity, which could be assets, processes, systems, etc. It was first introduced with a manufacturing operation focus [3], with the intention of using physical counterparts for system design, simulation, and risk management to reduce overall costs, improve productivity, and ensure product quality. The digital twin concept has evolved over time, and its deployment is highly IoT-oriented, currently. It is especially designed to connect things via wireless technology and networks, as well as to leverage the processing capability of the cloud. In the 6G era, the fusion of the physical and cyber worlds will be a major trend, which will lead to the digital twin evolving to the next level as the enabling technology, potentially contributing to the user-centric architecture design.

The term "reflection," as introduced in this chapter, incorporates the basis of the digital twin concept, but goes beyond the pure digital twin. As the virtual representative of any physical entity of the 6G system, reflection acts as a "smart digital agent," i.e., it does not only collect the behavior of physical entities, but also proactively triggers actions according to different circumstances. For instance, it could follow user's movement, get closer to users by migrating from a central cloud to the edge cloud, or allocate corresponding resources for users. It features functions such as identity management, authentication, access authorization, user digital asset management, mobility management, session management, and interaction with applications.

A reflection can represent entities at different scales, ranging from physical objects to an entire organization. It does not only involve digitally formulating or describing physical entities but also acts as an autonomous agent living within the network. On top of that, it operates the associated resource elements and interacts with other reflections. For these reasons, we can view reflection as an intelligent entity that manages all user resources in order to form, maintain, and control the user resource pool; and it also composes user services, which it intelligently projects and deploys onto the resource pool. Figure 28.4 illustrates this concept and depicts the functions that may exist in the reflection domain of User B. This domain is deployed in the mobile communication system according to the user profile and behavior in order to provide User B with a fully customized and user-centric, optimized service. This includes network functions, which can be composed for a specific user (i.e., USN), and instantiated as well as updated by the reflection of User B. Moreover, the AI capability of the 6G platform can be leveraged to trigger proactive network service provisioning.

Note that the 4G evolved packet core (EPC) or 5G core network can be seen as one specific reflection type, targeting the needs of traditional mobile network operators. As 5G steps on the scene, we are witnessing user type diversification. Many of these users are vertical industries, who have both resources and skills, yet also have distinctly different needs compared to mobile operators. Consequently, the reflection concept seeks to generalize the notion of core network functionality in an evolutionary way, supporting legacy needs while also allowing for (standard) support of new categories.

By using the reflection concept, services are controlled directly between reflections and deployed on corresponding resource elements in a fully decentralized manner. The distributed data provisioning system could be used to manage data for any format of digital assets, which are stored on the distributed database. As for the distributed mapping system, it could be used to provide the near-real-time update and enquire

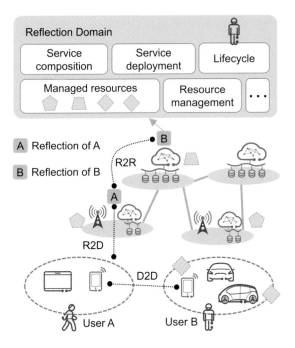

Figure 28.4 Reflective networking concept.

mechanism for the mapping between IDs, names, and addresses. In addition, the lifecycle management system could be used to manage and operate the reflections.

Architecture Consideration

From an architecture perspective, reflections may be supported by considering the following aspects.

- **Basic reflection services:** This mainly includes local resource management and reflection construction. A connection is provided between a resource and its reflection to allow for communication. This connection runs over the reflection-to-device (R2D) interface, as shown in Figure 28.4. The profiles and running statuses of the resource are sent over the connection to the reflection, and used to construct, update, and enrich the reflection itself. In the reverse direction, the reflection uses control signals to operate the resource that will actuate certain activities for realizing and running services.
- **Advanced reflection services:** This mainly includes service composition and deployment with policies. A reflection should be designed to respond to service invocation requests from the intra- and inter-reflection domains. The reflection checks resource availability and feasibility for the requested services. The outcome is that services are deployed to the resources (devices) over the interface between the physical entity and its reflection. Whenever an inter-reflection domain resource is required, the involved reflections will negotiate, for instance, over a

reflection-to-reflection (R2R) interface, as shown in Figure 28.4. As such, the R2R interface realizes a decentralized control plane, which is sustained by different reflection domains. In this case, services are cooperatively deployed through the individual controls of the respective reflection domains yet possibly on intersecting resource pools. In addition to the service composition, resource exposure and discovery services are also supported as part of the reflective services.

- **Reflection implementation:** The reflection domain will be established collaboratively with the data plane (to be introduced in Chapter 30) by relying on the data governance features provided by the 6G system.
- **Fine-grained autonomic resource management:** Given that the boundary is drawn in the service space and not in the physical resources, multiple independently operating reflections will typically provide services from a common resource pool constructed by all devices in the resource layer. Using device and infrastructure programmability, reflections can execute complex services over the infrastructure or even at a bigger scale. Without additional systemic mechanisms and coordination at the resource layer, this would lead to a classical split-brain problem, low service reliability, or huge overprovisioning, as individual reflections cannot optimize resource usage. The reason for this is the "known unknowns" with regard to the requests and operations of other reflections.

28.3.3   Digital Asset Management

On the basis of the UCN approach, a user's digital realm could be created with a clearly defined boundary, which contains abundant data resources, such as user profile, service profile, and resource profile (including both network as well as terminal resources). These are valuable digital assets that belong to the user, and they should be properly managed in full compliance with security and privacy regulatory requirements. The UCN approach could ensure that data ownership is returned to users in the design, which in turn means users would control how data are used and information is spread in the entire digital world. As such, the UCN design provides native support for digital asset management in order to reduce the risk of data abuse and privacy violation. Moreover, it enables users to monetize their own data. This mechanism can rely on DLT, among other technologies, which will be further discussed in Chapter 31.

Users in the digital society need a recognizable digital identity as an address for information exchange. By assigning a digital identity, the associated information such as status, behavior, and transaction will be comprehensible. The digitization of personally identifiable information brings forth numerous advantages. For instance, it allows for convenient and safe transactions, reduces costs, and improves work efficiency. The current lack of a reliable management scheme for digital identities has led to many issues, such as privacy leakage and data abuse. Therefore, we need to build the UCN foundation and design a digital identity scheme that meets the needs of identity authentication, authorization, digital asset management, etc.

References

[1] 3GPP, "System architecture for the 5G system," 3rd Generation Partnership Project (3GPP), Technical Specification (TS) 23.501, Aug. 2020, version 16.5.1. [Online]. Available: https://portal.3gpp.org/desktopmodules/Specifications/SpecificationDetails.aspx?specificationId=3144

[2] I. Stoica, R. Morris, D. Liben-Nowell, D. R. Karger, M. F. Kaashoek, F. Dabek, and H. Balakrishnan, "Chord: A scalable peer-to-peer lookup protocol for internet applications," *IEEE/ACM Transactions on Networking*, vol. 11, no. 1, pp. 17–32, 2003.

[3] M. Grieves, "Digital twin: Manufacturing excellence through virtual factory replication," White paper, vol. 1, pp. 1–7, 2014.

29 Native Trustworthiness

29.1 Background of Trustworthiness

29.1.1 From Philosophy to Society

The word "trustworthiness" immediately brings to mind philosophy. It serves as the foundation upon which efficient communities are built, for both humans and other beings. In today's day and age, the advanced societies we seek would be impossible without trust.

Over the years, trust has been defined in numerous ways. For instance, it is viewed as when "one party (the trustor) is willing to rely on the actions of another party (the trustee) [1]; the situation is directed to the future." "Trustworthiness is the characteristics or behaviors of one person that inspire positive expectations in another person" [2]. The Merriam–Webster dictionary refers to being trustworthy as being "dependable" and lists many synonyms, including calculable, dependable, reliable, responsible, safe, secure, solid, and true [3].

When recognizing the level of trust in something, more often than not, people try to define its characteristics with engineering language. Four common trust indices have been described in e-commerce: third-party privacy seals, privacy statements, third-party security seals, and security features [4]. Among them, security features play the biggest role in enabling consumers to trust e-commerce.

The study of risk analysis began after trustworthiness was introduced into the security field. When examining if something is trustable, the following three factors are considered: the source of information, the information itself, and the recipient. These three factors also include sub-factors, for example, accuracy, believability, and so on [5].

As indicated by the results of the above studies, trustworthiness can be defined and measured only if there is a specific scenario to rely on. Even when such a scenario exists, without a clear profile of trustworthiness being present, the relevant factors are subject to individuals. Moreover, "these factors can vary in importance depending on the individual making the trust decision and also on the situation; such is actually the subjective nature of trust" [6].

29.1.2 From Society to Industry

As technological development enters a phase where the physical and digital worlds co-exist, the trustworthiness of the information and communication technology (ICT) infrastructure becomes even more important. It greatly affects users' desire to sign contracts for the services provided by the ICT infrastructure. As such, all ICT stakeholders strive for trustworthiness.

The ITU-T recommendation of X.509 has given a definition of trust in the ICT field. "Generally, an entity can be said to 'trust' a second entity when it (the first entity) makes the assumption that the second entity will behave exactly as the first entity expects" [7]. The ITU has carried out ICT-related trust standardization since 2015, specifically in the IoT field, successively publishing the technical report, paper, and recommendation standards [8–12]. The documents propose the concept of trust, and classify it from both architectural and technical perspectives. After proposing a strategy for trust provisioning in the ICT system, the documents provide guidelines for analyzing trust in the network, cyber security, and IoT application fields.

Research on ICT trustworthiness began from IoT, and now includes the entire ecosystem. In 2018, a new work item, a security framework based on trust relationship in 5G ecosystem, gained the research approval of the ITU Telecommunication Standardization Sector (ITU-T). From the stakeholders' perspective, the time is ripe to identify the roles, relationships, and security-related responsibilities of stakeholders, and to reach a consensus on the 5G ecosystem [13].

In 2017, the National Institute of Standards and Technology (NIST) published a series of reports with the heading "Framework for cyber-physical systems" (CPS). The cyber-physical system public working group (CPS PWG) defines CPS trustworthiness as the "demonstrable likelihood that the system performs according to designed behavior under any set of conditions as evidenced by characteristics including, but not limited to, safety, security, privacy, reliability and resilience." This was the first time that a standards organization defined CPS trustworthiness [14].

The industries' discussions, definitions, and reference frameworks all aim to provide users with a trustworthy communication environment, and, as such, objective and subjective factors have both been repeatedly discussed. In terms of ICT research, the efforts geared towards achieving trustworthiness are ongoing.

Many models have been developed, including an ontology-based methodology proposed for analyzing mutual dependencies and conflicting requirements throughout the CPS lifecycle [15]. In addition, research has been conducted on the practical application of a game-theory-based trust measurement model, a radar graph tool describing communication trustworthiness, a cloud-model-based risk evaluation approach, and various other solutions.

Vendors and solution integrators have put a great deal of effort into meeting and even exceeding user needs, and are constantly adjusting their product credibility requirements to keep up with user preferences. For instance, requirements have expanded from only focusing on product development to also including security solutions in the initial stage of product design, and from implementing the best industry practices and standardization to actively participating in the construction of a healthy industry ecological environment.

From the communication perspective, how can we define trustworthiness? What scientific evidence should we present to help decision makers make the right call on whether 6G is trustworthy? In other words, what trustworthy qualities do we expect 6G to have?

By examining how 6G inherits the excellent capabilities and best practice of previous-generation networks, as well as taking into account multiple disciplines and societal issues, we can arrive at a view of 6G trustworthiness.

29.2   Complex Communication Trustworthiness

Without doubt, as a crucial infrastructure, mobile communication is a bridge connecting the physical and digital worlds. It continues to transmit information between people, between things, and between people and things. With the rapid development of communication technology, transmission capability is getting more advanced. As such, people have become accustomed to the information dividends brought by communication networks, family reunions, social gatherings, business, and other affairs closely related to people, as well as the exploration of the unknown in the universe.

People increasingly care about the trustworthiness of communication networks, and, as such, we believe that user acceptance is the only corresponding criterion. Individual users expect trustworthy networks to provide continued services, protect private information, and cause no physical injuries. From a different perspective, enterprise customers rate the trustworthiness of their networks in terms of stability, high-quality communication, no leakage of confidential information, no exploitation etc.

We should take all these factors into account for the 6G network. Put differently, where do we start, in defining trustworthiness in line with user expectations, and how do we achieve it? We believe that communication trustworthiness is a complex concept.

Trustworthiness involves abundant connotations with multiple properties.

The different dimensions include the use of technology, compliance with laws and regulations, consensus and collaborative actions, and positive beliefs. Each dimension has specific characteristics, which each have different facets.

- **Technology**: This serves as an anchor stabilizing the other dimensions. Any technology that can result in a reliable network should be used, such as cryptography, data analysis, machine learning, and security assessment.
- **Laws and regulations**: As "rules have been created and enforced through social or governmental institutions" [16], laws and regulations continuously affect market stability. From when 5G started gaining traction, countries began actively improving their laws and regulations on cyber security, communication security, and privacy protection. Relevant strategies and measures on testing, evaluating, and verifying equipment, devices, and entire networks have been introduced into the communication industry. The ongoing development of these laws defines the responsibilities of stakeholders, while also subtly leading the entire supply chain towards a more trustworthy network [17–19].
- **Collaboration**: Global standards organizations are an essential engine promoting collaboration. From 2G to 5G, the communication network has leaped from basic

phone calls and short message services to providing various industry applications. This would have been impossible to achieve without the collaboration of all ecosystem parties. ITU, 3GPP, the GSM Association (GSMA), and the Next Generation Mobile Networks (NGM-N) Alliance are well-known communication communities that provide stakeholders with an environment where they can freely discuss technology and business. Based on the knowledge and inspiration derived from these platforms, the communication network is gradually shifting from a closed to an open structure. By bringing network capabilities to light, cross-industry cooperation is beginning to emerge and industries are collaborating to deliver customized and higher-quality services to users.

A case in point is the 3GPP Release 16 specifications, which support two important vertical sectors (automotive and industrial automation). Release 16 is also assisting "other vertical industries such as transportation (e.g., Future Mobile Communication System for Railways) and media (e.g., 5G Mobile Broadband Media Distribution)" [20].

The security community is also striving to promote cooperation. This is because security sectors can withstand attacks more appropriately and efficiently through threat-intelligence and technology sharing, as well as joint work. Within this context, the theme of the world-renowned security event, RSA Conference 2014, was "share, learn, and secure." This would create opportunities for people to "capitalize on the ideas, insights, and relationships that may shape the future of information security" [21].

- **Belief**: This is the most elusive dimension. When users judge whether a communication network is trustworthy, what they believe to be the case dominates all other traits. We expect extensive scientific and technical knowledge that will help us reach a consensus on how to understand the technology and the corresponding practices. Phrased differently, if we listen and accept each other's views, we can reach resonance and establish collaboration. The convergence of more minds and hands will be more conducive to the development of enhanced communication innovations in the future.

Trustworthiness is a relative concept.

The judge, those judged, the method of judging, and uncertain or unexpected events are all factors that influence the trustworthiness of judgment.

Trustworthiness is vital.

With changes in human society, the connotations of trustworthiness and the methods adopted to achieve it are sustainable if they are optimized, evolved, and innovated.

29.3 Trustworthiness Design Rules

29.3.1 Principles

Two principles make up the core of 6G trustworthiness.

Principle 1: Nativeness

- **Trustworthiness capabilities must follow diverse 6G services:** The aspirations people have for 6G range from a sensing network to tactile healthcare with haptic feedback and low-orbit satellites. The diverse 6G network structure, services, and user requirements will make it an exceptional network in every sense of the word. To adapt to this trend, trustworthiness capabilities also need to be diverse. A 6G network that includes multiple technical and business domains may call for a set of trustworthiness capabilities which can be used to execute strict and centralized security access control for the centralized network part, and to customize authentication as well as authorization for the edge autonomy part.
- **Trustworthiness must be developed throughout 6G's lifecycle, including its design, development, and operations:** We can identify the trustworthiness requirements by sketching out their characteristics while at the stage of designing 6G. Along those lines, while developing 6G products, coding and manufacturing equipment will meet the design requirements of the previous stage. As such, 6G should be operated and deployed as well as configured with trust, which we must continuously evaluate and improve at each stage.

Principle 2: Balance

To achieve 6G trustworthiness, three factors should be considered: initial trust in visitors, cost of attacks, and recovery speed of operations.

Given the value of trustworthiness as a goal, if the designers of 6G have little initial trust in network visitors then, even if they hope that the cost of attacks to the attacker would be high and/or recovery would be fast, more countermeasures and strict plans will be needed.

- **6G visitors:** These can be applications for resource operations on the 6G network, including network access by user devices, database access by applications, function connection, log access, etc.
- **Cost of attacks:** Attackers only attack when they profit more than the cost of attacks.
- **Recovery:** The mechanisms used to resume normal operations or services should be capable of promptly mitigating attacks, both dynamically and continuously.

In reality, trustworthy architectures are usually designed along with the initial trust in visitors, following which access control solutions are proposed during the very early stages of the design process. In contrast, it is very difficult to predict the cost of attacks and recovery capability.

For example, some cases have critical requirements (e.g., high security under low-latency communications), which means that attackers have to put in more effort to succeed, whereas other cases are just normal phone calls. As such, these cases appeal differently to attackers which means the attack and profit ratio is also different. Furthermore, the speed with which normal service provisioning can be recovered when an attack occurs also differs between the two cases.

For these reasons, balanced trust might seem easy, but in reality, it is difficult to meet the requirements.

29.3.2   Objectives

From a technology dimension, security, privacy and resilience (which are established by cryptography and defense technologies) are the three primary characteristics of trustworthiness. As such, we refer to them as the Three Pillars of trustworthiness, which are underpinned by Ten Blocks (three facets of security, two facets of privacy, and five facets of resilience), as shown in Figure 29.1. The Three Pillars and Ten Blocks serve as the foundation upon which 6G trustworthiness can be achieved.

In line with the principles for achieving 6G trustworthiness, we need *a native trustworthiness architecture that meets the characteristics of security, privacy, and resilience, based on an inclusive trust model.*

This architecture should cover the entire 6G lifecycle to ensure no shortcoming. We have defined three objectives with regard to the Three Pillars and Ten Blocks.

Objective 1: Balanced security.

Availability, integrity, and confidentiality (the AIC triad) are the fundamental features of security. Balance is one of the principles of native trustworthiness, meaning the different protected asset/property may require a different level of protection or different weight in each facet, according to the different scenario.

Objective 2: Everlasting privacy protection.

"Privacy is commonly understood as the right of individuals to control or influence what information related to them may be collected and stored and by whom and to whom that information may be disclosed" [22]. To protect the identity and behavior of users, technologies like cryptography are supplied to guarantee that only parties authorized by users can interpret the content of the information transferred among them.

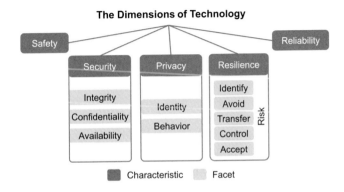

Figure 29.1 The dimensions of technology for trustworthiness.

Objective 3: Smart resilience.

According to the widely accepted definition provided by Wikipedia [23], "resilience is the ability of the network to provide and maintain an acceptable level of service in the face of various faults and challenges to normal operation." Facets are a series of abilities that Identify and measure risks. Situational awareness and big data analytics can be used to discover risks, and then activities that Avoid risks can be executed. We can Transfer all or some of the risks to other parties, thereby facilitating network recovery. Otherwise, we can Control the consequence to be as minimal as possible. Finally, we Accept residual risks that do not harm networks [16, 24, 25].

It is worth emphasizing that trustworthiness should be measured continuously to facilitate the prompt perception and distinguishing of threats and risks, respectively. As ITU-T stated in its report [25], the level of trust can be measured using a quantitative method similar to that used for quality of service (QoS) or quality of experience (QoE). Regardless of the approach used, the determination of a specific trust level depends on the associated services and applications. The trust level is a consistent, quantifiable measure of reliance on the character, ability, strength, or truth of someone or something [26].

We can approach risk analysis quantitatively or qualitatively. The former is used to assign monetary and numeric values to all elements of the risk analysis process. The latter uses a rating system to walk through different scenarios with different risk possibilities and rank the seriousness of threats and the validity of different possible countermeasures based on people's different subjective views [27].

From the risk analysis results, the effectiveness of current security mechanisms can be evaluated, and the necessary countermeasures must be implemented to reduce the overall risk to an acceptable level.

29.4   Trustworthiness Technologies

As mentioned in the previous sections, the Three Pillars and Ten Blocks are the base of 6G trustworthiness architecture, and a native trustworthiness architecture should be developed on the basis of an inclusive trust model. Starting from establishment of a trust model, and then studying appropriate technologies, is the way to build a trustworthy 6G network.

29.4.1   Multi-Lateral Trust Model

User trust in the existing network is endorsed by the corresponding subscribed mobile network operators (MNOs), who purchase and deploy network equipment that has already passed tests and verification. MNOs use centralized authentication and authorization entities to manage users.

This trust model may face many challenges on the path to achieving the three objectives of 6G trustworthiness, discussed at the end of the previous section. For instance, it's not easy to provide fine-grained centralized access control to fit both

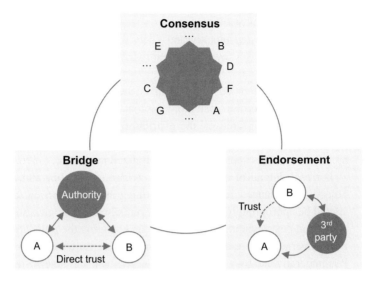

Figure 29.2 Multi-lateral trust model.

the high security level needed for centralized financial scenarios and a lightweight security mechanism for collecting localized sensing data. On top of that, instant trust is difficult to establish among cooperating partners, and we lack an effective solution for managing digital identities.

Since nativeness adheres to various 6G services, it is the first principle involved with designing trustworthiness. Within this context, a multi-lateral trust model is used to manifest the multiple possibilities of trust. It includes three modes: bridge, consensus, and endorsement, as shown in Figure 29.2, and its core feature is decentralized multi-party consensus, with co-existing centralized trust and third-party endorsement. The three modes are as follows.

- **Bridge**: This is the mode of establishing a trusted bridge between entities through the authorization framework with a central point, of authority such as a security policy center or security management of user profiles. Trust will be obtained from the current communication trust model.
- **Consensus**: The process of creating stronger and smarter communication systems all boils down to the trust between parties. These parties can be network components, different sides of the supply chain, or the roles of an industrial ecosystem. In this mode, transactions are attestable and responsibilities are shared by multiple parties. High efficiency and scalability will be provided as the main features of this mode, matching the agile and tailored access requirements of 6G.
- **Endorsement**: This refers to the mode in which an authoritative third party measures and evaluates network trustworthiness. For instance, as shown in Figure 29.2, party B could invite a third party to confirm whether party A is trustworthy, and the third party could endorse party A.

The data or transaction information required for third-party endorsement can be recorded and attested quickly and fairly through the consensus mode. In addition, consensus serves as a key supplement to the mutual trust in edge autonomous-service scenarios.

The above three modes are interlinked and may share key technologies, such as the cryptographic algorithms and protocols that should be adapted to the new 6G scenarios and architecture. Moreover, the distribution and full integration of the networking and computing resources also require an agile and lightweight zero-trust security architecture that is applicable to decentralized networks.

For each lateral of this model (i.e., bridge, consensus, and endorsement), all the objectives (balanced security, everlasting privacy protection, and smart resilience) should be appropriately attained. However, new technologies, such as artificial intelligence and post-quantum cryptography are changing the foundation of the trustworthiness constructed on traditional mechanisms. To ensure we achieve 6G trustworthiness, it is imperative that we understand the new attack and defense techniques.

29.4.2   Distributed Ledger Technology

When founded on cryptography, each block of the blockchain contains a cryptographic hash of the previous block, a timestamp, and transaction data. Given this, the distributed ledger does not need such a chain or proof of work. The distributed ledger technology (DLT) could be provided for many cases, network architectures, and applications (more details are provided in Chapter 31). This technology is essentially an asset database that can be shared across multiple sites, institutions, or geographic areas [28]. The blockchain is one type of distributed ledger, and non-blockchain distributed ledger tables, a distributed cryptocurrency, and a specific database architecture also exist [29].

Despite featuring certain benefits, today's blockchain still has limitations.

- **Lack of flexibility**: Blockchain is natively designed to be immutable, which means a high degree of both tamper evidence and resistance can be achieved. In many scenarios, this immutability provides stability for financial services; however, for numerous real-time communication use cases, some of the data needs to be changeable on various grounds, i.e., a modifiable or programmable parameter or parameters in a communication protocol or a virtual calculation circuit. Instead of being invalid in the blockchain ledger, under specialized circumstances, data would be deemed valid and integrated into the ledger.
- **Lack of performance**: In contrast to some legacy transaction processing systems, which are able to process tens of thousands of transactions per second, the Bitcoin blockchain can handle only three to seven transactions per second. The corresponding figure for the Ethereum blockchain [30] is as low as fifteen transactions per second, and as such, many observers do not consider blockchain technology to be viable for large-scale applications due to its relatively poor performance.

- **Incomplete security models**: Many of today's blockchain technologies are subject to various types of attacks. Some of their designs rely on certain consensus algorithms (recursive calls) that have proven to contain vulnerabilities, such as decentralized autonomous organization (DAO) attacks in the smart contract Ethereum [30].

  Another landscape of the blockchain security model is the emergence of quantum computing [31], which could potentially render the underlying cryptographic mechanisms and protocols vulnerable.

Blockchain-Like Technology for 6G

Along with the network architecture, consensus may be the most important mode in the multi-lateral trust model, as the 6G network architecture will trend towards a distributed nature. Currently, blockchain-like technology seems to have matured although various blockchain-based applications continue to emerge. The main challenge is how to ensure DLT meets the following 6G requirements.

- **Ultra-high throughput and ultra-low latency**: Unlike traditional blockchain, which achieves consensus slowly, new blockchain algorithms and architecture will be able to meet these requirements with the introduction of new consensus and cryptographic algorithms.
- **High availability and reliability**: 6G networks will require simplified management and operations to achieve high operational efficiency. That said, requirements for high availability and reliability should not be sacrificed but enhanced. In that regard, we can enhance availability by deploying blockchain technologies. In other words, the 6G network operations should be able to sustain a certain level of disasters or malicious attacks without requiring a large amount of human intervention.
- **High privacy protection and digital sovereignty**: When it comes to privacy protection, we must abide by privacy laws and regulations (e.g., GDPR) in different countries or regions. We can utilize DLT technologies to help us achieve the privacy objectives. Nonetheless, most of the privacy regulations address the fact that data owners have the right to delete and forget data, which contradicts the inherent immutability of blockchain [17]. Today's blockchain structures are based on hash chains, and the hash value is extremely difficult to change once a transaction is recorded on the blockchain. If changes are necessary, the current method involves starting a hard fork [32], which creates security vulnerabilities. The hard fork could impact many existing recorded blocks and it takes a substantial amount of time to revalidate the transaction. It is therefore necessary to conduct research on how to modify or edit transactions without impacting other blocks.

29.4.3 Post Quantum Cryptography

Quantum computers can efficiently solve hard mathematical problems (e.g., NP hard), i.e., integer factorization or the discrete log problem over various groups, "thereby rendering all public key cryptosystems based on such assumption impotent" [33].

Many vulnerabilities of the traditional schemes have to be considered prior to developing real quantum computers. For example, an adversary can store key exchange messages today and break them in 2035. Along those lines, key exchange using Diffie–Hellman (DH) key exchange is already vulnerable. When will large-scale quantum computers be built? How can we resist both quantum and classical computing adversaries [33]? NIST launched its post quantum standardization activity in December 2016. It aims to develop quantum-safe public key cryptography standards, including schemes for encryption/key establishment, and digital signatures [34].

Industrial Activities in Post Quantum Cryptography (PQC)

From an industrial perspective, the major activities are being led by technology giants. For example, Google experimented with an implementation of "new hope" [35] in "Canary" Chrome for a few months to evaluate the impact of hybrid schemes, and Thales implemented leading candidate algorithms and added them to popular open source crypto applications [36]. In addition, Microsoft is implementing PQC within an open VPN for an open source project [37], and this includes three different PQC protocols: Frodo-KEM, SIKE, and Picnic.

PQC for 6G

According to RAND Corporation, "quantum computers capable of cryptographic applications are expected, on average, to be approximately 15 years away – roughly 2033. However, experts assess that both earlier and much later development are possible" [38]. In the 6G era, where large-scale quantum computing could be used, the corresponding security architecture scheme should be considered. Therefore, the characteristics of PQC algorithms should be analyzed to adapt to 6G protocols. One possible way of going about this involves designing PQC algorithms to ensure that the models provide adequate flexibility for 6G and are applicable to multiple platforms. Moreover, PQC algorithms and protocols need to support a flexible 6G framework, such as flexible security levels, code sizes, and signatures that can be adapted to established protocols. These algorithms also need to provide more flexible key management to accommodate different key sizes for different protocols.

29.4.4   Autonomous Security

Autonomous security will be one of the key features for achieving a trustworthy 6G architecture. Instead of protecting systems and end users from malicious attacks as they occur, this architecture will adopt a more proactive approach. Combined with the system architecture design, autonomous security enabling technologies could perform dynamic intelligent defense, and such technologies include machine learning, artificial immune system [39], and more. Among all the enabling technologies, we can envision the wide adoption of AI/ML to resolve such challenges.

AI/ML can help detect threats and offer recommendations to security analysts, thereby shortening the response time from hundreds of hours to seconds and improving analysts' efficiency from one or two incidents to thousands per day. After being deployed in a simulated 6G network, an AI/ML model could continuously be trained with data, and improve its capabilities. Furthermore, by performing attack drills, autonomous security systems can help operators generate and customize effective security policies in a timely manner.

However, we should also keep the new challenges brought by new technologies in mind. The first challenge is technical reliability. The robustness of the deep neural network (DNN) directly affects the judgment of AI systems. Moreover, because the DNN is not transparent enough, it cause unfairness and inaccurate or unidentifiable results, and may even violate related laws or regulations (such as GDPR). AI training requires massive datasets, which may result in data leakage, tampering, theft, and misuse due to a lack of data security protection. The second challenge is societal applications. Security and privacy incidents may arise due to improper management and control of AI purposes, data quality issues, and insufficient knowledge of developers. The third challenge is legal requirements and responsibilities – no laws or regulations have clearly defined the rights and responsibilities of stakeholders [40].

Simply put, there is no standard definition of trustworthiness KPIs. But if we stop questioning subjective factors and set our sights on technical factors, which can be used for scientific reasoning, trustworthiness KPIs will be much simpler to define. Someday in the near future, we will figure it out, measure it, and finally achieve it.

References

[1] Wikipedia, "Trust (social science)," 2020. [Online]. Available: https://en.wikipedia.org/wiki/Trust_(social_science)

[2] Wikipedia, "Trustworthiness," 2020. [Online]. Available: https://simple.wikipedia.org/wiki/Trustworthiness

[3] Merriam–Webster dictionary, "trustworthy," 2020. [Online]. Available: https://www.merriam-webster.com/dictionary/trustworthy

[4] F. Belanger, J. S. Hiller, and W. J. Smith, "Trustworthiness in electronic commerce: The role of privacy, security, and site attributes," *Journal of Strategic Information Systems*, vol. 11, no. 3–4, pp. 245–270, 2002.

[5] J. R. Nurse, S. Creese, M. Goldsmith, and K. Lamberts, "Trustworthy and effective communication of cybersecurity risks: A review," in *Proc. 2011 1st Workshop on Socio-Technical Aspects in Security and Trust (STAST)*. IEEE, 2011, pp. 60–68.

[6] J. R. Nurse, I. Agrafiotis, S. Creese, M. Goldsmith, and K. Lamberts, "Communicating trustworthiness using radar graphs: A detailed look," in *Proc. 2013 11th Annual Conference on Privacy, Security and Trust*. IEEE, 2013, pp. 333–339.

[7] ITU-T Recommendation, "Information technology – open systems interconnection – the directory: Public-key and attribute certificate frameworks," 2000.

[8] G. Lee and H. Lee, "Standardization of trust provisioning study," Technical Report, International Telecommunication Union, 2005.

[9] ITU-T, "Future social media and knowledge society," 2015.

[10] ITU-T, "Trust provisioning for future ICT infrastructure and services," 2016.

[11] ITU-T Recommendation Y.3501, "The basic principles of trusted environment in ICT infrastructure," 2017.

[12] ITU-T Recommendation Y.3502, "Overview of trust provisioning in ICT infrastructures and services," 2017.

[13] ITU-T, Draft Recommendation X.5Gsec-t, "Security framework based on trust relationship for 5G ecosystem," 2019.

[14] E. R. Griffor, C. Greer, D. A. Wollman, and M. J. Burns, "Framework for cyber-physical systems: Vol. 2, working group reports," NIST Special Publication, 1500-202, 2017.

[15] M. Balduccini, E. Griffor, M. Huth, C. Vishik, M. Burns, and D. Wollman, "Ontology-based reasoning about the trustworthiness of cyber-physical systems," *IET Journal of IoT*, June 2018.

[16] Wikipedia, "Law," 2020. [Online]. Available: https://en.wikipedia.org/wiki/Law

[17] E. D. P. Board, "General data protection regulation," 2020. [Online]. Available: https://gdpr-info.eu/

[18] The European Network and Information Security Agency, "EU cyber security act, Europe," 2020. [Online]. Available: https://ec.europa.eu/digital-single-market/en/eu-cybersecurity-act

[19] The European Network and Information Security Agency, "A trust and cyber secure Europe," 2020. [Online]. Available: https://www.enisa.europa.eu/publications/corporate-documents/a-trusted-and-cyber-secure-europe-enisa-strategy

[20] Huawei, "Position paper – 5G applications," 2020. [Online]. Available: https://www-file.huawei.com/-/media/corporate/pdf/public-policy/position_paper_5g_applications.pdf

[21] Youtube, "RSA Conference. 2014: Share. Learn. Secure," 2014. [Online]. Available: https://www.youtube.com/watch?v=ti07UfL1gsk

[22] ITU-T, "Security in telecommunications and information technology," 2003.

[23] Wikipedia, "Resilience (network)," 2020. [Online]. Available: https://en.wikipedia.org/wiki/Resilience_(network)

[24] P. Trimintzios, "Measurement frameworks and metrics for resilient networks and services," Technical Report, European Network Information Security Agency, p. 109, 2011.

[25] ITU-T, "Trust in ICT," 2017. [Online]. Available: https://www.itu.int/pub/T-TUT-TRUST-2017

[26] ITU-T Recommendation, "Baseline identity management terms and definitions," 2010.

[27] S. Harris and F. Maymi, *Cissp all-in-one exam guide*. McGraw-Hill, 2016.

[28] UK Government Chief Scientific Adviser, "Distributed ledger technology: Beyond block chain," 2020. [Online]. Available: https://assets.publishing.service.gov.uk/government/uploads/system/uploads/attachment_data/file/492972/gs-16-1-distributed-ledger-technology.pdf

[29] Wikipedia, "Distributed ledger," 2020. [Online]. Available: https://en.wikipedia.org/wiki/Distributed_ledger#cite_note-UKscienceOffice201601-1

[30] V. Buterin *et al.*, "A next-generation smart contract and decentralized application platform," *White Paper*, vol. 3, no. 37, 2014.

[31] E. Rieffel and W. Polak, "An introduction to quantum computing for non-physicists," *ACM Computing Surveys (CSUR)*, vol. 32, no. 3, pp. 300–335, 2000.

[32] Ethereum, "The whole forking history," 2020. [Online]. Available: https://blockchain.news/news/ethereum-the-whole-forking-history

[33] L. Chen, S. Jordan, Y.-K. Liu, D. Moody, R. Peralta, R. Perlner, and D. Smith-Tone, "NISTIR 8105 report on post-quantum cryptography," National Institute of Standards and Technology, p. 10, 2016.

[34] NIST, "The whole forking history," 2020. [Online]. Available: https://csrc.nist.gov/Projects/post-quantum-cryptography/post-quantum-cryptography-standardization

[35] E. Alkim, L. Ducas, T. Pöppelmann, and P. Schwabe, "Post-quantum key exchange a new hope," in *Proc. 25th USENIX Security Symposium*, 2016, pp. 327–343.

[36] Thales, "Thales esecurity," 2020. [Online]. Available: https://github.com/thales-e-security?language=c

[37] Microsoft, "PQCrypto-VPN," 2020. [Online]. Available: https://github.com/Microsoft/PQCrypto-VPN

[38] M. J. Vermeer and E. D. Peet, "Securing communications in the quantum computing age: Managing the risks to encryption," RAND Corporation, 2020. [Online]. Available: https://www.rand.org/pubs/research_reports/RR3102.html

[39] E. Guillen and R. Paez, "Artificial immune systems – AIS as security network solution," in *Proc. International Conference on Bio-Inspired Models of Network, Information, and Computing Systems*. Springer, 2010, pp. 680–681.

[40] Huawei, "Thinking ahead about AI security and privacy protection," 2019. [Online]. Available: http://www-file.huawei.com/-/media/CORPORATE/PDF/trust-center/Huawei_AI_Security_and_Privacy_Protection_White_Paper_en.pdf

30 Data Governance Architecture Technologies

30.1 Background

Data are essential in today's digitalized society, and vast amounts of it are expected to be generated, collected, and exchanged in the 6G system in the future. Such data will be used to perform different operation and management tasks, including performance monitoring, configuration, and fault management. It will also be used as knowledge exchanged with other systems and business sectors to generate a broader value scope. Only in this way will mobile communications systems become a key enabler for further development of other industries, such as verticals.

Data governance itself may have varying connotations depending on its usage scope, which could vary from economy to technology. The concept of data governance refers to data management, maintenance and further development to obtain high-quality data as key organization assets through relevant processes and technology [1]. At present, the data generated in the mobile communication system of each mobile network operator (MNO) are isolated and stored separately according to technical domains, e.g., RAN, core network (CN), transport network (TN), OA&M, and terminals. The data owned by different network elements and players lack openness and transparency, resulting in data silos that form a major bottleneck for data collection and sharing. On the other hand, major OTT companies have built their expertise on data governance and monetization strategy (e.g., data storage, analytical services, rich application programming interfaces (APIs)), which are far more advanced than the telecommunications sector.

A data governance scheme in the 6G system will be essential to provide strong support for AI and sensing services; hence new approaches and system features are expected to emerge.

30.2 Design Considerations and Principles

Data governance goes far beyond conventional data collection and storage. In general, system design needs to consider four aspects, as shown in Figure 30.1.

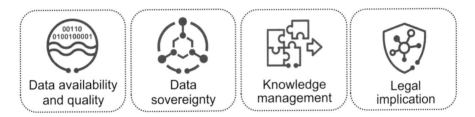

Figure 30.1 Design considerations for data governance.

Data Availability and Quality

Data availability and quality pose one of the biggest challenges to applying AI services in various industries. Increasing data availability involves collecting data not only from a single system's single domain, but also from multiple systems' multiple domains. This therefore raises a fundamental question: how can we break the physical boundaries (i.e., multi-vendor, multi-operator, and multi-industry) to let data flow into a heterogeneous data ocean?

Once the scattered and isolated data are collected and available for usage, it then becomes a question of how we control and improve the data quality. Obtaining vast amounts of data does not mean that the data are either high-quality or usable. As such, it is necessary to support efficient data processing while reducing computation complexity and energy consumption.

Data Sovereignty

As society transitions towards full digitization, the importance of data sovereignty, data security, and privacy is unprecedented. In terms of governance, many countries have enacted privacy protection laws or regulations. Service providers are also updating their privacy protection schemes. Furthermore, major governments around the world are developing or have issued data management regulations. For example, the EU officially released the GDPR in 2018, regulating data usage at the EU level [2]. Then in 2019, China released the Data Security Management Regulations, which together with the Cybersecurity Law released in 2016, constitute the Chinese equivalent of the GDPR [3]. The United States is also implementing privacy-related laws, such as California's Consumer Privacy Act, which came into effect in January 2020 [4]. How can we fully explore the intrinsic value of data to provide precise support for services while also protecting privacy in order to respect data sovereignty? This has become a hot topic in recent years. The design of the 6G system should take into account regulatory uncertainties, especially in different geographical regions.

Knowledge Management

In general, knowledge can be considered as processed data with a specific purpose or value; it can be directly used by physical or virtual entities in different technical and business domains. Knowledge management covers the generation, update, and exposure of knowledge. For knowledge generation and update, the source and quality of data should be carefully considered, and precautions should be taken against unreliable or even malicious sources of low-quality or harmful data. For knowledge exposure, this relies on an appropriate platform and interface design.

Legal Implication

Data collection and usage are becoming more sophisticated and sensitive, often incorporating live feeds of information from sensors and various other technologies. These enhanced capabilities have yielded new streams of data and new types of content that raise policy and legal concerns over possible abuse: malicious institutes or governments may repurpose these capabilities for reasons of social control. Similarly, new technology capabilities make it difficult for ordinary people to discern the difference between legitimate and fraudulent technology content, such as an authentic video versus a "deep fake." As such, it will become more critical to maintain the fragile balance between preserving the social benefits of technology and preventing undesirable repurposing of these new technology capabilities for social control and liberty deprivation. In order to detect fraud and prevent abuse of these enhanced technology capabilities, more drastic legal and policy tools are needed.

30.3   Architecture Features

The key feature in the systematic design of data governance is an independent data plane (as shown in Figure 30.2), which will deliver general data-relevant capabilities in order to provide transparent, efficient, endogenous security and privacy protection for internal and external functions within 6G. The following sections introduce the basic concepts and relevant network functions and services.

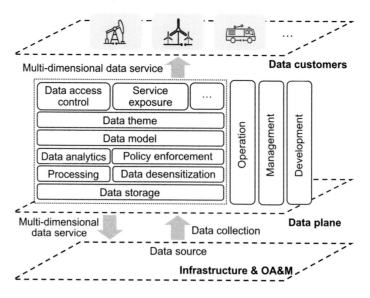

Figure 30.2 Independent data plane to enable complete data governance.

30.3.1   Independent Data Plane

The independent data plane aims to implement data governance scheme for the 6G system. It handles data from different business entities and processes it regardless of the data source for its entire lifecycle, which covers the following aspects: data generation and collection, data processing and analytics, data service provisioning. Hence, an independent data plane can provide data services for external business entities such as vertical industries (e.g., automotive, manufacturing, and eHealth), as well as for the 6G system itself (e.g., control plane, user plane, and management plane), for network automation and optimization purposes. Targeted data may include configurations, states, and logs related to network operations, users' private data, sensor data, and service data provided by other players.

The collected data will form a rich data resource, which can be organized in a distributed manner. The data usually need to be pre-processed (e.g., data anonymization, format shaping, denoising, transformation, and feature extraction) before it can be used, because directly using raw data is problematic for applications such as AI and sensing.

To ensure data integrity and legitimacy during data handling, regulatory and non-regulatory data policies will need to be enforced by default, covering geographic restrictions, national or regional privacy regulations, and other rules. The delivery of data to the data plane should comply with the data usage rights and obligations specified in the data contract. Furthermore, the data plane should provide data desensitization, which is a key service, in order to achieve data privacy protection.

All the services provided by the data plane mentioned above are operated and managed by a self-contained OA&M system.

Another important aspect of the data plane is generating knowledge based on data collection, processing, and orchestration. This should be performed in accordance with contract requirements in order to orchestrate the processing and transmission of relevant multi-source data.

The data governance framework itself could be constantly evolving and enriching due to new data sources, data models, and data themes that are tailed and utilized by data customers. Hence, it supports real-time framework development in parallel with framework operation and management.

Because the data plane is a logical concept, it can be implemented in a centralized manner with hierarchical architecture. Alternatively, it can be implemented as a logical function and distributed across edge or deep-edge nodes. The following sections discuss some key components of the data plane.

30.3.2   Data Governance Multi-Player Roles

The data governance ecosystem includes two roles: from data customer to data provider and from data owner to data steward. These could be taken by the same or different business entities. Hence, data governance in 6G is a typical scenario that involves multiple players, which could be potential data customers that consume the data or knowledge provided by the 6G system, or could be data providers for the 6G system. 6G could have its own data governance framework, but on the other hand, it is also possible for 6G to implement the data governance framework together with other industry players on the basis of its own domain knowledge. Different evolution or revolution paths may exist. As such, it is essential to establish how data rights are negotiated among different business entities during the operation phase. This could be achieved using decentralized technologies such as blockchain. Chapter 31 discusses this in detail. Nevertheless, as a new component of mobile communications systems, independent data planes may trigger new functional and interface requirements for standardization activities.

30.3.3 Data Resource

The data resource contains rich data content, comprising structured or unstructured data and pre-processed, post-processed or raw data.

Efficient collection of data (for example network status and user behavior such as mobility patterns) from the wireless environment is the prerequisite for data governance. Intelligent methods can then be employed to analyze the data and deliver the derived knowledge to internal or external customers. It is therefore important to understand the source of data.

Figure 30.3 shows some of the main data source categories in the 6G system.

- **Infrastructure:** This refers to the communication system and includes all types of physical and virtual resources, such as the RAN, TN, and CN. It also includes computing resources, covering the cloud, edge, and deep edge. Infrastructure-related data are mainly generated within the infrastructure, including computing resources, communication resources (e.g., status of a network function), sensing information

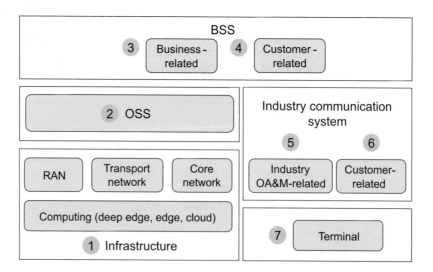

Figure 30.3 Main data source categories.

(e.g., from the RAN), and certain user profiles (e.g., mobility information, location, and associated context).

- **Operation supporting system (OSS):** This layer contains all OA&M-related data, such as physical equipment status, system operation information, and service provisioning information.
- **Business supporting system (BSS):** This layer contains all data related to business logic, such as information about customer and partner relationship management. More importantly, it also contains subscription data for both consumer and enterprise customers. Customers should have full ownership of such data and should therefore have full control over it.
- **Industry communication system:** When the industrial sector adopts 6G, the collected data may include information related to industrial use case OA&M. It may also contain data such as industrial user profiles (e.g., traffic patterns and mobility) and business/service data stored in the cloud. Industrial customers should have full ownership over this type of data.
- **Terminal:** Data from the terminal side includes terminals' computing and communication resources, service usage profiles, and sensing knowledge. End users should have full ownership over this type of data.

30.3.4 Data Collection

In 6G, one of the major roles of data governance is to provide suitable methods for creating data resources, which requires an appropriate architecture as well as network function support. The first step to achieving this is data collection. The key factors involved are as follows.

- Establish an agreement (e.g., data authorization) and a secure connection with the data source.
- Receive data collection requirements; determine what data to collect; and decide where, when, and how to collect the data according to the requirements.
- Notify the data source of the data attributes.
- Collect data from the data source and save it to the database.
- Operate and maintain the data in the database.

30.3.5 Data Analytics

Managing a data resource opens up the possibility of providing data analytics as a service for different types of customers. Four types of analytics can be applied.

- **Descriptive analytics** mines statistical information about historical data to provide network insights, such as network performance, traffic profile, channel conditions, and user perspectives.
- **Diagnostic analytics** enables autonomous detection of network faults and service impairments, identifies the root causes of network anomalies, and ultimately improves network reliability and security.

Table 30.1 Examples of multi-dimensional data services provided by the data plane.

Customer	Example services
Infrastructure network manager	Infrastructure network performance supervision, analysis, prediction, and assurance Infrastructure network resource utilization supervision and optimization Infrastructure network configuration, supervision, and optimization
Communication service manager	Service performance (e.g., QoS) supervision, analysis, prediction, and assurance Service charging optimization Service security supervision, analysis, prediction, and assurance
Device connectivity manager	Device location tracking and resolution Device activity status configuration and tracking Device mobility mode supervision, analysis, and prediction
Content forwarding manager	Data buffering and delivery scheme optimization
Third-party customer	User behavior analysis (e.g., degree of interest in specific services), user location and distribution, etc. Social context analysis (e.g., users' social neighborhood and sentiments, and users' surrounding location and mobility)

- **Predictive analytics** uses data to predict future events such as traffic patterns, user locations, user behavior and preference, resource availability, and even faults.
- **Prescriptive analytics** leverages predictive analytics to provide suggestions for resource allocation, content placement, and more.

Data analytic services will generate knowledge derived from the data plane. This may include proactive knowledge (e.g., recommendations on actions) and passive knowledge (e.g., information sharing and actions decided by customers).

Such data analytics can be requested by customers and tailored to meet their requirements. The data plane should provide on-demand and multi-dimensional service exposure and data exposure. Table 30.1 lists some examples of the types of services that could be provided to customers. It could be expected that the types of customers will be much more diverse than the examples listed in the table, and that their requirements and use cases for data analytics will differ.

30.3.6   Data Desensitization

Collecting and storing sensitive data involves privacy risks and the responsibility to protect privacy. Data desensitization is essential for handling privacy concerns as well as legal guidelines, especially for supporting AI and sensing tasks in the 6G design.

For AI tasks in particular, a cross-domain design could be considered. There has recently been a significant amount of research into differential privacy in the AI domain [5, 6], focusing on how to anonymize the training data of individual devices.

Data desensitization during model training and AI inference is an essential criterion in the 6G design. Approaches that can be taken to enable differential privacy include injecting noise into the training data without jeopardizing its statistical properties so that the trained model still captures features in the original dataset [7], and applying cryptographic techniques so that learning is based on encrypted (rather than decrypted) data [8]. An alternative approach is for devices to send model parameters rather than training data – two examples of this approach are federated learning [9] and split learning [10].

One key problem is that a malicious insider with complete knowledge of the learning mode can construct information similar to the training data by exploiting the gradual course of model convergence [11]. In federated learning, for example, this may result in information being leaked to malicious devices. Consequently, it is vitally important to consider how heterogeneous types of learning methods should be handled, as well as their limitations, without losing the generality of data desensitization.

References

[1] "Data governance," wikipedia. [Online]. Available: https://en.wikipedia.org/wiki/Data_governance#Data_governance_organizations

[2] European Parliament, "Regulation (eu) 2016/679 of the European Parliament and of the council of 27 April 2016 on the protection of natural persons with regard to the processing of personal data and on the free movement of such data, and repealing directive 95/46/ec (general data protection regulation)," *Official Journal of the European Union L*, vol. 119, pp. 1–88, 2016.

[3] "Measures for data security management (draft for comments)," May 2019. [Online]. Available: http://www.cac.gov.cn/2019-05/28/c_1124546022.htm

[4] "AB-375 Privacy: Personal information: businesses." [Online]. Available: https://leginfo.legislature.ca.gov/faces/billTextClient.xhtml?bill_id=201720180AB375

[5] P. Vepakomma, T. Swedish, R. Raskar, O. Gupta, and A. Dubey, "No peek: A survey of private distributed deep learning," *arXiv preprint arXiv:1812.03288*, 2018.

[6] C. Dwork, "Differential privacy: A survey of results," *Proc. International Conference on Theory and Applications of Models of Computation*. Springer, 2008, pp. 1–19.

[7] C. Dwork, F. McSherry, K. Nissim, and A. Smith, "Calibrating noise to sensitivity in private data analysis," in *Proc. Theory of Cryptography Conference*. Springer, 2006, pp. 265–284.

[8] M. Minelli, "Fully homomorphic encryption for machine learning," Ph.D. dissertation, 2018.

[9] B. McMahan, E. Moore, D. Ramage, S. Hampson, and B. A. Y. Arcas, "Communication-efficient learning of deep networks from decentralized data," in *Proc. Conference on Artificial Intelligence and Statistics*. PMLR, 2017, pp. 1273–1282.

[10] O. Gupta and R. Raskar, "Distributed learning of deep neural network over multiple agents," *Journal of Network and Computer Applications*, vol. 116, pp. 1–8, 2018.

[11] B. Hitaj, G. Ateniese, and F. Perez-Cruz, "Deep models under the gan: Information leakage from collaborative deep learning," in *Proc. 2017 ACM SIGSAC Conference on Computer and Communications Security*, 2017, pp. 603–618.

31 Multi-Player Ecosystem Architecture Technologies

31.1 Background

Prior to 5G, the earlier generations of mobile communication had a relatively closed ecosystem, where different business entities cooperated with each other only if necessary. For instance, roaming required the cooperation of multiple MNOs to address technical constraints. Such cooperation normally involved laborious closed negotiation of contracts that covered a wide range of technical, business, and legal aspects. The aim for 5G was to expand the ecosystem towards vertical industries. As such, it was decided to construct a system that appealed to different types of customers via network capability exposure and corresponding interfaces, for example, focusing exclusively on network functions.

6G is expected to be the next-generation digital platform, which will offer ubiquitous digital services. In order to unify the diverse range of services, it will be necessary to attract players from different domains, including the ICT industry as well as all other vertical industries. This unification process is already at play, albeit in its infancy, and will evolve in the future.

New players, such as verticals, along with conventional MNOs, will form a diverse 6G ecosystem. This ecosystem should be open – a fundamental architecture design requirement – in order to enable technical and business collaboration in the multi-player environment. Furthermore, such collaboration should be efficient, transparent, and trustworthy. An open ecosystem is also likely to drive new business models for multi-player collaboration.

6G should ensure openness for a wide variety of players, so that they are equally involved in certain types of network OA&M, and can generate business revenue. This means that the existing basic multi-player paradigm, which is based on contracts or network function interactions, will not be suitable for 6G. Instead, 6G will evolve to a deeper participatory paradigm, and it will be presented as the next-generation digital platform for the entire society. But in order to accomplish this, several key challenges that hinder the entire ICT industry must be addressed.

- **Natural boundary between systems:** A network system, by design, is currently a closed system that has a private database, control policy, and service operation logic. Taking user identity management as an example, user credentials are private data that service providers store in data silos. This impedes the ability to share information.

- **Expensive collaboration and interoperation:** Collaboration is possible only after both parties negotiate mutually agreeable terms and sign a contract, which is a laborious and time-consuming process. Even with signed contracts, however, many practical issues still arise due to unclear criteria and misunderstanding.
- **Lack of trust between participating players:** Although direct interaction is possible, potential risks are involved, meaning that participating players are reluctant to trust each other. As such, a mutually trusted third party is needed for coordination. The dependence on a third party incurs operation costs, degrades efficiency, and also grants excessive power to the third party.

6G will inherently be a multi-player network environment and business ecosystem. It will be necessary to ensure that interactions between different players are trustworthy and secure, and that secure alliances among different players are established and terminated flexibly. As such, overcoming the above challenges in the 6G architecture design is a fundamental requirement in shifting to a participatory paradigm and achieving a truly open ecosystem.

31.2   Design Considerations and Principles

New design principles will be needed to tackle new challenges that arise. It is therefore necessary to consider the following three design principles.

Principle 1 – Openness

6G should be more open in terms of sharing network information and knowledge, network operations, and capability exposure for multi-player collaboration.

As alluded to in Chapter 30, data isolation refers to current systems' segmenting data into isolated domains, where data are stored privately and not shared. Conversely, openness redefines how data (both private and public) from the mobile communication domain should be handled. Data should be stored and accessed by multiple players from different technical and business domains, facilitating the flow of information across different system boundaries.

Making network operation more open will ensure that the flow and treatment of information in one or more domains is more transparent. Today's network system is considered a black box, where external players have access to only output results. Such a closed system raises concerns about potential privacy and security issues. This not only undermines customer confidence in the network system, but also compromises the willingness of different players to participate due to a lack of information necessary for mutual interactions.

Openness preserves the boundary between two players' realms rather than eliminating it along with related control policies, but the way in which information is transferred will be redesigned. Additionally, openness creates mutual benefits via publicly available information without making the system fully transparent.

Principle 2 – High Interoperability

A high degree of interoperability among multiple players is needed in 6G. Although a mobile application may currently involve services provided by multiple service providers, interoperation is either cumbersome or limited due to insufficient openness.

In 6G, a higher degree of interoperation is needed in order to ensure E2E performance for most use cases in the future, as many of them will likely have stringent service requirements such as ultra-low latency. Such interoperability will require different players to work together in order to dynamically optimize multiple domains, meaning that cross-domain operations will become the norm.

The 6G network will reconsider the concept of interoperability. Interoperability in 6G will enable seamless cross-domain utilization under the direct control of different players, rather than today's implementation, which exposes interfaces for parameter exchange.

Principle 3 – Trustworthiness

6G will be a decentralized system with an open ecosystem where different players can join and leave freely. The trustworthiness of the 6G system has been elaborated in Chapter 29. From a multi-player perspective, a trustworthy 6G system should:

- provide a multi-player cross-domain data management framework, where fine-grained data access control can be implemented;
- be the trust anchor when providing information for multiple players; and
- enable auditable service provisioning, operations, and management, for instance, via system logs that must be authentic and undeniable.

These three design principles correlate with each other.

31.3   Architecture Features

This section focuses on one of the key technologies involved in creating a collaborative multi-player ecosystem – distributed ledger technology (DLT). DLT is expected to provide native support for the proposed design principles. Because the collaborative multi-player ecosystem will extend to cover both business- and technology-related concepts, it incorporates all relevant network features presented in this book, namely, network AI, user-centric network, native trustworthiness, and integrated non-terrestrial networks.

DLT delivers the key features necessary to rebuild the mobile network system on top of an open, interoperable, and trustworthy platform. Such a platform will attract players from various sectors, ranging from government and education to public health and service providers such as finance and transportation. This section also discusses the key network functions and entities that DLT impacts.

31.3.1   Distributed Ledger Technology

Over the past decade, blockchain technology has become incredibly popular due to cryptocurrency and the profound impact it has had on today's digital society. The core concepts behind blockchain technology emerged after the Paxos protocol [1], setting the stage for the development of Bitcoin. Bitcoin, released pseudonymously by Satoshi Nakamoto in 2008, is a peer-to-peer electronic cash system that aims to function as a public ledger for cryptocurrency transactions. The following year, the Bitcoin cryptocurrency blockchain network was established. By using a blockchain, Bitcoin is decentralized so that no single user can control the electronic cash and no single point of failure exists – this promotes the use of Bitcoin. Its primary benefit is enabling direct transactions between users without needing a trusted third party.

In addition to its use in digital currencies such as Bitcoin, since 2014 blockchain technology has been applied to other applications through smart contracts. Blockchain 1.0, used for cryptocurrency applications, has evolved to blockchain 2.0 and 3.0. The evolution of blockchain technology has enabled significantly different markets and opportunities, for example, decentralized applications (Dapps) were not possible with blockchain 1.0. Blockchain with smart contracts enables business logic and business process mechanisms to be built into the chain, and makes it possible to exchange common digital assets besides Bitcoins. The main benefits of blockchain technology were first realized in the digital economy industry. But as it evolved to become a general-purpose technology, blockchain offered a number of advantages in terms of handling actions transparently among various business entities. Such advantages could also be utilized in the design of 6G architecture, for example, to create an open and collaborative ecosystem and for security purposes.

Blockchain could be used as a digital system for recording asset transactions in multiple places at the same time [2, 3]. Unlike traditional databases, distributed ledgers have no centralized data store or administration functionality. Although the terms blockchain and DLT are often used interchangeably, they are distinctly different in that blockchain is merely a subset of distributed ledger technology.

The key features of DLT are as follows.

Distributed Consensus

DLT inherently employs a decentralized architecture, where multiple participants collectively form a distributed system. Distributed consensus is the core of a DLT-based system. In a distributed consensus procedure, distributed ledger records can transit from one state to the next without centralized control. Each state change is subject to a democratic competition procedure: after the majority of nodes reach agreement, the new state is accepted and globally synchronized on every node. This ensures that all nodes in the DLT-based system always have a global view of the system.

Immutability and Auditability

The distributed consensus procedure ensures immutability in the DLT-based system [4]. Because of the distributed consensus protocol (e.g., Proof-of-Work in

Bitcoin), it is computationally infeasible to forge ledger records unless a majority of nodes are compromised. Depending on which distributed protocol is used, the proportion of compromised nodes needed to make forgery computationally feasible ranges from 30% to 51% [5, 6].

The distributed consensus procedure also ensures auditability. The ledger contains all historical state records because each new record is appended to it – this is the only way to update the ledger. Due to the immutability feature, every node has a complete copy of the records, enabling local auditing.

Smart Contract

A smart contract is an executable binary code that applies predefined processing logic to the ledger records. One of the main differences between a smart contract and a normal application is that the execution of a smart contract is securely guaranteed and fully automated. In other words, if predefined conditions are met, the execution of a smart contract cannot be stopped.

In addition to defining the processing logic against data records stored in the ledger, a smart contract also exposes a set of *APIs*. Users can call these APIs to trigger the smart contract by sending a transaction to the smart contract. Once a smart contract is published in the DLT-based system, it is committed into the distributed ledger of every node in the system, after which it becomes public and immutable.

The key features of DLT can be leveraged to help achieve the three main principles in designing an open multi-player ecosystem, namely, openness, interoperability, and trustworthiness.

- **Openness** aims to provide an easy way to share data and information across multiple domains in 6G. The features of distributed consensus make it possible to share data and information globally without a centralized third party. At the same time, the immutability feature helps to ensure the integrity of shared data and information. Because any attempt to access data and information via DLT will be logged in the ledger, the auditability feature is central to realizing the openness principle.
- **Interoperability** aims to transform 6G into a multi-player participatory platform that enables the cooperative provisioning of services. The smart contract feature enables external players to access the data shared from another domain through specific APIs according to the conditions defined by the data owner. In addition, the triggered smart contract (APIs) will be automatically executed, which improves interoperability.
- **Trustworthiness** aims to make 6G a trust anchor of data storage, information sharing, and accountability. The immutability feature helps to improve the security and trustworthiness of data and information shared across multiple domains, while the distributed consensus feature enhances trustworthiness, as different players need to reach agreement over the shared data and information. In other words, system states are determined through a democratic competition procedure, which is securely guarded by benign participants.

In the 6G ecosystem, the mobile network is expected to act as a digital platform where multiple players from other domains cooperate with each other to provide services. This requires the new design principles described earlier to be implemented. DLT has a number of promising features that may help realize the proposed new design principles. Integrating DLT into the 6G network can help to achieve transparent and decentralized network functions, aligning with the new design principles. However, despite its numerous advantages, DLT also has limitations. For example, in cases where large amounts of data and information need to be shared, DLT may cause bottlenecks such as low throughput, high latency, and high power consumption. Although there is more scope for the technology to be improved, further commitment from researchers is required.

The following sections explore how DLT can transform the next-generation mobile communication system.

31.3.2   Multi-Player Platform

DLT can be used to create a new identity management scheme and digital asset management paradigm, where users control their own digital assets and the user scope ranges from consumers to enterprises. This will help to simplify network control and OA&M. Furthermore, DLT-based operation supporting systems (OSSs) and business supporting systems (BSSs) will be simplified, as no third-party coordinator is needed. Such capabilities are essential in a multi-player environment.

Figure 31.1 shows the basic design of DLT-integrated architecture for 6G. DLT is located at the foundation layer in the next-generation mobile communication system to support the open ecosystem. DLT is the basis on which a new identity management layer is built, providing a self-sovereign identity scheme so that authentication can be performed seamlessly across different domains. A data management and data access (DM/DA) layer is built for managing network and user data, supporting fine-grained data sharing while preserving native privacy across multiple domains. Based on these two layers, specific layers can be built for network control and the OSS/BSS.

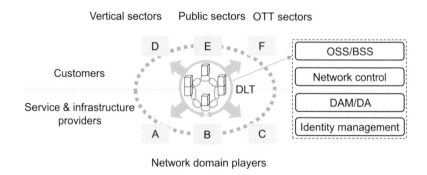

Figure 31.1 A general illustration of DLT-integrated architecture for 6G.

Other players from different domains can also join in the mobile network coalition by creating DLT-based shared data planes. Each player defines its own data sharing policies and exposes a smart contract to implement autonomous processing logic regarding the shared data in a distributed ledger or across different ledgers.

31.3.3   Identity Management

Users want to control their data and communicate anonymously or pseudonymously to protect their identity and privacy. This requires the use of advanced technologies such as biometrics and DLT in addition to cryptographic mechanisms.

In a conventional mobile communication system, a user requires a SIM card in order to use a mobile operator's network. When applying for a subscriber identity module (i.e., a SIM card), the user provides personal details such as a date of birth, gender, address, and billing method. This is equivalent to creating an account (i.e., user credentials) in the operator's database. Whenever the user accesses the network, the user's information is verified against that stored in the database. The user is granted network access only if the verification succeeds. Every operator stores such credentials in a central user data repository (UDR), which is isolated and belongs to only one domain; however, this is expected to change in the 6G era.

As shown in Figure 31.2, the purpose of using DLT is to realize a self-sovereign identity scheme [7]. Specifically, an authority (e.g., an operator) issues a certificate to a user after approving the user's profile. The certificate (UE-Cert) contains the user's public key and the authority's signature. The authority (e.g., from network domain A) publishes its own certificate (Domain-Cert), which contains the authority's public key to the ledger. This certificate is publicly accessible to everyone. In order for another authority (e.g., from network domain B) to perform authentication, it only needs to obtain the Domain-Cert issued by network domain A. If the signature of the

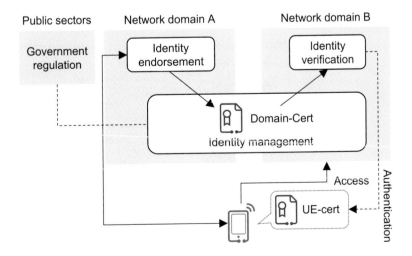

Figure 31.2 DLT-based identity management.

user's certificate can be verified with the retrieved public key, the user's certificate was indeed issued by its claimed originator authority and the user is therefore authenticated successfully.

A self-sovereign identity mechanism opens up new possibilities for identity management in the mobile network. First, the private key can be generated by the user rather than being created in the operator's database. Because only the private key can be used for decryption, the sovereignty of the identity is given back to the identity owner (i.e., the user). Second, the operator no longer functions as an identity creator, but rather as an identity endorser. This frees the operator from maintaining a large amount of sensitive user information, such as usernames and passwords. Third, any operator can verify an identity in a certificate by obtaining the corresponding public key, meaning that authentication is no longer restricted to a per-operator basis.

A DLT-based identity management scheme has the potential to create an open platform for identity endorsement. In some cases, it may be more advantageous for national- or city-level government departments to endorse identities, rather than (or in addition to) operators performing this role. The more trustworthy the endorser is, the more authoritative the identity system will be.

31.3.4   Data Management

DLT is a key technology in building a unified, trustworthy data plane for the 6G mobile network. With DLT, users can gain more control over their data instead of letting other players (such as OTT service providers) have exclusive control over it. In this way, all users will hold the keys to access their own data.

DLT-based data management and data access of users' digital assets enables users to control their own assets while also opening up an opportunity for a new business model. As shown in Figure 31.3, data can be easily and securely shared between different players [8]. Specifically, DLT can help in two ways: (1) establishing a distributed

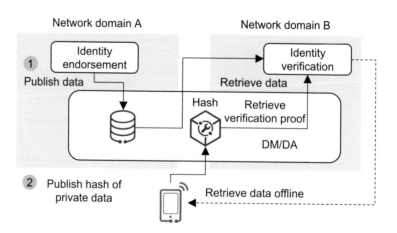

Figure 31.3 DLT-based data management.

consensus on published data, and (2) helping to directly share data offline while also guaranteeing the integrity of the original data.

To establish a distributed consensus on published data, data are directly published and validated by the entire network with a distributed consensus protocol. Every part of the network shares the same view of the ledger data, meaning that data are shared in the distributed consensus process. Due to the append-only manner, however, one challenge is that directly publishing data to the ledger causes a sustainability issue of the network data layer. Another challenge is that, because the published data are replicated on every node and publicly accessible, data confidentiality cannot be ensured. As a result, this approach is suitable for only a small amount of data that is relatively static and insensitive.

To directly share data offline while also guaranteeing the integrity of the original data, the verification information (typically the hash value) of the original data rather than the original data itself is published to the ledger. In order to verify data integrity, a recipient can refer to the published verification information. This is because the immutability feature of DLT guarantees that such proof originated from the data publisher. Furthermore, data are shared offline between the sender and recipient (e.g., network domain B receives data directly from the UE).

DLT-based DM provides a new scheme to manage and share data in the 6G mobile network. First, data do not need to be privately stored in one domain. Instead, the data can be encrypted and spread across the entire network, because DLT guarantees its integrity. Second, fine-grained access control can be implemented via smart contracts. Any request to access the published data by triggering a smart contract will be recorded, facilitating future auditing. This is especially important in regard to data that are used for collaborative control, such as network states, session context, and QoS requirements.

DLT-based DM may transform the data plane into a decentralized data marketplace, where digital assets are exchanged directly between the original asset owners and asset buyers. DLT will help ensure that data access, transfer, and auditing can be fully monitored in a decentralized manner.

31.3.5   Network Control

DLT can benefit two key aspects of network control, namely, session management and access and mobility control.

Session Management

In session management, a user plane forwarding path is established for a UE according to the UE's session request. At present, session management computes the forwarding path based on global information about the network layer or predefined simple forwarding rules. In 6G, however, session management may be handled by many distributed control entities across different domains. This may make it difficult to obtain global information, and may require multiple session control entities to share network information with each other regardless of whether they reside in the same domain.

DLT can help address this issue by creating a shared distributed ledger, on which different control entities can publish network information in their own domains. Publishing policies can be implemented to ensure that sensitive information is not exposed. Sessions can be established across multiple domains by utilizing smart contracts that are published by different session control entities. Smart contracts define the conditions and inputs necessary for a forwarding path to be created in a specific domain. After a smart contract is called, DLT guarantees its execution and a segment of a forwarding path is deployed accordingly. Whenever a smart contract is called, all operations are recorded by the ledger for auditability. These benefits make DLT an ideal choice for enabling data sharing across multiple domains.

Access and Mobility Control

Access and mobility control (session and service continuity) face a similar problem to that faced by session management when the number of control entities increases and the entities are distributed in the 6G mobile network.

- **Access control:** In current implementations of access control, a UE is authenticated by requesting an authentication service, which interacts with the UDR. In the UDR, a shared key challenge is generated by the authentication service for the UE. If the UE passes the challenge, access control grants the UE access permission. The mechanism of the shared key challenge is generated on the basis of the provided previously conceived UE identity, which is decrypted by the authentication service in order to reveal the UE's permanent identity.

  DLT can be used even if a traditional shared key authentication scheme is implemented. The shared key challenge can be prepared in advance according to the UE's permanent identity stored in the UDR. The authentication service prepares a verification smart contract and publishes it to a distributed ledger accessed by a set of relevant access control entities. Any access control entity can retrieve the prepared challenge and send it to a UE. The UE then solves the challenge and returns its solution to the access control entity, which submits the solution to the smart contract. If the smart contract confirms that the solution is correct, the UE authentication is successful. This eliminates the need for multiple control entities to perform sequential on-demand signaling, especially when a network function (NF) needs to be selected among multiple candidate NFs.

  This approach also works in a cross-domain scenario, where authentication services in different domains prepare the challenges and the corresponding verification smart contracts. The authentication process is the same as that in the single-domain scenario, without any inter-domain signaling.

  An alternative and better approach is to adopt an endorsement verification paradigm based on self-sovereign identity management. Assuming that a UE has an operator-authorized certificate, the operator publishes its certificate to a distributed ledger, which is propagated across different domains. In this case, the UE connects to an access control entity and requests network access, providing its certificate at the same time. The access control entity parses the domain identifier (i.e., who originally issued the UE's certificate) and retrieves the operator's certificate from the distributed ledger. The access control entity then verifies the UE's certificate. Authentication is completed if verification is successful.

Endorsement verification simplifies the authentication process significantly, although careful consideration must be given to how frequently a UE should apply for a new certificate. Of particular note is the fact that the authentication/authorization function and the operator who issued the certificate do not need to be in one domain.

- **Mobility control:** During a UE handover, a mobility control entity has to share UE context information and possibly RAN states with session management. This event triggers sequential interaction and signaling among multiple mobility control and session management entities. In a highly distributed environment, this approach is inefficient, especially if there are frequent handover events (due to dense cell deployment in the 6G network).

  DLT can be used to establish a proactive handover strategy, in which a local handover NF cluster is created so that UE context in the NF cluster can be synchronized in advance. This will shorten the handover time needed to send and create new UE context at the target NF (RAN). Efficient synchronization can be achieved because such clusters contain only a limited number of local UEs.

Many other mobile NFs share similar problems with session management and access and mobility control. Such NFs can also be transformed to work on the basis of DLT.

31.3.6 Operation and Business Support

In order to drive the digital transformation of society and industry in the future, the business environment and ecosystem need to be opened up further in terms of the development, deployment, operation, control, and management of wireless networks. A wide variety of players will be involved, playing different roles and working together to establish a joint ecosystem and develop new business models. This mainly influences the design of the conventional OSSs and BSSs in the 6G mobile communication system.

Multi-Player OSS

In the conventional OSS, information must be amassed in a central location before decisions and adjustments can be implemented on the network. Latency, network failures, asynchronous network states, and other reasons may influence the control decision. Even worse, different control entities may implement conflicting decisions, further compromising the effectiveness of network control.

The contradiction between the distributed nature of the network and a globally optimal operation decision is a key challenge that needs to be overcome. In order to make better decisions, the OSS needs to have more detailed information about its operated system. However, establishing a global view of the entire system is challenging (e.g., information needs to be up to date, and decision-making is complex). On the other hand, if we make a distributed decision, the information available locally may be insufficient (e.g., local decisions from isolated workers may conflict). Furthermore, signaling synchronization is always cumbersome and costly.

DLT, thanks to its distributed consensus mechanism, can be used to overcome the above contradiction. A DLT-enabled network data layer will allow collected network running states (or verification information if confidentiality is required) to be shared and synchronized directly from distributed NFs. More importantly, this data layer will verify published data, achieving distributed consensus for each piece of information. In this way, a DLT-enabled network data layer provides the OSS with a reliable data source that offers a global view of the network for many operation purposes, such as failure detection, data analysis (machine learning), and QoS monitoring.

More importantly, such a DLT-based OSS eliminates the boundaries between different players, where *information/states* of other domains can be shared smoothly so that every domain can have abundant, trustworthy, and synchronized information for their own operations.

Multi-Player BSS

In the conventional BSS, DLT is particularly useful because it enables a cross-domain pricing scheme and simplifies cross-domain settlement. Both pricing and settlement terms can be formed as smart contracts, which will completely replace paper-based contracts. The contractual terms can be agreed between two participants either online or offline, and once smart contracts are published to a distributed ledger, the full contracts are immutably committed.

In the pricing scheme, a smart contract is automatically executed and becomes effective immediately once a buyer triggers it, resulting in a predefined outcome. In the settlement scheme, relevant operations (e.g., billing calculation and transactions) will be executed automatically when the conditions are met, without any human intervention, according to the contractual terms. No third-party coordinator is needed throughout the entire process.

References

[1] L. Lamport, "The part-time parliament," *ACM Transactions on Computer Systems*, vol. 16, no. 2, pp. 133–169, 1998.

[2] M. Pilkington, "Blockchain technology: Principles and applications," in *Research handbook on digital transformations*. Edward Elgar Publishing, 2016.

[3] Z. Zheng, S. Xie, H. Dai, X. Chen, and H. Wang, "An overview of blockchain technology: Architecture, consensus, and future trends," in *Proc. 2017 IEEE International Congress on Big Data*. IEEE, 2017, pp. 557–564.

[4] F. Hofmann, S. Wurster, E. Ron, and M. Böhmecke-Schwafert, "The immutability concept of blockchains and benefits of early standardization," in *Proc. 2017 ITU Kaleidoscope: Challenges for a Data-Driven Society (ITU K)*. IEEE, 2017, pp. 1–8.

[5] M. Castro, B. Liskov *et al.*, "Practical byzantine fault tolerance," in *Proc. Third Symposium on Operating Systems Design and Implementation (OSDI)*, vol. 99, pp. 173–186, 1999.

[6] A. Gervais, G. O. Karame, K. Wüst, V. Glykantzis, H. Ritzdorf, and S. Capkun, "On the security and performance of proof of work blockchains," in *Proc. 2016 ACM SIGSAC Conference on Computer and Communications Security*, 2016, pp. 3–16.

[7] O. Jacobovitz, "Blockchain for identity management," Technical Report, The Lynne and William Frankel Center for Computer Science Department of Computer Science. Ben-Gurion University, Beer Sheva, 2016.

[8] Y. Zhu, Y. Qin, Z. Zhou, X. Song, G. Liu, and W. C.-C. Chu, "Digital asset management with distributed permission over blockchain and attribute-based access control," in *Proc. 2018 IEEE International Conference on Services Computing (SCC)*. IEEE, 2018, pp. 193–200.

32 Non-Terrestrial Network Integrated Architecture Technologies

32.1 Background

Overview

Within the industry, and in particular with regard to beyond-5G networks, there is renewed interest in integrating non-terrestrial network systems, such as satellites, HAPSs, and drones with land-based networks. Such integration aims to achieve a ubiquitous wireless network, covering presently uncovered areas and bringing true 3D coverage alive as described in Chapter 6. While wireless deployments have continued to grow with each successive generation of wireless technology, service gaps may still exist in some geographic areas, particularly where deploying a wireline-connected base station is either uneconomical or infeasible.

Although non-terrestrial networking technologies are often labeled as competing with their terrestrial counterparts, they can be seen as complementary in some network deployments, and should be considered on the basis of whether they are orbital or otherwise. For example, HAPS systems, drones, and balloons are non-orbiting platforms and, in terms of architecture, are generally consistent with existing network architectures. These platforms are specifically deployed to cover a particular geographic region. Orbiting platforms, on the other hand, are deployed in a constellation; consequently, they can provide ubiquitous coverage. In this chapter, we explore how to provide universal, ubiquitous coverage on a global scale, focusing specifically on satellite technology.

Research is currently ongoing in a number of areas related to satellite networking. This research will hopefully produce functionality specific to satellite networks, enabling network operators to create new service offerings and revenue streams. By leveraging the advantages of satellite networks, 6G capabilities will exceed many of those offered by other wireless and wireline technologies.

Satellite Networks

Satellite technology has been used for telecom and broadcast applications since the mid 1960s. This section focuses on satellite networks used for interactive services, such as voice and data. Broadcast applications (e.g., radio and television) are omitted, as they are predominantly offered over GEO satellites; likewise, special data relay networks for military and space exploration are omitted.

Reusable launch systems and satellite production are currently witnessing technical advancements that are driving down the cost for satellite deployment. It is expected that technical innovations and cost reductions will continue, and that low-cost satellite technology and related launch systems will be fully mature for 6G deployment.

Satellite technology has attracted more and more interest as an enabler for building high-performance networks. However, to minimize the amount of latency involved in satellite-based communication, LEO or VLEO satellites are required. But because these satellites orbit at a relatively low altitude, each one can provide only limited coverage. Consequently, larger networks or mega-constellations are required. Companies such as SpaceX and Amazon, to name just a few, are in the process of applying or acquiring licenses to deploy constellations that consist of multiple thousands of satellites for their Starlink [1, 2] or Kuiper networks [3].

Table 32.1 lists representative examples of satellite networks that are being planned, providing services, or currently being deployed. These networks focus on providing broadband services, aiming to offer enhanced performance compared with services provided over geosynchronous satellites.

In addition to providing select details about each satellite network, Table 32.1 lists the corresponding configuration. Note that Starlink plans to use approximately 40,000 satellites [1, 2] in both polar and inclined orbit configurations to form a complex

Table 32.1 Example satellite networks (the details provided in this table are correct as of September 2020, but are subject to change).

Satellite network	Status	Number of satellites	Altitude	Configuration
Irridium	Operational (voice and low-speed data)	66 (six orbits)	780 km	Polar (86 degrees)
Teledesic	Bankrupt	840 (initial)	700 km (initial)	Polar
		288 (revised plan)	1400 km (revised)	
SpaceX-Starlink	Currently being deployed (testing)	4408 Generation 1 Phase 1 (being deployed)	540–570 km 328–640 km	Polar and Inclined
One-web	Pending as of September 2020 (initially backed by SoftBank)	648 (planned, 74 deployed as of March 2020)	1200 km	Polar
Telesat-LEO-SAT	Currently being	292	1000 km (polar)	Polar and
	deployed (testing)	(117) minimum	1200 km (inclined)	Inclined
Amazon-Kuiper	In planning stage	3236	590 km, 610 km, 630 km	Inclined (multi-shell)
O3b/SES	"The Other 3 Billion," currently being planned	20	8062 km (medium Earth orbit)	Equatorial

Table 32.2 Details of SpaceX Starlink constellation (Generation 1, Phase 1) [1].

Group	Altitude	Inclination	Number of planes	Satellites per plane	Total satellites
1	550	53	72	22	1584
2	540	53.2	72	22	1584
3	570	70	36	20	720
4	560	97.6	6	58	348
5	560	97.6	4	43	172
Constellation total					4408

Table 32.3 Details of SpaceX Gen2 Starlink constellation (Generation 2) [2].

Group	Altitude	Inclination	Number of planes	Satellites per plane	Total satellites
1	328	30	1	7178	7178
2	334	40	1	7178	7178
3	345	53	1	7178	7178
4	360	96.9	40	50	2000
5	373	75	1	1998	1998
6	499	53	1	4000	4000
7	604	148	12	12	144
8	614	115.7	18	18	324
Constellation total					30000

Table 32.4 Details of Project Kuiper (Amazon) constellation [3].

Group	Altitude	Inclination	Number of planes	Satellites per plane	Total satellites
1	630	51.9	34	34	1156
2	610	42	36	36	1296
3	590	33	28	28	784
Constellation total					3236

constellation. Phase 1 is currently being deployed and includes only inclined orbit configurations (see Table 32.2). The second-generation network (see Table 32.3) plans to employ an additional 30,000 satellites, including highly inclined (e.g., near polar) orbit configurations. Conversely, Amazon plans to use only inclined orbit configurations (see Table 32.4), but will operate in three orbital shells.

32.2   Design Considerations and Principles

3GPP has shown significant interest in the use of satellites in mobile communication systems. For example, a general architecture and related issues are described in 3GPP TR38.811 [4, 5], while some potential solutions are documented in 3GPP TR38.821 [5]. The solutions currently address direct connection between UE and satellite, and connection to a remote radio access and core node, with varying levels of radio protocol integration directly within the satellites. While this is only one approach, it is applicable when satellites are involved in the transmission of wireless traffic to only the core network of the RAN.

The full value of satellite networks can be brought into play only when satellites support connectivity to the RAN, the core network, and even the Internet at large (Internet connectivity is often described as "broadband access" in this case). Such ubiquitous connectivity is necessary for satellite networks to become integral in the 6G system, rather than being sidelined as an independent system requiring interconnection.

32.2.1   Satellite Constellation

The satellite network is in continuous motion dictated by the fundamental laws of physics. A single satellite orbits the earth in an elliptical path, with the center of the earth as one focus. Most satellites used for communication purposes are deployed in a near-circular orbit. That is, the inter-foci distance is close to zero. A satellite position is generally defined as a box surrounding a nominal point in a circular orbit. The box provides limits on the eccentricity of the orbit. An orbit is typically characterized by the nominal altitude of the satellite and the inclination of the orbit. One that crosses directly over the poles is termed a polar orbit and has an inclination of 90 degrees; otherwise, it is termed an inclined orbit.

The inclination of the orbit defines the northern and southern extremes covered by the satellite. The inclination and the ground-facing antenna properties define the maximum and minimum latitudes that can be covered by a specific orbit.

A collection of satellites is termed a constellation. When multiple satellites form a network, they may travel along the same or different orbits.

The geometry of a constellation affects its ability to provide global coverage. For circular and near-circular orbits, the most important constellation geometries are the Polar and Walker–Delta constellations, as shown in Figure 32.1.

Constellations that consist of polar orbits will provide complete global coverage. However, while most satellites will travel in the same direction, there are two cases where satellites in one orbit will travel in an opposite direction to those in the adjacent orbit. In such cases, inter-satellite communication links are constantly being reconfigured. An area termed the orbital seam is formed and, because traffic intended to be passed over the seam has to be indirectly routed, this area has an impact on performance. Additionally, interface tracking is difficult as the satellites cross the poles, meaning that inter-satellite links (ISLs) between orbits are generally shut down over the poles.

Walker–Delta constellations are formed when multiple inclined orbits are used. Such constellations avoid the formation of a seam because the satellites remain relatively fixed between immediately adjacent orbits until the satellites reach the top of

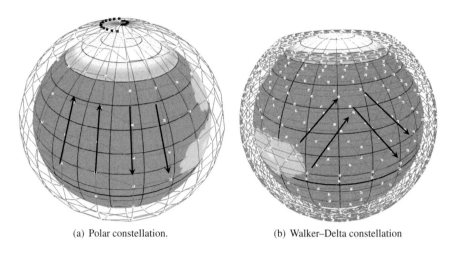

(a) Polar constellation.   (b) Walker–Delta constellation

Figure 32.1 The two most important constellations for circular and near-circular orbits.

their orbits (e.g., when a satellite traveling northward begins to travel southward). As satellites travel southward, they will cross other orbits, opening up the possibility for satellites to form an additional adjacency. In order to utilize this adjacency for data routing, an additional interface (called a "fifth link") is required with rapid tracking capabilities. However, it may not be possible to use the fifth link because it can result in seam-like behavior. (The fifth link refers to the link that connects satellites traveling in opposite directions to each other in the Walker–Delta constellation.)

Because coverage of a Walker–Delta constellation is limited by the inclination, it may be unsuitable when low-earth-orbit satellites are required to provide coverage to high latitudes. To overcome this limitation, companies such as SpaceX (Starlink) are deploying networks that combine the characteristics of polar and Walker–Delta constellations covering various altitudes to provide global coverage. For example, users in polar regions will access satellites in highly inclined orbits (e.g., near-polar orbits) in conjunction with terminals capable of operating at low elevation angles. Users in higher latitudes will also utilize similar terminals.

Figure 32.2 shows an example constellation whose density is similar to what SpaceX is planning. Although this example only shows a single orbital shell and omits the additional polar orbits providing coverage to the polar regions, it facilitates understanding the scope of these new networks.

32.2.2   Low Latency at Global Scale

In certain scenarios – especially over longer distances – a LEO satellite constellation can provide comparable or even lower delay than a ground-based fiber network. This is because radio waves and light travel faster in free space than light travels in a fiber.

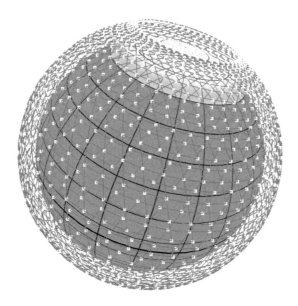

Figure 32.2 High-density constellation.

32.2.3   Connectivity Provisioning

To integrate with the 6G system, both traditional bent-pipe connectivity (where the satellite acts as a simple relay node) and the more advanced networked connectivity (where traffic is relayed between a source and destination at longer distances involving multiple satellites) could be used. This can be implemented by using ISLs or by configuring satellites to relay traffic among each other via ground stations.

By using a multi-hop satellite relay, we can cover users who would normally be outside the reach of the bent-pipe architecture (e.g., users on ships or in aircraft) as well as terrestrial users. This not only allows traditional services to reach almost all users worldwide, but also brings benefits for services such as global emergency notification and disaster response.

To achieve global coverage and low latency, constellations will require hundreds or even thousands of satellites – depending on the satellite altitude – and may include more complex multi-shell or space-relay networks. Consequently, it is expected that 6G will build upon this with critical new routing and addressing methods. This in turn will allow satellite constellations to function as a generic routing platform, capable of being deployed as a core technology that extends mobile communication system to the sky.

32.2.4   Multi-Service Capability

Satellite technology will extend coverage to areas underserved by existing wireless networks. A LEO satellite may cover a radius of about 300 km. But by using ISLs or ground-based relay stations, this coverage can be extended further, ultimately covering most of the globe. As a result, operators will be able to cover a much larger service area and realize new service opportunities.

In addition to acting as simple routers or relays in the sky, satellite networks are expected to carry different types of services. Such networks, while primarily targeted at expanding mobile communication systems, should allow multiple services to be carried on over the common but ubiquitous core. In order to achieve this, services need to be mapped at a common layer – typically the IP layer – before being tunneled over the satellite network.

32.3   Architecture Features

The integrated satellite network in the 6G system will differ from that of terrestrial networks, largely due to the continuous satellite motion. For this reason, it is essential that we know what the key architecture features are in order to realize the full 3D coverage vision. In that regard, the terrestrial mobile communication system could be truly extended to the sky, especially when a certain part of it is placed on satellite systems.

32.3.1   Latency

The use of LEO satellites makes it possible to achieve lower latency than that provided by existing fiber-based networks. In RF transmission, such as microwave, latency through the transmission system is approximated by the speed of light in a vacuum. For fiber-based systems, the index of refraction reduces the transmission speed to approximately 68% of the speed of light in a vacuum. Over long distances, this reduction can be significant compared to RF transmission, and, despite the longer path length over a LEO satellite, the satellite route can be faster. Point-to-point microwave systems typically provide low latency, but in many cases, terrestrial deployment is difficult (e.g., over mountains) or impossible (e.g., over oceans). Satellites may be used in these cases to provide optimum service quality in terms of low latency.

We need to consider two cases in particular. The first is the multi-hop case, which involves carrying traffic over multiple satellites and long distances (e.g., around the earth). In this case, multiple ISLs are concatenated. The second is the single-hop case, which involves the use of a single satellite to relay ground traffic. This is sometimes referred to as "bent-pipe" and also applicable to the HAPS configuration. The following analysis is based on propagation delay without including equipment delay and the assumption that the fiber path directly connects the two ground points without deviation. In reality, there will always be some deviation, resulting in a fiber path length that is longer than the point-to-point link, which in turn leads to a conservative analysis.

Multi-Hop Case

The general geometry of a satellite path compared to a ground-based path is shown in Figure 32.3(a). Here, the path includes an uplink and downlink connection, as well

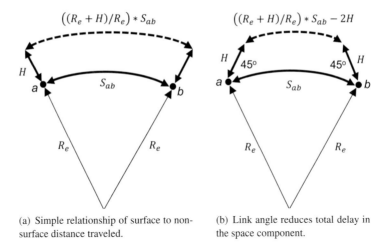

(a) Simple relationship of surface to non-surface distance traveled.

(b) Link angle reduces total delay in the space component.

Figure 32.3 Geometry of satellite link.

as multiple connections between satellites. For simplicity, the first and last satellites are shown immediately above the terrestrial users and the satellite path is shown as a curved path. In reality, the satellite may not be directly above the terrestrial users and the transmission path shown will be composed of many straight line segments between multiple satellites. The distance between two points a and b on the earth is usually calculated using the *haversine* formula, and indicated on the figure as S_{ab}. The distance of the satellite path is increased by a factor related to the altitude, as shown in Figure 32.3(a).

The main factors impacting the difference in propagation between the ground-based fiber and the space-based satellite link include reduced fiber speed due to the index of refraction in the fiber, and the increased length of the satellite path, which now also includes the two ground-satellite links. For the ground-space link, which is RF-based, and the ISLs, the speed propagation delay is given by the speed of light in a vacuum. We assume that the index of refraction in the fiber is 1.47; therefore, the propagation speed is approximately 68% of the speed of light. The path lengths of the ground signal between the two points a and b is usually calculated as a great-circle route between those two points on the surface of the earth, and this is shown as S_{ab} in Figure 32.3(a). The satellite path represents a series of point-to-point free-space optical links. Provided that there is a suitable satellite density, concatenated paths are approximated by a chord on a sphere with radius equal to the radius of the earth plus the satellite altitude. The total propagation delay of the satellite path can be calculated from this distance plus the distance of the two ground-space access links. Therefore, given any two points on the earth, a and b, and the satellite altitude, the propagation delay of the satellite path can be compared to a fiber-based terrestrial path. A useful measure is the altitude, which is equal for satellite or ground transmission for a given distance between the two points. This can, for example, help us determine whether

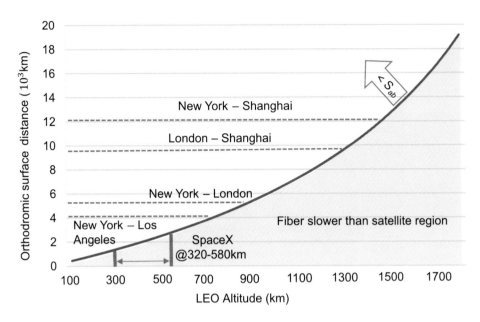

Figure 32.4 Effective region where LEO provides lower propagation delay, assuming straight paths and no per-hop equipment delays.

using satellites is beneficial. Figure 32.4 illustrates this with regard to addressing financial trading needs by understanding the benefits of reduced latency.

We can see that to be useful for financial trading between LA and New York, for example, the maximum orbit of the LEO satellite would need to be approximately 700 km, assuming zero delay-forwarding between satellites, which also are assumed to lie on a perfect orthodrome between LA and New York. We can also see that the upper bound occurs where the satellite is in an approximately 2000 km orbit, and the distance that must be covered to justify a satellite this high has to be nearly half the earth's circumference. The implication is that, to achieve a lower delay than fiber orbits, using LEO in the 600 km and below range will be necessary, and the assumption is that the paths are relatively straight (both on the ground and in space).

The preceding analysis represents scenarios where a satellite is directly above a ground station or terminal. It is worth noting that, in reality, the total path may be shorter than that shown in the illustration. The best-case scenario is when the satellite path is on the great-circle route between the ground stations or terminals, which have visibility to a satellite that is moderately inclined, as shown in Figure 32.3(b). However, this depends on the relative locations of the ground stations or terminals and the constellation density. Due to the motion of the satellites relative to the ground stations or terminals, this minimum delay will be a transient effect. The user is likely to experience propagation-delay variation with a variability period related to the altitude of the satellites. Nevertheless, this provides good estimates of the lower bound on latency.

Single-Hop Case

In the preceding case, we assumed that the satellite network is relaying traffic over multiple satellites. However, in certain scenarios, user traffic can be directly relayed to another ground station or terminal via a single hop, as opposed to being relayed over multiple satellites, and this is also referred to as a bent-pipe configuration. For the LEO constellation case, the bent-pipe configuration can play a large role in addressing traffic access at the edge of a network. It is worth noting, however, that the use of services over a GEO satellite for this configuration will result in excessive delay.

The propagation delay through a satellite in the single-hop case between two users can be via a ground-connected fiber or two ground-to-space links. For the purposes of this analysis, we are neglecting the effect of the earth's curvature, and assuming the terrestrial fiber connects the two users directly.

The propagation over the space portion will be continually varying due to the satellite motion. The minimum delay will occur when the satellite is equidistant from the users. We assume that the satellites will handoff to provide optimum network performance and, as such, the handover maximum occurs when the satellite is directly above one user. Note that both the maximum and minimum propagation delays will also vary with the satellite altitude.

Figure 32.5(a) is an example of both the minimum and maximum propagation delay for a LEO satellite operating at an altitude of 395 km. For short distances, we can see that the terrestrial fiber path provides lower delay. Note that fiber delay increases at a faster rate than the delay of the space portion, owing to the different speed of light in fiber. For the minimum delay described above, we see that at a point-to-point fiber distance of approximately 750 km, the delay over satellite is better than that provided by fiber. At 750 km, there will still be cases where the delay of the satellite exceeds that of the ground fiber due to satellite motion. That said, we can observe that the maximum delay crosses the terrestrial fiber delay when the fiber distance is approximately 1000 km.

A similar relationship can be seen for HAPSs that operate at a lower altitude. For example, as shown in Figure 32.5(b), for a HAPS operating at an altitude of 20 km, point-to-point propagation delay exceeds what fiber can provide for distances of approximately 50 km. Again, we assume that fiber paths are line-of-sight distances. In reality, fiber lengths will be longer, which in turn reduces the distance between users, while satellite provides better performance over fiber.

For both the HAPS and satellite cases, the relationship where the minimum and maximum delays exceed the fiber delay is fixed. This can be seen in Figure 32.6, which covers the regions used for both HAPS and LEO satellites. The region indicated by the dotted line is the area between conventional and orbital flight, and HAPS typically operates from 20 km to 25 km. On the other hand, while the specific boundary is not defined, LEO typically operates in the region above 300 km.

Even when the satellite or HAPS performance is only on par with fiber-optics cable, the advantages of HAPSs and satellites lie in the ubiquitous coverage they provide. As such, fiber deployment is not needed, which means services can be deployed more rapidly.

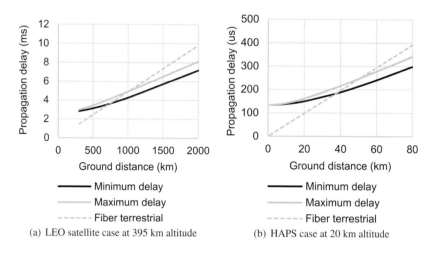

(a) LEO satellite case at 395 km altitude   (b) HAPS case at 20 km altitude

Figure 32.5 Propagation delay for the LEO satellite and HAPS cases.

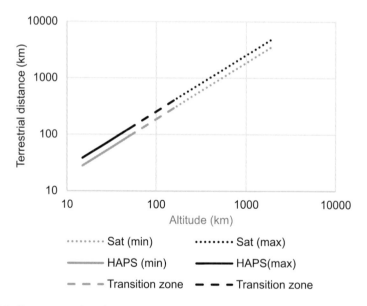

Figure 32.6 Cross-over point where HAPS or satellite delay is equivalent to point-to-point terrestrial fiber.

Short Summary

The overall delay for HAPSs and satellites, including both the single-hop and multi-hop cases, shows that latency through the non-terrestrial portion of the network can provide better latency performance than that achieved by the current terrestrial fiber. In all cases, the advantage gained is dependent on the altitude of the non-terrestrial network (NTN) platform and the distance between the two communicating ground terminals. For satellites, there will be some variation in performance due to satellite motion, which will bound the worst-case latency that a user experiences.

With regard to the single-hop case, for the typical LEO altitudes of approximately 400 km, performance begins to equal that of terrestrial fiber when the distance between users is approximately 750 km. At distances of about 1000 km, the satellite starts to outperform a direct fiber link between two users. This gain continues to increase and will only be limited by the radio coverage area, which depends on the altitude. It is important to take into account the fact that terrestrial users rarely have direct fiber links, and, in most cases, the fiber path exceeds the geographic distance between users. Terrestrial connections are also expected to include multiple service nodes between geographically distant users. A case in point is broadband services, where users can expect multiple routing elements to be present, each adding further latency. This is network-specific, but typically reduces the distance at which NTN provides superior latency performance over fiber.

For the multi-hop satellite case, the performance gains are similar to the preceding observations. The main difference lies in the fact that the NTN portion is only limited by the constellation design, as opposed to the coverage area of one satellite. Assuming proper network element design, the space portion of the NTN can be designed to constrain per-node latency. In terms of individual user performance, satellite motion results in some variation of latency. In the single-hop case, the variation is due to the motion of the single satellite relative to the users. However, in the multi-hop case, each user accesses a different satellite, and each satellite may contribute to additional path delay. This can ultimately result in a wider range of latency variation, but since multiple satellites are used the equivalent terrestrial delay would be substantial and gains can still be achieved with NTN multi-hop satellite.

The performance gains provided by NTN are achieved through the use of current fiber technology. It is worth mentioning that the industry has been conducting research on hollow-core fiber development, which will reduce the delay of terrestrial fibers. Even though we are still in the early stages, if fiber systems are developed using hollow-core fibers, the resulting latency advantage provided by NTN will reduce substantially. That said, this is not seen as an impediment to NTN, which mainly aims to provide benefits in terms of ubiquitous coverage. It is also worth pointing out that our analysis only addresses latency.

While HAPSs also exhibit delay variation and a similar performance as LEO satellites, they have a much lower altitude, and provide better performance at a subsequently shorter point. Since HAPSs are stationary, they deliver performance that is equivalent to that provided by fibers for users between 35 km and 50 km apart. The fact that HAPs are stationary (even considering minor motion variation) also leads to a fixed path length between two connected users, who do not witness a significant variation in latency. This performance is based on the assumption that the HAPS is providing simple connectivity between two users. As in the satellite cases above, we have not factored in the equipment processing delay. As such, if service processing can be provided on the HAPS platform rather than being backhauled to a terrestrial location, further advantages can be achieved.

Although both HAPSs and satellites can provide better latency, we also need to consider other factors that impact delay on a per-network basis. These include the elevation angle, constellation density, constellation type, and routing mechanisms. Performance (latency) can be improved if the ground station (or user) can see multiple satellites and subsequently select the one that minimizes delay. This may have a bearing on which RF interface is used.

32.3.2   Connectivity Models

Services carried by satellite networks in the 6G system are expected to provide ubiquitous connectivity to ground-based users. Seamless integration with the ground-based networks to provide ubiquitous access services is paramount. These services thus will include both direct services (e.g. direct broadband over dedicated satellite receivers) as well as indirect services such as mobile backhaul or other relay type services (e.g., data services in commercial aircraft). 6G will also require capability for direct user-to-user and machine communications, including vehicle-to-vehicle (i.e., V2X). To handle the diversified use cases mentioned in Chapter 6, different types of architecture support could be provisioned.

In order to provide 3D coverage, 6G ground-based stations may require new capabilities in order to provide different integration schemes with satellite networks. For instance, satellite networks can act as backhaul for the extension of terrestrial wireless service, as shown in Figure 32.7. Satellite networks can also serve as direct connections with 6G terminals, as shown in Figure 32.8. This covers the cases for both fixed and mobile services. In Figure 32.8, the satellite network has a direct connectivity to a UE. The satellite network carries both user data and control in the same tunnel. It is possible that processing can be performed in the satellite network directly. This is a form of processed payload. In this case, the satellite network may provide separate tunnels.

The mobile case can also be extended to allow the aggregation of multiple users by using a device with one satellite channel. Since each user may transfer data to different

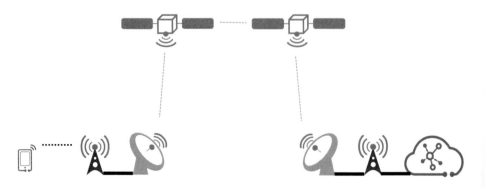

Figure 32.7 Satellite carrying a mobile service (service extension case).

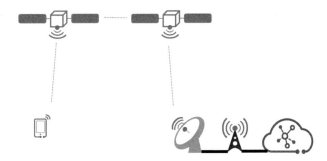

Figure 32.8 Satellite carrying a mobile service (direct connection case).

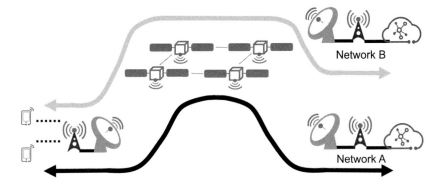

Figure 32.9 Aggregated mobile services on the ground.

locations, the device should participate in the satellite routing system to process traffic accordingly (see Figure 32.9).

For simplicity, the architecture of Figure 32.9 shows processing on the ground. In some cases, this may also be delivered by a stationary airborne vehicle (e.g., HAPS) with backhaul provided by the satellite network. From the perspective of the satellite network architecture, the HAPS and ground aggregation terminals are the same.

Three-dimensional coverage is especially essential for the IoT technology when it is applied to remote areas with limited or no ground base station deployment (e.g., sensors used to monitor forest fires and climate change). Satellite networks could support IoT or sensor network connectivity using the model shown in Figure 32.7. Direct connectivity similar to Figure 32.7 is also possible with IoT and sensor networks over satellites. The IoT sensor devices are anticipated to consume a low amount of power and may require special support from the satellite network, for instance, special control commands to power on devices.

32.3.3 Routing in Space

Although the routing protocols can appear similar to the protocols used in terrestrial networks, the differences are critical and therefore the satellite network should be treated as a separate layer in the entire 6G system. As is the case for all layer networks, the association between the networks forms a client–server relationship. In this case, the satellite acts as a server layer to the data provided by the client, either a terrestrial network or a direct user. The methods for carrying data across a network vary according to the specific technology employed in the client network.

The satellite network aims to provide the ability to carry data in packet form. Since this network will mostly connect with IP-based networks, the satellite network architecture will carry client signals using a tunnel model similar to the one used in terrestrial data networks. User connections will therefore represent tunnels between user access points to the satellite networks (or within the satellites themselves) and terrestrial gateways or other users as appropriate. Since it is not possible to support the Internet or even the service provider routing state, the tunnel model is required to encapsulate and carry the user data with addresses that are appropriate for the satellite network.

In terrestrial networks, the routing system routes the packet to the final destination, which is typically close to a specific gateway. Routing to this gateway may be considered optimal if based on the longest prefix match. However, in certain satellite network scenarios (e.g., those due to power issues), it may be more appropriate to route to an alternative intermediate gateway and then perform final routing to the destination over the terrestrial network, as shown in Figure 32.10.

The routing system will need to accommodate the preceding function, and, as such, routing in space mechanisms may need further improvement for 6G. For instance, in terrestrial networks and existing routing protocols, source and destination addresses are subnet-based and do not include direct information regarding the geographic location of devices. As this fails to address the mobility of network elements, the new architecture design should consider mobility support through a number of mechanisms. We therefore need to conduct research on how the identifier and location of each end station coupled with geographic routing can address mobility within this context.

In 6G satellite networks, the addressing scheme could be based on geographic addresses, which means the source and destination are based on corresponding geo-

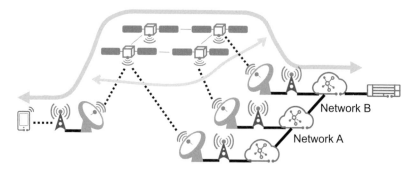

Figure 32.10 Data delivered to the closest ground station (when multiple path options may exist).

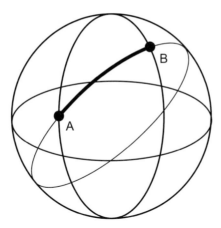

Figure 32.11 Orthodrome relative to the great circle that includes it.

graphic locations, as opposed to IP addresses. If the route could be determined through the geographic distance between the destination and many known intermediate satellites, it would become conceptually easier to determine routing costs. The shortest path between two points on the globe is referred to as a great-circle route, or orthodrome, as shown in Figure 32.11. Therefore, routing based on minimizing this distance is referred to as orthodromic routing, and further research on this subject may be worthwhile.

32.3.4   Operations, Administration, and Maintenance

An operational model that considers the movement of satellites over different administrative regions or countries does not exist in terrestrial networks and may require accommodation by various OA&M and routing systems.

In light of this, the communication scheme for the 6G-integrated satellite network OA&M should be considered in the corresponding design scope, which in turn will enable ground control centers to manage any satellite in the constellation, while those satellites send OA&M updates and responses accordingly. In other words, requirements will not be met by OA&M that is only possible when a satellite is over or within the range of a specific ground station.

This means that a packet network must also allow OA&M data packets to be forwarded or terminated and originated within the satellites' OA&M control software. We can achieve this through a wide variety of methods, among which the most common method for terrestrial packet networks involves carrying both control and data planes over common inter-router links. For satellite OA&M networking, the most cost-effective method will likely include a common routing control plane but separated data planes, and possibly a pre-emptive QoS controlled sub-channel on the ISLs. For reasons previously discussed, it is not possible to use existing OA&M control planes such as OSPF/ISIS or existing OA&M addressing.

Moreover, it is also important to define the interface between the satellite OA&M system and the terrestrial OA&M system if they belong to different business entities, and this is imperative to achieving end-to-end service operation and management. As this may be similar to the multi-player ecosystem scenarios, reference can be made to Chapter 31 for the corresponding research scope.

Due to the complex nature of satellite constellations, machine learning will likely be beneficial to the operation and administration of satellite-based networks.

References

[1] "Space exploration holdings, LLC, SpaceX non-geostationary satellite system, attachment A, technical information to supplement schedule S." [Online]. Available: https://fcc.report/IBFS/SAT-MOD-20181108-00083/1569860

[2] "Space exploration holdings, LLC, SpaceX non-geostationary satellite system, attachment A, technical information to supplement schedule S." [Online]. Available: https://fcc.report/IBFS/SAT-LOA-20200526-00055/2378671

[3] "Kuiper systems LLC, technical appendix, application of Kuiper systems LLC for authority to launch and operate a non-geostationary satellite orbit system in Ka-band frequencies." [Online]. Available: https://fcc.report/IBFS/SAT-LOA-20190704-00057/1773885

[4] 3GPP, "Study on new radio (nr) to support non-terrestrial networks," 3rd Generation Partnership Project (3GPP), Technical Report (TR) 38.811, July 2020, version 15.3.0. [Online]. Available: https://portal.3gpp.org/desktopmodules/Specifications/SpecificationDetails.aspx?specificationId=3234

[5] 3GPP, "Solutions for NR to support non-terrestrial networks (ntn)," 3rd Generation Partnership Project (3GPP), Technical Report (TR) 38.821, Dec. 2019, version 16.0.0. [Online]. Available: https://portal.3gpp.org/desktopmodules/Specifications/SpecificationDetails.aspx?specificationId=3525

Summary of Part VI

In Part VI we explored six key architectural aspects that together may define the structure of the 6G network. 6G will truly embrace and integrate communication, computing, and storage. The mobile communication system should not just focus on sending data to the right place for processing. Instead, the mobile communication system itself is the optimal place to process relevant data given its ubiquitous connectivity and computing resources at the edge. However, this requires deep convergence of all available resources from all stakeholders in order to create a controllable open environment. This is necessary if we wish to establish universal and ultra-high performance ICT infrastructure.

As the structured and layered network architecture becomes more distributed, it will eventually lead to a new type of architecture that leverages deep intelligence at the edge. It will be self-contained and self-evolved but apply to large-scale networking. Delivering ultra-high performance, 6G will be more than just wireless connectivity. It will involve communication elements – such as base stations and terminals – that are capable of performing universal sensing. These features will lay a solid foundation for connecting all intelligent devices via an intelligent and trustworthy platform. In this way, 6G will become the foundation of all vertical infrastructure, redefining the boundaries between terminals, connectivity, and the cloud.

In addition, it is important to work together with vertical industry customers on 6G from the very beginning. Our experience in the development of 5G industries shows that 5GAA (for the automotive industry) and 5G-ACIA (for the manufacturing industry) play essential roles in advocating 5G adoption and penetrating potential new markets. Early feedback from potential customers – not only from the business aspect, but also from the technology aspect – is critical to determine the direction of future 5G releases. The lessons learned in 5G will be adopted for 6G industry development. Suitable industry platforms should be formed at the right time with the right partners for the right purposes.

Part VII

Summary and Future Work

33 6G Ecosystem and Roadmap

33.1 6G Initiatives and Ecosystem

In 2019, we witnessed 5G becoming reality as global rollout began. While continued efforts aim to achieve better service quality and more mature applications in vertical industries by leveraging the NR standardization framework, 6G is on the horizon, aiming to solve many more challenges that society faces by delivering a wealth of new potentials. Over the coming decade, we expect the industry to continue technological research, ecosystem alignment, and global standardization related to 6G.

Since 2018, a number of initiatives into 6G research have been launched. For example, industry and academic circles in Europe, China, Japan, South Korea, and the USA have been engaged in identifying the typical application scenarios, key capabilities, and potential technologies for the next-generation wireless network, which is expected to be commercialized in the 2030s.

33.1.1 ITU-R Initiatives

For the last 30 years, the radio communication division of the International Telecommunication Union, known as ITU-R, has been coordinating government and industry efforts in developing a global broadband multimedia IMT system. It has successfully led the way in developing IMT-2000 (3G) [1], IMT-Advanced (4G) [2], and IMT-2020 (5G) [3], and is now initiating a new cycle towards 2030 and beyond.

In February 2020, ITU-R WP 5D agreed on the work plan for a preliminary draft Report ITU-R M.[IMT.FUTURE TECHNOLOGY TRENDS], focusing on radio technologies. This plan is expected to be completed in June 2022 and will provide a broad technical overview of terrestrial IMT systems emerging in 2030 and beyond. The ITU-R is also expected to launch a vision study for next-generation IMT technologies around June 2021 and complete it before the World Radiocommunication Conference in 2023 (WRC-23).

33.1.2 Regional Activities

Europe: Framework Programme and Horizon Europe
Framework Programmes (FPs) are funding projects created by the European Union (EU) and European Commission to support and promote research, and have been

providing financial support for implementing European research and innovation policies since 1984. For example, the WINNER [4] project in FP6 and its follow-up defined the bases for 4G LTE technology innovation, while the METIS [5] project in FP7 and its follow-up laid a solid foundation for the success of 5G technologies. Horizon 2020 is the eighth FP, running from 2014 to 2020 and focusing on innovative science, industrial leadership, and tackling societal challenges. In terms of research, Horizon Europe will be the new research and innovation FP, establishing a new partnership between the EU, industry, small and medium-sized enterprises (SMEs), and research institutes.

In May 2020, NetWorld2020 – the European Technology Platform for communications networks and services – published a white paper [6] on the European Union Strategic Research and Innovation Agenda (SRIA) for 2021–2027. The white paper was delivered to the European Commission as part of preparations for establishing the FP Horizon Europe. Created by more than 130 authors from industry and academic circles around the world, the white paper offers general guidance and discusses various aspects of communications towards 2030 at great length. It covers: global megatrends; policy frameworks, and KPIs towards 2030; discussions about human-centric and vertical services; technological aspects of system architecture, edge computing, radio technology, optical network, security, satellite, and devices; and emerging trends.

China: 6G Initiatives

In 2019, the China IMT-2030 Promotion Group was established [7]. Several working groups have been formed according to various aspects of 6G research for mobile communications towards 2030 and beyond, covering 6G requirements and vision, spectrum, potential radio access technologies, networking technologies, and so on. The IMT-2030 Promotion Group brings together all those involved in exploring 6G study, including universities, research institutes, operators, vendors, and verticals.

In November 2019, China's Ministry of Science and Technology together with five other ministries jointly held a kick-off meeting to officially announce the start of research and development into 6G [8]. At the meeting, a national 6G technology R&D working group and an overall expert group were established – these groups will work together with the ministries to study a strategy for 6G R&D. The China Communications Standards Association (CCSA) TC5 WG6 [9] is responsible for studying advanced wireless technology standards. It is currently working on the report "Research on vision and requirement of the B5G generation mobile communication systems."

Japan: Beyond 5G Promotion Strategy

Japan's Ministry of Internal Affairs and Communications (MIC) held a "Beyond 5G Promotion Strategy Roundtable" [10] in January 2020. The roundtable aims to formulate a comprehensive strategy to deal with the needs and technological progress expected when Beyond 5G – the next generation to 5G – is introduced.

MIC also announced the outline of its Beyond 5G strategy [11] in April 2020 and started a public consultation. Two months later, in June 2020, MIC published a "Beyond 5G Promotion Strategy-Roadmap towards 6G" [12].

South Korea: R&D Strategy for Future Mobile Communications

In August 2020, South Korea's Ministry of Science and ICT (MSIT) published a report titled "The research and development strategy for future mobile communications of 6G" [13]. The report states that MSIT will invest 200 billion won (South Korea currency unit) over five years, starting in 2021, to encourage technological innovations in 6G as well as promoting international cooperation, strengthening the industry ecosystem, and leading the 6G global market.

USA: Promoting 6G Leadership

In March 2019, the Federal Communications Commission (FCC) opened the spectrum range from 95 GHz to 3 THz for experimental purposes [14], paving the way for research and testing of beyond-5G wireless technologies. The spectrum range opened up by the FCC has significant potential to achieve extremely wide bandwidth wireless communications in the future.

In May 2020, the Alliance for Telecommunications Industry Solutions (ATIS) announced a call "Promoting US leadership on the path to 6G" [15]. ATIS explained that 6G development had already begun, and shared its vision for collaboration across government, academic, and industry sectors to ensure that the USA remains at the forefront of technology leadership for the next decade.

33.1.3   Views from Industry and Academia

Looking ahead to the next 10 years, industry and academic circles have presented and exchanged their views at IEEE conferences and via numerous white papers.

China Mobile published the white paper "The vision and requirements for 2030+" in November 2019. The paper provides the initial KPIs for 6G, and describes the five main characteristics of 6G networks, namely, "on demand service, extreme simplicity, soft network, endogenous intelligence, and endogenous security" [16].

NTT DOCOMO published the white paper "5G Evolution and 6G" in January 2020. The paper discusses how mobile communication technologies will evolve towards 6G as well as describing the related requirements, use cases, and potential technologies. 6G was considered to expand "The Third Wave: Resolution of social issues; Human-centered value creation," which was initiated by 5G in the 20-year cycle of changing services of mobile communications [17].

Samsung published its white paper on 6G in July 2020. Its vision of 6G is to "bring the next hyper-connected experience to every corner of life." The white paper discusses 6G at great length, covering related megatrends, services, requirements, and candidate technologies, and predicts the timeline for completing 6G standardization and commercialization [18].

6G is currently a hot topic in academia, where it has been the subject of extensive research since 2019. It has featured prominently in the agenda of almost all relevant conferences, seminars, and workshops, including the International Conference on Acoustics, Speech, & Signal Processing (ICASSP) 2020, the IEEE Wireless Communications and Networking Conference (WCNC) 2020, the European Conference on

Networks and Communications (EuCNC) 2020, the IEEE International Conference on Communications (ICC) 2020, the IEEE International Symposium on Personal, Indoor, and Mobile Radio Communications (PIMRC) 2020, the IEEE Vehicular Technology Conference (VTC-Fall) 2020, and the IEEE Global Communications Conference (GLOBECOM) 2020. For example, at ICASSP in May 2020, 6G was extensively discussed in keynotes [19] and also the industry panel themed on "The B5G/6G challenge – a race of enabling technologies and design principles" [20]. Furthermore, numerous research papers have been published in various subject areas, covering predictions and requirements to use cases and potential technologies.

6Genesis is one of the flagship initiatives in the Academy of Finland's national research funding program led by the University of Oulu. Starting in 2018, it is an eight-year research initiative to develop, implement, and test key enabling technologies for 6G [21]. The 6Genesis Flagship program sponsors the 6G Wireless Summit, which is an annual conference hosted by the University of Oulu. The first summit was held during March 24–26, 2019, at Levi in Finnish Lapland, where it released its first 6G white paper titled "Key drivers and research challenges for 6G ubiquitous wireless intelligence" [22]. The second summit was held as a virtual event in February 2020 due to the outbreak of COVID-19. In total, the 6G Flagship program has published 11 new 6G white papers [23].

Another paper of note is "Communications in the 6G era," published by researchers at Nokia Bell Laboratories in March 2020 [24]. The paper discusses various aspects of 6G, including new use cases, key requirements and performance indicators, and the fundamental technology transformations most likely to be defined for 6G systems. Other noteworthy discussions about 6G include the following.

- [25] identifies the key drivers of 6G, including multi-sensory XR applications, connected robotics and autonomous systems, wireless brain–computer interactions, and blockchain and distributed ledger technologies. It also provides five recommendations in terms of the trends, challenges, and associated research in 6G.
- [26] depicts 6G as a multi-dimensional and autonomous network architecture that integrates space, air, ground, and underwater networks to provide full coverage and unlimited wireless connectivity. It identifies several promising technologies for the 6G ecosystem, including terahertz communication, ultra-massive MIMO, large intelligent surfaces and holographic beamforming, orbital angular momentum multiplexing, laser and visible-light communication, blockchain-based spectrum sharing, quantum communications and computing, molecular communications, and the Internet of Nano-Things.
- [27] presents a range of guidelines for future development, with a diagram depicting the time, frequency, and space resource utilization of 6G. The authors analyzed various research findings related to 6G design, including multi-band ultra-fast transmission techniques, super-flexible integrated network designs, and multi-mode multi-domain joint transmission, as well as machine learning and big-data-assisted intelligent approaches. The authors also explored three open issues – power supply, network security, and hardware design – and proposed potential solutions.

- [28] summarizes the main challenges and potential of each enabling technology, and associates each of them with specific use cases. It identifies a number of 6G use cases, including VR, holographic telepresence (teleportation), eHealth, pervasive connectivity, Industry 4.0 and robotics, and unmanned mobility, and discusses the related enabling technologies. The analysis suggests that, in order to meet requirements, novel disruptive communication technologies, innovative network architectures, and integration of intelligence in the network will be needed.
- [29] provides insight into 6G communications from a human-centric perspective. It considers that 6G should be human-centric in addition to extending 5G in terms of higher bandwidth and lower latency.

There are also many insightful contributions that focus on the application scenarios of 6G, such as tactile Internet, above-100 GHz spectrum, AI-enabled communications, and the related potential technologies. These contributions include the following.

- [30] proposes that sub-terahertz and VLC technologies are powerful enablers of 6G.
- [31] outlines promising research that involves massive MIMO, digital beamforming, and antenna arrays.
- [32] introduces the concept of a semantic-effectiveness (SE) plane as a core part of future communication architectures.
- [33] presents an overarching vision for machine type communication in 6G, including relevant performance indicators and key enabling technologies.
- [34] delves into the fundamental opportunities, challenges, and approaches associated with creating future wireless, sensing, and position location systems for spectrum bands above 100 GHz.

AI/ML for wireless has the potential to redefine wireless communication systems and is an attractive topic in 6G research. The article [35] provides a comprehensive tutorial on the use of machine learning, with a particular focus on enabling a variety of applications in future wireless networks. It explores a number of emerging applications, such as unmanned aerial vehicles (UAVs), wireless VR, mobile edge caching and computing, IoT, and spectrum management. It also discusses how AI/ML can be used to address a variety of wireless communication problems. In [36], the authors posit that 6G will go beyond mobile Internet – 6G will be required to support ubiquitous AI services, spanning from the core network to connected user devices. Furthermore, they predict that AI will play a critical role in designing and optimizing 6G architectures, protocols, and operations.

33.2   Roadmap to 2030

While 5G is continuously evolving and new releases emerge, ITU-R WP 5D has launched a research project to study the technology trends of IMT systems up to

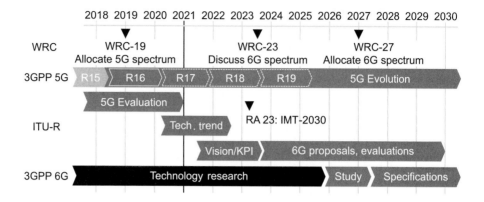

Figure 33.1 Timeline for 6G research and standardization in ITU-R and 3GPP.

2030 and beyond. A preliminary timetable for 6G research expects this study to be completed in June 2022, and indicates that another project centering on the future vision of IMT systems will start in mid-2021 with an expected completion date of mid-2023, before WRC-23. This vision report will provide recommendations for the use cases, KPIs, and spectrum bands of 6G.

Once the vision report has been published, it is expected that research on 6G technologies will continue, because in-depth research into the business values and models highlighting the unique competitive advantages of 6G is as important to the overall ecosystem as discussions about the enabling technologies. Learning from the experience gained in 5G, we can see that different vertical industries have very different requirements while one vertical industry spanning multiple regions has extremely diverse requirements. It takes time to fully analyze how these requirements will affect the design of wireless communication systems, so 3GPP may start an overall study into the new radio interfaces and core networks around the end of 2025 (R20), while starting research into specifications around the end of 2027. At present, we expect the first version of 6G standardization to be released sometime around 2030, as shown in Figure 33.1.

Over the next decade, we will enter an era where people, things, and intelligence are all connected. The physical world will be sensed in real time and mapped to interconnected digital signals in order to achieve large-scale intelligence. Offering enhanced communication services and native AI capabilities, the next generation of wireless communication will empower society to take its next step forward in addressing the challenges we currently face.

To build the vision of 6G and identify the technologies that will make 6G a reality requires that industry, academic, and ecosystem players work together, as in 4G and 5G. The standardization of 6G worldwide will undoubtedly be the path to success for decades to come.

References

[1] "Framework for the radio interface(s) and radio sub-system functionality for international mobile telecommunications-2000 (IMT-2000)," Recommendation ITU-R M.1035, Mar. 1994.

[2] "Framework and overall objectives of the future development of IMT 2000 and systems beyond IMT-2000," Recommendation ITU-R M.1645, June 2003.

[3] "IMT Vision – framework and overall objectives of the future development of IMT for 2020 and beyond," Recommendation ITU-R M.2083-0, Sept. 2015.

[4] "WINNER," May. 2020. [Online]. Available: http://www.ist-winner.org/

[5] "METIS," Sept. 2020. [Online]. Available: https://metis2020.com/

[6] "Smart networks in the context of NGI," Strategic Research and Innovation Agenda 2021–27, Sept. 2020. [Online]. Available: https://bscw.5g-ppp.eu/pub/bscw.cgi/d367342/Networld2020%20SRIA%202020%20Final%20Version%202.2%20.pdf

[7] "IMT-2030," Sept. 2020. [Online]. Available: http://www.cbdio.com/BigData/2020-01/20/content_6154253.htm

[8] "MST 6G," Sept. 2020. [Online]. Available: http://www.chinanews.com/sh/2019/11-07/9001283.shtml

[9] "CCSA TC5 WG6," Sept. 2020. [Online]. Available: http://www.ccsa.org.cn/navMore?title=TC5%3A%20%E6%97%A0%E7%BA%BF%E9%80%9A%E4%BF%A1&index=0&menuIndex=3

[10] "Appeal for opinions on strategy outline of beyond 5G promotion," Apr. 2020. [Online]. Available: https://www.soumu.go.jp/main_sosiki/joho_tsusin/eng/pressrelease/2020/4/14_1.html

[11] "Beyond 5G promoting strategy (overview) (ver.1.0)," Sept. 2020. [Online]. Available: https://www.soumu.go.jp/main_sosiki/joho_tsusin/eng/presentation/pdf/200414_B5G_ENG_v01.pdf

[12] "Beyond 5G promotion strategy," Sept. 2020. [Online]. Available: https://www.soumu.go.jp/main_sosiki/joho_tsusin/eng/presentation/pdf/Beyond_5G_Promotion_Strategy-Roadmap_towards_6G-.pdf

[13] "The research and development strategy for future mobile communications of 6G," Sept. 2020. [Online]. Available: https://www.msit.go.kr/web/msipContents/contentsView.do?cateId=_policycom2&artId=3015098

[14] "FCC opens spectrum Horizons for new services & technologies," Sept. 2020. [Online]. Available: https://www.fcc.gov/document/fcc-opens-spectrum-horizons-new-services-technologies

[15] "Promoting U.S. leadership on the path to 6G," July 2020. [Online]. Available: https://www.atis.org/wp-content/uploads/2020/07/Promoting-US-Leadership-on-Path-to-6G.pdf

[16] "The vision and requirements for 2030+," Nov. 2019. [Online]. Available: https://www.shangyexinzhi.com/article/372578.html

[17] NTT DOCOMO, inc. "White Paper 5G evolution and 6G," Jan. 2020. [Online]. Available: https://www.nttdocomo.co.jp/english/binary/pdf/corporate/technology/whitepaper_6g/DOCOMO_6G_White_PaperEN_20200124.pdf

[18] Samsung, "6G vision: The next hyper-connected experience for ALL," July 2020. [Online]. Available: https://cdn.codeground.org/nsr/downloads/researchareas/6G%20Vision.pdf

[19] "6G: From connected everything to connected intelligence," May 2020. [Online]. Available: https://2020.ieeeicassp.org/industry-program/keynotes/6g-from-connected-everything-to-connected-intelligence/

[20] "The B5G/6G Challenge – a race of enabling technologies and design principles," May 2020. [Online]. Available: https://2020.ieeeicassp.org/industry-program/panels/the-b5g-6g-challenge-a-race-of-enabling-technologies-and-design-principles/

[21] M. Katz, M. Matinmikko-Blue, and M. Latva-Aho, "6Genesis flagship program: Building the bridges towards 6G-enabled wireless smart society and ecosystem," in *Proc. 2018 IEEE 10th Latin-American Conference on Communications (LATINCOM)*. IEEE, 2018, pp. 1–9.

[22] M. Latva-Aho and K. Leppänen, "Key drivers and research challenges for 6G ubiquitous wireless intelligence (white paper)," 6G Flagship, Oulu, Finland, 2019.

[23] "11 new 6G research white papers published," Sept. 2020. [Online]. Available: https://www.6gchannel.com/

[24] H. Viswanathan and P. E. Mogensen, "Communications in the 6G era," *IEEE Access*, vol. 8, pp. 57 063–57 074, 2020.

[25] W. Saad, M. Bennis, and M. Chen, "A vision of 6G wireless systems: Applications, trends, technologies, and open research problems," *IEEE Network*, vol. 34, no. 3, pp. 134–142, 2019.

[26] Z. Zhang, Y. Xiao, Z. Ma, M. Xiao, Z. Ding, X. Lei, G. K. Karagiannidis, and P. Fan, "6G wireless networks: Vision, requirements, architecture, and key technologies," *IEEE Vehicular Technology Magazine*, vol. 14, no. 3, pp. 28–41, 2019.

[27] P. Yang, Y. Xiao, M. Xiao, and S. Li, "6G wireless communications: Vision and potential techniques," *IEEE Network*, vol. 33, no. 4, pp. 70–75, 2019.

[28] M. Giordani, M. Polese, M. Mezzavilla, S. Rangan, and M. Zorzi, "Toward 6G networks: Use cases and technologies," *IEEE Communications Magazine*, vol. 58, no. 3, pp. 55–61, 2020.

[29] S. Dang, O. Amin, B. Shihada, and M.-S. Alouini, "From a human-centric perspective: What might 6G be?" *arXiv preprint arXiv:1906.00741*, 2019.

[30] E. C. Strinati, S. Barbarossa, J. L. Gonzalez-Jimenez, D. Ktenas, N. Cassiau, L. Maret, and C. Dehos, "6G, the next frontier: From holographic messaging to artificial intelligence using subterahertz and visible light communication," *IEEE Vehicular Technology Magazine*, vol. 14, no. 3, pp. 42–50, 2019.

[31] E. Björnson, L. Sanguinetti, H. Wymeersch, J. Hoydis, and T. L. Marzetta, "Massive MIMO is a reality; what is next?: Five promising research directions for antenna arrays," *Digital Signal Processing*, vol. 94, pp. 3–20, 2019.

[32] P. Popovski, O. Simeone, F. Boccardi, D. Gündüz, and O. Sahin, "Semantic-effectiveness filtering and control for post-5G wireless connectivity," *Journal of the Indian Institute of Science*, vol. 100, no. 2, pp. 435–443, 2020.

[33] N. H. Mahmood, H. Alves, O. A. López, M. Shehab, D. P. M. Osorio, and M. Latva-aho, "Six key enablers for machine type communication in 6G," *arXiv preprint arXiv:1903.05406*, 2019.

[34] T. S. Rappaport, Y. Xing, O. Kanhere, S. Ju, A. Madanayake, S. Mandal, A. Alkhateeb, and G. C. Trichopoulos, "Wireless communications and applications above 100 GHz: Opportunities and challenges for 6G and beyond," *IEEE Access*, vol. 7, pp. 78 729–78 757, 2019.

[35] M. Chen, U. Challita, W. Saad, C. Yin, and M. Debbah, "Artificial neural networks-based machine learning for wireless networks: A tutorial," *IEEE Communications Surveys & Tutorials*, vol. 21, no. 4, pp. 3039–3071, 2019.

[36] K. B. Letaief, W. Chen, Y. Shi, J. Zhang, and Y.-J. A. Zhang, "The roadmap to 6G: AI empowered wireless networks," *IEEE Communications Magazine*, vol. 57, no. 8, pp. 84–90, 2019.

Index